D1016200

TAKING SIDES

Clashing Views in

Energy and Society

SECOND EDITION

Thomas A. Easton
Thomas College

The McGraw·Hill Companies

TAKING SIDES: CLASHING VIEWS IN ENERGY AND SOCIETY,
SECOND EDITION

Published by McGraw-Hill, a business unit of The McGraw-Hill Companies, Inc., 1221 Avenue of the Americas, New York, NY 10020.
Taking Sides is published by the **Contemporary Learning Series** group within the McGraw-Hill Higher Education division.

1 2 3 4 5 6 7 8 9 0 DOC/DOC 1 0 9 8 7 6 5 4 3 2 1

MHID: 0-07-351449-7
ISBN: 978-0-07-351449-9
ISSN: 1944-7469

Managing Editor: *Larry Loeppke*
Senior Developmental Editor: *Jill Meloy*
Senior Permissions Coordinator: *Shirley Lanners*
Marketing Specialist: *Alice Link*
Lead Project Manager: *Jane Mohr*
Design Coordinator: *Brenda A. Rolwes*
Cover Graphics: *Rick D. Noel*
Buyer: *Nicole Baumgartner*
Media Project Manager: *Sridevi Palani*

Compositor: MPS Limited, a Macmillan Company
Cover Image: © Ingram Publishing RF

www.mhhe.com

Editors/Academic Advisory Board

Members of the Academic Advisory Board are instrumental in the final selection of articles for each edition of TAKING SIDES. Their review of articles for content, level, and appropriateness provides critical direction to the editors and staff. We think that you will find their careful consideration well reflected in this volume.

TAKING SIDES: Clashing Views in Energy and Society

Second Edition

EDITOR

Thomas A. Easton
Thomas College

ACADEMIC ADVISORY BOARD MEMBERS

Preface

Most fields of academic study evolve over time. Some evolve in turmoil, for they deal in issues of political, social, and economic concern. That is, they involve controversy.

It is the mission of the Taking Sides series to capture current, ongoing controversies and make the opposing sides available to students. This book focuses on issues related to society's use of energy. These issues are political, social, economic, environmental, and technical. The book does not pretend to cover all such issues, for not all provoke controversy or provoke it in suitable fashion. But there is never any shortage of issues that can be expressed as pairs of opposing essays that make their points clearly and understandably.

The basic technique—presenting an issue as a pair of opposing essays—has risks. Students often display a tendency to remember best those essays that agree with the attitudes they bring to the discussion. They also want to know what the "right" answers are, and it can be difficult for teachers to refrain from taking a side, or from revealing their own attitudes. Should teachers so refrain? Some do not, but of course they must still cover the spectrum of opinion if they wish to do justice to the scientific method and the complexity of an issue. Some do, although rarely so successfully that students cannot see through the attempt.

For any Taking Sides volume, the issues are always phrased as yes/no questions. Which answer—yes or no—is the correct answer? Perhaps neither. Perhaps both. Perhaps we will not be able to tell for many years to come. Students should read, think about, and discuss the readings. They should carefully consider the "Is There Common Ground?" sections at the end of each issue. Then they should come to their own conclusions without letting my or their instructor's opinions dictate theirs. The additional readings mentioned in the *introductions* should prove helpful. Most should be available in the school's library or through the school's access to online magazine and journal databases.

This second edition of *Taking Sides: Clashing Views in Energy and Society* contains 38 readings arranged in pro and con pairs to form 19 issues. For each issue, an *introduction* provides background, a brief description of the debate, and related material. Each part is preceded by an *Internet References* page that lists several links that are appropriate for further exploring the issues in that part.

Changes to this edition Four of the issues from the first edition have been retitled and refocused; they are Issue 11, Is It Time to Think Seriously About "Climate Engineering"?; Issue 13, Should the United States Stop Planning for Permanent Nuclear Waste Disposal at Yucca Mountain?; Issue 16, Is Wind Enough?; and Issue 17, Are Biofuels a Reasonable Substitute for Fossil Fuels?

Nine issues are entirely new; they are Issue 3, Is America Ready for the Electric Car?; Issue 4, Is Shale Gas the Solution to Our Energy Woes?; Issue 5, Should We Drill for Offshore Oil?; Issue 7, Is Human Activity Responsible for Global Warming?; Issue 8, Is Global Warming a Catastrophe That Warrants Immediate Action?; Issue 9, Will Restricting Carbon Emissions Damage the Economy?; Issue 12, Is It Time to Revive Nuclear Power?; Issue 15, Is Renewable Energy Green?; and Issue 19, Is It Time to Put Geothermal Energy Development on the Fast Track? Four of the six issues retained from the first edition have one new essay each. There are 28 new essays altogether.

A word to the instructor An *Instructor's Resource Guide with Test Questions* (multiple-choice and essay) is available through the publisher for the instructor using Taking Sides in the classroom. Also available is a general guidebook, *Using Taking Sides in the Classroom,* which offers suggestions for adapting the pro-con approach in any classroom setting. An online version of *Using Taking Sides in the Classroom* and a correspondence service for Taking Sides adopters can be found at www.mhhe.com/cls.

Taking Sides: Clashing Views on Issues in Energy Science is only one title in the Taking Sides series. If you are interested in seeing the table of contents for any of the other titles, please visit the Taking Sides Web site at www.mhhe.com/cls.

Thomas A. Easton
Thomas College

Contents In Brief

Contents

Topic Guide

This topic guide suggests how the selections in this book relate to the subjects covered in your course. You may want to use the topics listed on these pages to search the Web more easily.

On the following pages, a number of Web sites have been gathered specifically for this book. They are arranged to reflect the units of this Taking Sides edition. You can link to these sites by going to www.mhhe.com/cls.

All the articles that relate to each topic are listed below the bold-faced term.

(Continued)

Introduction

The history of human civilization is in large part the story of how people have increased their use of energy. Ten millennia ago, when people were changing from hunting and gathering to more settled lives centered on herding and farming, each person was limited to the energy of their own body plus whatever could be gained from fire for cooking, heating, pottery making, and a few other uses. A plain-English definition of "energy" is "the capacity to do work," and at that time no one had any way to magnify the work they could do with their own muscles and handheld tools.

J. R. McNeill, in *Something New Under the Sun: An Environmental History of the Twentieth-Century World* (Norton, 2000), notes that when people adopted shifting agriculture, they increased the amount of available energy by a factor of 10, largely because agriculture increases the amount of sunshine turned into food for people and thus the number of people and the work they can do. Settled agriculture gave another tenfold increase by increasing productivity. Domesticating large animals such as horses, cattle, elephants, and camels gave another large boost by greatly increasing the amount of work even a well-fed person can accomplish. By the beginning of the Industrial Revolution (about 1800), however, human muscle still supplied a large portion of all the energy used in civilization (estimates vary from 70 percent to 85 percent). Since then, per capita energy use has climbed very rapidly. David Price, "Energy and Human Evolution," *Population and Environment: A Journal of Interdisciplinary Studies* (March 1995), notes that in 1860, humans used about 100 kilograms of coal equivalent per capita. In 2000, the figure was about 2000 kilograms. The amount of work done around the world, using energy other than that of the human body alone, was the same as what could be done by 280 billion human beings. Price translates that as meaning that every human being on Earth has the equivalent of 50 invisible slaves working for them and adds that in highly technological countries such as the United States, every person has more than 200 such slaves.

The historical increase in energy use has been paralleled by growth in the human population. It has also required changes in technology. Wind was harnessed very early to drive ships, and later windmills. Water power also has a long history of being used to grind grain and saw wood. The Industrial Revolution truly began when the steam engine was invented and the energy content of fossil fuels (coal only at first) could be released. The first major uses were industrial, but coal furnaces made central heating available for homes, coal gas made gas lighting available, and when electrical generators were developed and installed, the use of energy in the home and in industry climbed very rapidly. To supply that energy, the scale of water power (hydroelectric dams) grew, fossil fuel use grew, and new energy sources such as nuclear power were devised. The internal combustion engine and the automobile also greatly increased the per capita use of energy.

This brief account glosses over an immense number of details. History is more than changes in energy use and more than the development of technology. But many of the other aspects of history—religion, trade, war, even art—are enabled by energy use and technology, for only when people need not spend all their time meeting essential needs such as food and shelter do they have time to devote to other pursuits. When agriculture first made it possible to spend less time getting food, it also made it possible to be priests and generals, sculptors and musicians, masons and plumbers, merchants and lawyers, scientists and engineers (among others). Modern agriculture requires large amounts of energy for fuel, fertilizers, pesticides, irrigation, and more and is so productive that a very small portion of the population can feed the rest. On-farm employment has dropped from 90 percent in 1800 to 2 percent today.

It also ignores environmental impacts of energy use. Those impacts were small when per capita energy use was small, as well as when population was small (population is about 7 billion today, but it did not reach 1 billion until 1800). The account above makes it clear that agriculture is an energy technology, and even thousands of years ago it led to deforestation, erosion, desertification, and soil salinization. An energy technology with surprising effects is the use of wind to power ships. The British Empire dominated the world with sailing ships built of wood. The need for raw materials for building ships played a major role in exhausting Britain's forests, which sparked the nineteenth-century shift from wood to steel as ship-building material. But steel production needs coal, which means coal mines, which have a tendency to flood with groundwater. The first steam engines (in the eighteenth century) were built to pump the water out of the mines. Once they had been sufficiently improved, they powered the steel ships as well as a vast amount of industry on land.

The use of coal—and later oil and natural gas—produces its own environmental problems, including pollution of air and water and global warming (see Issues 6–8), which prompt us to find other ways to meet our energy needs. We must also consider that the amount of fossil fuel buried in the ground is finite (see Issue 1). "Finite" means that eventually we must use it all. To this thought we must add that demand is increasing as other countries strive to improve standards of living and levels of technology. According to the U.S. Energy Information Administration's *International Energy Outlook 2010* (www.eia.doe.gov/oiaf/ieo/highlights.html), world energy consumption is projected to increase by 49 percent between 2007 and 2035. Problems thus threaten to grow worse and the time when we have in fact used all the planet's stock of fossil fuels grows ever closer. Given modern civilization's dependence on energy, it seems clear that we must find replacements for fossil fuels before that day. Fortunately, fossil fuels are by no means the only way to generate energy for human use.

How Do We Use Energy?

One of the most revealing portraits of how modern civilization uses energy can be seen at http://svs.gsfc.nasa.gov/vis/a000000/a002900/a002916/. This image shows outdoor electric lighting around the world, as seen from space. The lights are brightest in urban centers and dimmest in rural areas and in urban areas with less technological infrastructure. As a corollary, they are also brightest in areas where population is dense and numerous. If we reflect on the past, we must recognize that the lights have brightened as population and urbanization have grown, but that before the adoption of electric lighting urban and population centers must have been nearly as dim as the countryside. If we project the situation into the future, with world population projected to be increasingly more than 50 percent urban after 2008, we can see the lights growing brighter and the bright zones spreading. Yet it is worth noting that any light that can be seen from space represents wasted energy (wasted in the sense that it is not going where it is wanted, on the ground). This waste is reduced when cities adopt directional lighting, parking garages, and enclosed shopping districts (such as malls); if cities do so widely enough, future space cameras may actually see a less brilliant picture of the Earth. Cameras sensitive to heat (infra-red) radiation today map hot spots that correspond well to the bright spots. Heat too is waste, and this waste too can be reduced. But the laws of thermodynamics dictate that any use of energy will unavoidably emit waste heat.

We do not use energy only for electric lighting. Electricity itself powers some trains, appliances, computers, communications, and industrial machinery. Fossil fuels power steel mills, factories, cars, trucks, trains, airplanes, ships, farm equipment, electrical power plants, and home heating, and provide the energy needed to make the fertilizers (and other chemicals) that maintain our food supply. They also provide the energy used to manufacture the rocket fuel that puts communication and weather satellites in orbit and carries people to and from the International Space Station. Many of these uses benefit the consumer. Much also goes to military uses. Either way, they provide jobs and incomes, boost the economy, and raise the modern standard of living far above what was attainable in the past.

How Do We Obtain Energy?

In the United States, 35 percent of the energy used comes from petroleum, of which 72 percent goes to transportation (and accounts for 94 percent of the energy used by transportation); 23.4 percent comes from natural gas, with roughly a third going to residential uses and a third going to industrial uses; and 19.7 percent comes from coal, 93 percent of which is used for generating electricity. A mere 8 percent is nuclear, all of which goes to generating electricity. Alternative or renewable energy sources (e.g., sun, wind, biomass, hydroelectric, tidal, geothermal) provide only 7.7 percent of the mix, mostly as electricity. (See the Energy Information Administration's *Annual Energy Review 2009* [Report No. DOE/EIA-0384(2009), www.eia.doe.gov/aer/].)

Heat and motion are at present the twin keys to making energy useful to humans. Most of the energy we use today—80 percent—comes from burning fossil fuels in internal combustion engines, heating furnaces, and power plants. By itself, burning releases heat, and sometimes, as is the case with home heating, that is the desired form of the energy. Heat also causes gases to expand,

and in an internal combustion engine that phenomenon is harnessed to cause movement of pistons, crankshafts, and wheels. In an electrical power plant, the heat is used to boil water and produce steam under pressure, which is used to spin the generators that make electricity.

Nuclear power (Issues 12–14) also generates heat to make steam and spin generators to produce electricity. Geothermal power plants (Issue 19) use underground heat to do the same thing. Hydroelectric and tidal power plants use the force of falling water to spin the generators (see Issue 18). Wind power relies on the force of moving air to do the same job (see Issue 16). Only photovoltaic power (solar cells) uses an entirely different phenomenon—the interaction of light with the electrons in a semiconductor—to generate electricity. At present, it accounts for only a small fraction of total electricity production (roughly a tenth of wind power use), but its use has been increasing rapidly. Ken Zweibel, James Mason, and Vasilis Fthenakis, in "A Solar Grand Plan," *Scientific American* (January 2008), describe how solar power could be developed, with the aid of government funding, to supply 69 percent of the United States' electricity and 35 percent of its total energy by 2050. By 2100, it could supply all the nation's electricity and 90 percent of its energy.

The statistics are both illuminating and frightening. Eighty percent of the energy used in the United States comes from fossil fuels. Transportation is almost totally dependent on petroleum, most of which is imported (see Issue 2) and all of which could be used more efficiently. If we run out of petroleum—as we will, since it is finite in supply—we will need other liquid or gaseous fuels to power cars, trucks, trains, ships, and airplanes. This may mean converting coal to liquid form, as has been done in the past. It may also mean increasing use of biofuels (see Issue 17). Each of these alternatives has problems, for coal too is finite in supply and biofuels compete with the human need for farmland and crops. If the alternatives cannot meet the need for liquid fuels, and if we cannot find other ways to power cars, trucks, trains, ships, and planes, transportation must be much reduced. One "other way" is electricity, but electric vehicles (see Issue 3) need improved methods of storing electricity; better batteries and "super-capacitors" are under development.

Environmental Issues Related to Energy

Historically, energy use did not strike people as causing many significant problems. It was seen as a source of benefits. But by the 1600s, a few people had begun to comment on the smoke that plagued cities and suggest moving workshops and factories to the countryside (see, e.g., John Evelyn, *Fumifugium: Or the Inconvenience of the Aer and Smoake of London Dissipated* (1661)). Since then, the human population has grown and industry has come to occupy more of the landscape, provide more jobs, and meet more needs. It has also grown much more energy-intensive.

During the twentieth century, people began to recognize the damage that energy use could do to the environment. People protested the building of dams that would flood scenic landscapes (John Muir, founder of the Sierra

Club, led and lost the fight to stop the damming of the Hetch Hetchy Valley). By the 1960s, air pollution was an issue, and the U.S. Clean Air Act was passed. Nuclear power provoked concern about leakage of radioactive materials from both power and reprocessing plants (see Issue 14) and about the hazards of nuclear wastes (see Issue 13). Concern over coastal flooding by tidal power projects has blocked wider implementation of this energy source. People object to the noise associated with windmills, as well as the threat they pose to migratory birds (see Issue 16). People argue against the damming of rivers in part because of the damage done to fish populations (see Issue 18). Today, prime concerns are biofuels (Issue 17) and global warming (Issues 7 and 8), an inevitable consequence of adding carbon dioxide to the atmosphere by burning fossil fuels and a potential threat to wildlife, forests, coastlines, and human communities. Fossil fuel use also poses other problems, including oil spills, landscape destruction, and disruption of wildlife, and attempts to increase the supply are leading the fossil fuel industry into new areas such as the deep ocean (see Issue 5). Coal is plentiful, and the coal industry believes it can be used cleanly, without continuing its infamous problems (see Issue 6).

Energy—in large quantities—is essential to modern civilization. Since it seems unlikely that we can ever find a source of energy whose use does not affect the environment, we may have to accept environmental problems as a necessary part of the price (in addition to money and political turmoil) we must pay for energy. However, this is not to say that we should ignore those problems. It serves human interests to be aware of the problems and to seek to minimize them and their consequences. Among other things, this must mean choosing energy sources that create fewer problems as well as repairing damage already done and finding work-arounds (such as fish-ladders to help migratory fish get past dams).

Fortunately, there are alternatives to the energy sources (fossil fuels) that dominate the current market. They include nuclear, solar, wind, tidal, geothermal, and hydroelectric power, and they do have less environmental impact. But at present they play only a small role. If fossil fuels vanished overnight, it would take many years to expand that role to the point where civilization could function as it does today. Fossil fuels are not about to vanish that quickly, but as noted above, their supply is finite. There will come a day when they will no longer be able to meet our needs. It thus behooves us to develop and implement alternatives before they are needed.

Global Warming

The energy-related environmental problem of most concern to governments and environmentalists today is of course global warming (see Issue 7). In 2007, the Intergovernmental Panel on Climate Change (IPCC; www.ipcc.ch/) issued its Fourth Assessment Report, saying in no uncertain terms that "Warming of the climate system is unequivocal, as is now evident from observations of increases in global average air and ocean temperatures, widespread melting of snow and ice, and rising global average sea level." The warming is due to human activities, notably including the burning of fossil fuels, which over the course of a few decades releases carbon dioxide originally removed from the atmosphere over millions of years (many millions of years ago). Since carbon dioxide is a greenhouse gas, meaning that it slows the release of heat to space, it causes warming. The impacts on ecosystems and human well-being (especially in developing nations) are expected—despite the difficulty of making precise predictions—to be serious (see Issue 8). Indeed, analysts are saying that even though measures to prevent or minimize global warming will be expensive, they will be much less expensive than just letting global warming run its course. It is thus not surprising to see momentum gathering for action, although the U.S. government remains opposed to mandatory controls on carbon emissions and the pro-action and anti-action sides continue to accuse each other of deceptive practices.

What sorts of steps might be taken to fight global warming? Burning less fossil fuel is an obvious answer to the question, but it is also a difficult one because of civilization's dependence on oil, natural gas, and coal. Researchers are developing ways to keep carbon emissions from reaching the atmosphere (see Issue 10) and to keep solar heat from arriving (see Issue 11), while economists are developing ways to encourage compensatory actions (see Issue 9). Some people—even some environmentalists—are arguing that part of the answer must be increased use of nuclear power (see Issues 12 and 15). We will also surely need wind power (Issue 16), biofuels (Issue 17), hydroelectric power (Issue 18), and geothermal power (Issue 19).

Coping with Consequences

The twin issues of global warming and the finiteness of fossil fuels force us to face some hard questions. Because we are not about to run out of or stop using fossil fuels in the near future, we need to say only a little about what the sudden loss of 80 percent of the current energy supply would mean for civilization. If we resist the temptation to be pessimistic, we can say that it would probably *not* mean the end of civilization. There would be a period of adjustment while alternatives were rapidly deployed and people found ways to live using less energy (as by moving closer to work and decreasing their standard of living), but over time we would adapt. Many people would find the necessary changes difficult and painful. They will not be impossible, but knowing that they may lie ahead should motivate us to deploy alternative energy sources before the crisis is upon us.

Global warming is a more urgent matter. Changes in climate can mean changes in our ability to produce food, and late in 2007, the UN Food and Agricultural Organization reported that the world food supply is dwindling and the world's poor face a serious crisis. Both the supply and the price of food were soon affected by high fuel prices, diversion of land and crops to biofuels, and global warming. Residents of poor countries face the prospect of not being able to afford adequate food even if it is available. In addition, high prices for food and fuel mean that aid agencies are finding that their budgets cannot supply as much aid as in the past. Millions of people face the prospect of starvation. Although this does not threaten civilization itself, it would be a major disaster.

Climate change also means that sea levels will rise. Some South Pacific island nations are already protesting that this will inundate their islands. Others, such as Bangladesh, face the loss of low-lying coastal zones that are currently home to millions of people. However, the Netherlands (Holland) is already preparing to cope. David Talbot, in "Saving Holland," *Technology Review* (July 2007), describes how planners there are carefully studying potential impacts of global warming on a country much of whose land is already below sea level and kept habitable by extensive dikes. Among other measures, they are contemplating building houses, greenhouses, and even roads that can float on the rising waters.

Other potential impacts of climate change include the loss of arctic ice (and the gain of new shipping routes), increases in hurricane frequency and strength, spreading droughts, and migration of tropical diseases such as malaria into temperate zones. Environmentalists are also concerned that many species will die out. However, the other impacts mentioned are of more immediate concern to humans, for they threaten drastic change to people, landscapes, and economies. Many of these impacts will be measured in terms of money; one report says that the prospective monetary impact of global warming is so great that it justifies spending a great deal of money now to forestall it.

Can we forestall such impacts? Michael M. Crow, in "None Dare Call It Hubris: The Limits of Knowledge," *Issues in Science and Technology* (Winter 2007), argues that "we seem to be operating beyond our ability to plan and implement effectively, or even to identify conditions where action is needed and can succeed." Furthermore, he says, current issues "are child's play compared to the looming problems of global terrorism, climate change, or possible ecosystem collapse; problems that are not only maddeningly complex but also potentially inconceivably destructive." There are limits rooted in human psychology, politics, economics and more that define "the boundary conditions that we face in learning how to manage our accelerating impact on Earth," and an important question is "How can we create knowledge and foster institutions that are sensitive to these boundary conditions?" It is hubris (overweening pride that defies the gods) to think that we can always "predict, manage, and control nature," while at the same time devoting "little effort to the apparently modest yet absolutely essential question of how, given our unavoidable limits, we can manage to live in harmony with the world that we have inherited and are continually remaking."

Yet, no one finds it acceptable to say "we cannot cope." We *must* cope, for the alternative is unthinkable. Fortunately, powerful voices say we *can* cope. The International Energy Agency's *Energy Technology Perspectives 2010* (www.iea.org/Textbase/techno/etp/index.asp) makes the point (among others) that in our use of energy, we are not living sustainably. In order "to achieve the 50% CO_2 emissions reduction, government funding for RD&D [research, development, and demonstration] in low-carbon technologies will need to be two to five times higher than current levels." The technologies needed already exist, and many of them are discussed in this book.

Why Should We Care?

To some people, what is at stake in connection with energy is wealth. Enormous sums of money flow to energy companies, who see changing from fossil fuels to something else as an unmitigated economic disaster. The governments of oil-rich nations will also suffer if the oil runs out or people stop using it. The governments of oil-using countries gain through taxes on fuel use, and in late 2007, states were complaining that because of high gasoline and diesel prices, people were driving less. The movement toward more efficient cars also promises to decrease fuel use, and therefore the tax revenue from fuel use. Since that revenue is used to maintain roads and bridges, some people are already expecting problems unless other sources of funds are found.

Some people—notably countries that supply fossil fuels to the rest of the world—believe that what is at stake is political power. They too see changing away from fossil fuels as disaster. But many see continuing with business as usual—using fossil fuels, adding to global warming, delaying the shift to alternative energy sources—as the disaster. Civilization may need energy, but it does not need mass starvation, rising sea levels, and the many other problems that accompany energy use.

Still others dream of a better future. At present, this means a rising standard of living defined by improving incomes and the ability to afford cars, computers, refrigerators, air-conditioners, and other energy-intensive possessions that the people of industrialized nations (the United States, Europe, Japan, Australia, and a few others) take for granted. China is rapidly industrializing and increasing its consumption of energy, largely in the form of fossil fuels. The nation is very conscious of the accompanying problems, for Chinese cities are infamous for their air pollution. Fortunately, China is investing heavily in renewable energy (see Lisa Mastny, ed., *Renewable Energy in China: Current Status and Prospects for 2020* (Worldwatch Institute, 2010)). Reducing the use of fossil fuels, lacking a viable alternative, would be seen as a giant step backward. To countries that have not yet taken even China's steps forward, reducing fossil fuel use would be seen as oppression. Indeed, there are already complaints that trying to prevent global warming by restricting the use of fossil fuels amounts to a conspiracy to keep poor peoples poor.

Why should we care? Some people argue that we should protect the environment from the effects of human actions. Some say that we should protect *people* from the effects of human actions (which can also protect the environment). If we fail to do so, people will suffer. The basic question is one of priorities. We must make choices about what matters to us most: people or the environment; the present or the future; and convenience or responsibility? Most people would agree that it is important to meet human needs, to see to it that everyone has enough food, clean water, and at least basic health care and education. These are among the UN's eight Millennium Development Goals:

1. Eradicate Extreme Hunger and Poverty
2. Achieve Universal Primary Education
3. Promote Gender Equality and Empower Women

4. Reduce Child Mortality
5. Improve Maternal Health
6. Combat HIV/AIDS, Malaria, and Other Diseases
7. Ensure Environmental Sustainability
8. Develop a Global Partnership for Development

The use of energy appears here in Goal 8. Development means improving standards of living. Unfortunately, there is a conflict between development and sustainability that some feel is irreconcilable. Can we increase global energy use without destroying the environment, and in a way that can be kept up for the long-term future? If we can, it will not be easy.

False Hopes

We must cope, or at least hope that we can do so. Lacking hope, it makes little sense to attempt to foresee problems and devise ways of warding them off or repairing them. Yet, it does not seem to be a lack of hope that leads some people to say that we do not face energy-related problems. Instead, it is commitment to the status quo, to business as usual, to careers rooted in the customary ways of doing things, to investments, and to political backers who make their money from established energy industries (such as oil and coal companies). It is refusal to contemplate any need to change standard of living by driving a smaller car or giving up summer air-conditioning and vacation travel. It is even just plain denial; people are often reluctant to see problems before they are imminent, and politicians too often reinforce that reluctance in order to win votes. It is hard for a bearer of bad news to win an election.

Hope clearly marks those who spend their careers developing workable alternative energy sources, identifying future problems, and looking for solutions to those problems. It also marks those who wax enthusiastic over "fringe" ideas such as "cold fusion" or "free energy," both of which look at the moment like false hopes, for neither one can be demonstrated and physicists are highly skeptical of the theory behind them.

Are there other false hopes? The greatest may be the hope that all the problems will just go away, that the scientists will prove to be wrong about finite oil or global warming, or that a technological breakthrough will make all our worry a waste of time. A lesser false hope is that conservation can solve all our energy-related problems. It can help, and it is even essential if we are serious about minimizing environmental impacts, but it is not enough.

The Need for Critical Thinking

Many energy-related issues, both in this book and elsewhere, prompt vigorous and noisy debate. Students and voters alike must struggle to make sense of the arguments and decide which side seems most likely to be right. It is not enough to take a politician's or a newscaster's word. People must examine and assess the data themselves, in a process often called "critical thinking."

Taking Sides books like this one are designed to help develop the necessary skills by presenting opposing sides, providing background and context information, and pointing users toward additional information.

To assess opposing views on energy issues, it can help to have had a course or two in physics and chemistry, but even without that it can help to identify vested interests, if any. Vested interests can take many forms. For instance, government projects are infamous for their inertia, with critics of excess funding saying that they have only two stages: "too soon to tell" (whether they will work) and "too late to stop" (even if they do not work) because politicians have staked their reputations on them, bureaucracies have extended power structures into them, and jobs have been based on them. Should we believe anyone who argues in favor of a project in the "too late to stop" stage?

We can identify other vested interests by asking who pays the debaters. Can they be trusted if they support the views of those who pay them? A representative of a fossil fuel company, or a researcher funded by the fossil fuel industry, who argues against restraints on the burning of fossil fuels may deserve some skepticism. So may a politician whose party is known to favor business interests, arguing against measures that would reduce corporate profits, or a politician whose party is known to favor the environment, arguing *for* such measures. A proponent of a vested interest who argues for measures that would harm that interest may seem more trustworthy.

"May" is a key word here. It can be helpful to identify a debater's biases, but bias alone need not invalidate an argument. What is their attitude toward the data? Do they cite only data that support their view? Do they recognize conflicting data and try to fit it into an overall understanding of the situation? Are they open to other interpretations of the data?

What are their values or their priorities? This is another kind of bias. Some debaters value the natural world over the human world. Some put economic values over spiritual or aesthetic values. Some may seem too willing—or too unwilling—to accept risk. And although all values have their places in human life, it can be extraordinarily difficult to reconcile them in these debates, where so much is at stake.

Thomas A. Easton
Thomas College

Internet References . . .

Association for the Study of Peak Oil & Gas (ASPO)

ASPO is a network of scientists and others having an interest in determining the date and impact of the peak and decline of the world's production of oil and gas.

http://www.peakoil.net

The Energy Bulletin

The Energy Bulletin's "Peak oil primer" is a more sober—but also sobering—look at the situation.

http://www.energybulletin.net/primer.php

Energy Independence Now

Energy Independence Now is a California-centered nonprofit organization dedicated to catalyzing a rapid transition to a clean, renewable energy and transportation economy.

http://www.energyindependencenow.org/

Energy Tomorrow

Energy Tomorrow is an industry-sponsored Web site that provides information on many energy issues. For information on shale gas, see the Web site.

http://energytomorrow.org/issues/energy-security/throughout-the-world/shale-gas/

Marcellus Shale Coalition

The Marcellus Shale Coalition is an industry association "committed to the responsible development of natural gas from the Marcellus Shale geological formation and the enhancement of the region's economy that can be realized by this clean-burning energy source."

http://marcelluscoalition.org/

Hydraulic Fracturing at the EPA

The Environmental Protection Agency is embarking on a study of hydraulic fracturing ("fracking"), used to obtain natural gas from deep shale beds, and its potential impact on drinking water, human health and the environment.

http://www.epa.gov/ogwdw000/uic/wells_hydrofrac.html

U.S. Department of Energy—Clean Coal

The Department of Energy's Clean Coal information page outlines work being done to make coal a cleaner fuel.

http://www.fossil.energy.gov/programs/powersystems/cleancoal/

UNIT 1

Fossil Fuels

Modern civilization is powered by fossil fuels—coal, oil, and natural gas. Unfortunately, the supply of these fuels is finite, they are not evenly distributed around the world, and their waste products pose problems of pollution and climate change. What can we do if and when an important fuel runs out? How do we deal with the vulnerability inherent in depending on other nations for supplies of essential resources? What is the role of those vested interests that have grown rich on the status quo?

Understanding these controversies is essential preparation for understanding the debates that will shape politics and headlines, as well as the changes in energy use that must come over the next few years.

- Can the United States Continue to Rely on Oil as a Major Source of Energy?
- Is It Realistic for the United States to Move Toward Greater Energy Independence?
- Is America Ready for the Electric Car?
- Is Shale Gas the Solution to Our Energy Woes?
- Should We Drill for Offshore Oil?
- Should Utilities Burn More Coal?

ISSUE 1

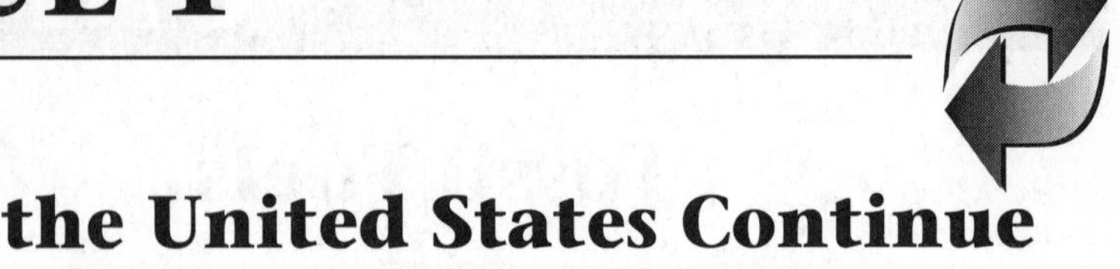

Can the United States Continue to Rely on Oil as a Major Source of Energy?

YES: Eric Gholz and Daryl G. Press, from "U.S. Oil Supplies Are Not at Risk," *USA Today Magazine* (November 2007)

NO: Tom Whipple, from "Peak Oil," *Bulletin of the Atomic Scientists* (November/December 2008)

Learning Outcomes

After studying this issue, students will be able to:

- Explain what "peak oil" means.
- Explain what the consequences of "peak oil" are likely to be.
- Describe at least three factors that will retard "peak oil" or lessen its impact.

ISSUE SUMMARY

YES: Eric Gholz and Daryl Press argue that predictions that global oil production must slow are based on scant evidence and dubious models of how the oil market responds to scarcity.

NO: Tom Whipple argues that the coming peak in global oil production and the subsequent decline will decrease standards of living worldwide for decades, until new energy technologies can be brought into production.

The world's supply of oil is finite. The amount oil companies can take out of the ground depends on how many oil deposits have been found and how easy it is to get the oil out of the ground. How long the oil companies can keep pumping oil depends on how quickly they discover new oil deposits to replace the ones they have exhausted and on improvements in methods of getting oil out of the ground (essential once the easy-to-pump oil has been removed). If the rate of pumping slows, the price of oil rises, giving the oil companies more money. With more money, they can fund improvements in technology and searches for new deposits, which then increase the rate of pumping. Optimists say the cycle can continue for the foreseeable future. Pessimists say there must come a point where the amount of oil that can be pumped in a year reaches a maximum; this is the "peak oil" moment. After that, the yearly supply must decline.

The January/February 2006 issue of *World Watch* magazine was devoted to the debate over whether "peak oil" poses a real problem. Red Cavaney, president and chief executive officer of the American Petroleum Institute, a trade group, titled his contribution "Global Oil Production About to Peak? A Recurring Myth." Christopher Flavin, president of the Worldwatch Institute, argued in "Over the Peak" that it is no myth. Oil extraction from the ground has exceeded new discoveries for two decades. Flavin also noted oil industry studies that predict peak production within a decade and called continuing to expand use of oil and expecting supply to continue to meet demand "irresponsible and reckless." Robert K. Kaufmann, professor in the Center for Energy and Environmental Studies at Boston University, said in "Planning for the Peak in World Oil Production," that we will not run out of oil overnight. The decline will be gradual, but if society does not begin taking steps now—for instance, increasing energy taxes to reduce demand and stimulate development of alternate energy technologies—"the effects could be disastrous."

Jim Motavalli, "The Outlook on Oil," *E Magazine* (January/February 2006), says that "one conclusion is irrefutable: The age of cheap oil is definitely over, and even as our appetite for it seems insatiable (with world demand likely to grow 50 percent by 2025), petroleum itself will end up downsizing." Peak oil may in fact have already happened, in 2005; if it hasn't, it will soon; see Robert L. Hirsch, Roger H. Bezdek, and Robert M. Wendling, "Peaking Oil Production: Sooner Rather than Later?" *Issues in Science and Technology* (Spring 2005). Talk against the problem may be motivated by the enriching effects of high oil prices on corporate and investor bank accounts. If we do nothing, tough times—even disaster—loom ahead. Some peak oil believers, says Bryant Urstadt in "Imagine There's No Oil," *Harper's* (August 2006), go so far as to say that as oil supply declines, "The economy will begin an endless contraction, a prelude to the 'grid crash.' Cars will revert to being a luxury item, isolating the suburban millions from food and goods. Industrial agriculture will wither, addicted as it is to natural gas for fertilizer and to crude oil for flying, shipping, and trucking its produce. International trade will halt, leaving the Wal-Marts empty. In the United States, Northern homes will be too expensive to heat and Southern homes will roast. Dirty alternatives such as coal and tar sands will act as a bellows to the furnace of global warming. In response to all of this, extreme political movements will form, and the world will devolve into a fight to control the last of the resources. Whom the wars do not kill starvation will. Man, if he survives, will do so in agrarian villages. It is a terrible scenario . . ."

Even though new technologies are getting more oil from existing wells (see Leonardo Maugeri, "Squeezing More Oil from the Ground," *Scientific American* (October 2009)), by May 2008, skeptics of "peak oil" were having a harder time defending their views. Oil was not running out, but conflicts between supply and demand had already driven oil prices well above 2006's $60 a barrel (which had seemed alarming at the time), and much higher prices were being forecast. "Peak oil" was in the news, with the Reuters agency saying on May 28, in "Peak Oil: Fact or Fallacy?" that "The price of crude oil has hit record levels above $135 a barrel, pushing industry analysts to take more seriously peak oil, or the idea that global production is near an apex after which it will decline sharply." The global economic crisis brought prices back down, but the basic problem remains. Mark Hertsgaard, "Running on Empty," *Nation* (May 12, 2008), notes that peak oil is imminent and is likely to trigger an economic crisis of its own. Richard A. Kerr, "World Oil Crunch Looming?" *Science* (November 21, 2008), notes that "Unless oil-consuming countries enact crash programs to slash demand . . . 2030 could bring on a permanent global oil crunch that will make the recent squeeze look like a picnic," and the crunch could come much sooner. "Oil Production 'Will Peak in 5 Years,'" *Professional Engineering* (November 12, 2008), says the date may be as early as 2013, after which "high oil prices will become a long-term trend." Mark Rowe, "When Will the Oil Flow Slow?" *Geographical* (June 2010), notes that in more than 100 countries, peak production is already past. Dates for a global peak extend all the way to 2040, but Terry Macalister, "Key Oil Figures Were Distorted by US Pressure, Says Whistleblower," *The Guardian* (November 9, 2009), says the real date is much closer. One effort to reach a realistic date concluded that it may be before 2020; see Richard A. Kerr, "Splitting the Difference Between Oil Pessimists and Optimists," *Science* (November 20, 2009).

Paul Roberts, "Tapped Out," *National Geographic* (June 2008), regrets that so far discussion of changing our energy-hungry lifestyles is "off the table," but behavioral change was already being detected in some reports. According to the U.S. Department of Transportation, thanks to high gas prices Americans drove fewer miles in March 2008 than in any March since 1979. CNN called the driving decline more rapid than at any time since the government started keeping records in 1942 (when gas rationing went into effect). It is fair to say that many people were feeling the pinch and taking what steps they could to adapt. Yet, a great deal more can be done. Charles H. Eccleston, "Climbing Hubbert's Peak: The Looming World Oil Crisis," *Environmental Quality Management* (Spring 2008), calls for an international program to make energy alternatives more available. Timothy J. Considine and Maurice Dalton, "Peak Oil in a Carbon Constrained World," *International Review of Environmental & Resource Economics* (2007), note that continuing use of oil conflicts with efforts to reduce greenhouse gas (carbon) emissions. "A clear policy direction for carbon regulation that encourages technological innovation is imperative as peak oil approaches."

Roger Howard, "Peak Oil and Strategic Resource Wars," *The Futurist* (September–October 2009), warns that the coming decline in oil production may lead to wars between producer states. Fred Curtis, "Peak Globalization: Climate Change, Oil Depletion and Global Trade," *Ecological Economics* (December 2009), projects that because global trade depends on cheap transportation of freight, getting away from fossil fuels, whether by choice (to deal with climate change) or not (oil depletion), must shift production of goods closer to consumers. Despite warnings that peak oil is a looming disaster, it seems unlikely that modern civilization will grind to a halt. Oil is not the only fossil fuel available. Vast amounts of coal are already used for electricity production, and coal can be used to make liquid fuels (see Zhang Ruihe, "Should China Develop Coal to Oil?" *China Chemical Reporter* (December 6, 2007)). We are thus unlikely to run out of gasoline for our cars or heating oil for our homes. Fuel will, however, be more expensive. We can also use natural gas and biofuels. If we wish to get away from fossil fuels, anything that generates electricity (such as nuclear, solar, and wind power, among others) will help keep people on the road, either in electric cars or in hydrogen-fueled cars. Developing these alternatives will, however, require rapid action and extensive funding to minimize the severity of the crisis so many people are prophesying.

In the YES selection, professors Eric Gholz and Daryl G. Press argue that predictions of "peak oil"—the moment when global oil production must slow and begin to decline—are based on scant evidence and dubious models of how the oil market responds to scarcity. Oil supply depends more on economics than on geology, and there is plenty of oil to go around. Tom Whipple, editor of *Peak Oil Review*, argues that the coming peak in global oil production and the subsequent decline is largely ignored by decision-makers, but it is coming and it will decrease standards of living worldwide for decades, until new energy technologies can be brought into production.

YES

Eric Gholz and Daryl G. Press

U.S. Oil Supplies Are Not at Risk

It is past time for political and ecological alarmists to stop spreading unfounded fears that America's energy security somehow is endangered.

Many Americans have lost confidence in their country's "energy security" over the past several years. Oil prices already were high by historic standards in 2005 when Hurricane Katrina ravaged the Gulf Coast and temporarily shut down the refineries, pipelines, and offload terminals at that large port complex, highlighting the apparent vulnerability of U.S. oil infrastructure. Furthermore, growing chaos in Iraq reminds Americans of their country's limited ability to control events in the oil-rich Persian Gulf region. Finally, the reliability of even America's domestic oil supplies was called into question last year when poor maintenance temporarily closed the pipelines that carry oil from Alaska to the contiguous 48 states. That a foreign company (British Petroleum) manages the Alaska pipeline only reinforces the overarching feeling that the U.S. has little control over the energy supplies it vitally needs.

Because the U.S. is a net oil importer, and a substantial one at that, concerns about energy security naturally raise foreign policy questions. One set of arguments is based on fears about dwindling global oil reserves and their increasing concentration in politically unstable regions. Those so-called peak oil worries have led some foreign policy analysts to call for increased U.S. efforts to stabilize—or, alternatively, democratize—the politically tumultuous oil-producing regions. A second concern focuses on the rise of China and Beijing's alleged strategy for "locking up" the world's remaining oil supplies through long-term purchase agreements and aggressive diplomacy. According to a number of analysts, the U.S. must respond to China's energy policy, outmaneuvering Beijing in the "geopolitics of oil," or else American consumers will find themselves shut out from global energy markets. Finally, many analysts suggest that even the "normal" political disruptions that sometimes occur in oil-producing regions—occasional wars and revolutions—hurt Americans by disrupting supply and creating price spikes. U.S. military forces, those analysts claim, are needed to enhance peace and stability in crucial oil-producing regions, particularly the Persian Gulf.

Each of those fears about oil supplies is exaggerated. Peak oil predictions about the impending decline in global rates of oil production are based on scant evidence and dubious models of how the oil market responds to scarcity. In fact, even though oil supplies increasingly will come from unstable regions,

From *USA Today Magazine,* vol. 136, issue 2750, November 2007 (excerpts).

the ongoing investments designed to reduce the costs of finding and extracting oil are a more effective response to that instability than trying to fix the political problems of faraway countries. Furthermore, fears of China are overstated as well. Chinese efforts to lock up supplies with long-term contracts will, at worst, be economically neutral for the U.S. and even may be advantageous. The main danger stemming from China's energy policy is that current U.S. fears may create a self-fulfilling prophecy of Sino-U.S. conflict. Finally, instability in the Persian Gulf poses surprisingly few energy security dangers, and the U.S. military presence there actually exacerbates problems rather than helping to solve them.

Those arguments do not mean that the U.S. can ignore energy concerns. Global demand for energy is soaring and shows no sign of relenting. Furthermore, oil supplies, though currently abundant, eventually will begin to run low, and the world will need to develop other energy sources. Yet, neither of those problems requires the sort of activist military policies that many foreign policy analysts suggest. . . .

Geologic features determine the location and quantity of oil deposits, but they do not determine "oil supply" in any meaningful sense. Supply depends on the difficulty (and hence cost) of oil exploration and production and on companies' economic decisions about how much money to spend looking for new oil fields, developing pumping capacity from the fields they find, and filling pipelines with oil. In any given region, geologic factors, such as the porosity of the rock, determine whether meaningful oil deposits exist and how expensive they are to discover and tap. However, geology merely creates the playing field for oil exploration and extraction. The amount of oil that actually can be "produced" at any given time, that is, extracted from the ground, transported to refineries, refined, and then transported in various forms to end users, depends on how much money oil companies have invested in a given field.

Prices drive fluctuations in oil supply. High prices encourage producers to pump their working fields at a higher rate to maximize profits before prices drop; lower prices lead them to reduce production. Companies with large inventories of oil generally respond to high prices by selling their stocks, unless they expect prices to rise even higher in the future. Price troughs encourage them to hold (or expand) their inventories, reducing supply in the short term.

Similarly, expectations about future petroleum prices shape long-term trends in oil supply. Oil companies, some of which are owned by the governments of countries with large reserves, decide how much to invest in exploration, new extraction technologies, and refining and transportation infrastructure and whether to pay large up-front costs to tap difficult-to-reach fields (such as those under deep water). Those major decisions, far more than geologic constraints, determine how much oil can be produced in the coming decades. Moreover, in the oil industry, like all others, investment decisions are driven by expectations about future prices—if the companies expect oil prices to be high, they will invest more heavily today since the enormous up-front expenditures will be recouped by high per barrel prices in the future but, if they expect prices to be low, they will trim investment, reducing future supplies.

Supply and Demand

Oil prices do not merely affect supply; they also play a key role in determining global demand. In the short term, demand does not change much in response to price fluctuations. People need to drive to work and heat their houses even if oil prices soar, so they tend to cut expenses elsewhere rather than go without oil. Higher prices, though, will reduce long-term demand. As prices increase, companies spend money on more efficient equipment and production processes, and individuals buy more efficient cars and improve the insulation in their homes. Finally, high prices spur investment in equipment that uses non-petroleum energy sources, reducing the demand for oil.

Although rising prices generally dampen demand, in the short term, climbing prices actually may spark additional demand. If the factors pushing up prices seem likely to continue, then consumers, brokers, and producers may decide to fill their inventories so that they can profit from the even higher price they expect in the future. Such speculation is the principal mechanism at work when fears of war or political instability drive up oil prices. Yet, this dynamic occurs only in the short term. Eventually, inventories become full or the price rises sufficiently that speculators start to sell their inventories. Demand returns to a level commensurate with actual consumption, and the price temporarily is depressed because the market draws supply from ongoing extraction and from the excess inventory. Day-to-day prices may bounce around quite a bit as consumption, extraction, and inventory strategies adjust, but that volatility is centered on a price level determined by "real" supply and demand.

The overall point is that the oil market has its idiosyncrasies and arcane details, but it generally functions like other markets. Rising prices increase supply, stimulate investment, and reduce demand. Price fluctuations match up the amount of supply on the market at any given time with the amount of demand, such that there are no "gaps" between supply and demand on a day-to-day basis.

Market forces shape oil prices, but they do not act alone. More than in most other industries, political risk tempers companies' enthusiasm for making expensive investments because many oil-producing regions are politically volatile. Will local governments nationalize companies' investments or raise taxes and fees for future extraction? Will terrorists destroy key equipment, or will a war disrupt the flow of oil to markets? In essence, companies explore and drill less intensively in unstable regions than they would otherwise because the expected costs due to political risks must be added to the purely economic costs. Companies must expect oil prices to rise by an extra margin before they are willing to invest in volatile regions.

Oil companies understand political risk; they have made their profits by dealing with it for their entire history. The big corporations manage portfolios of investments in different parts of the world, increasing the likelihood that at least one of their investments will be affected by political events at any given time, but reducing the probability that a substantial fraction of their oil revenue will be disrupted all at once. Because oil companies' investments account for a baseline level of political risk, that baseline is built into the overall level of today's available oil supply. In especially "lucky" times, when little goes wrong politically, an unexpectedly high level of oil will be available on world markets, and oil prices may fall; conversely, in especially "unlucky" times, oil prices temporarily may rise. In sum, political risk affects the overall level and geographic location of investments in the oil industry, but it does not change the fundamental supply dynamic. The quantity of oil available today depends on the investment decisions made in previous decades. Future levels of supply hinge on current investments.

Supply disruptions and political risk are not the only necessary adjustments to the basic supply-demand framework in oil markets. The world's major oil exporters have formed a cartel, the Organization of Petroleum Exporting Countries, to try to affect prices by controlling supply. The cartel members negotiate agreements to mute the normal, competitive market pressure to produce up to the point where price equals marginal cost. Although the logic is simple, making a cartel work is difficult. First, even monopolists are uncertain about the actual strength of demand for their product, and OPEC members often disagree about how much to restrict supply. They also often are at odds about how much production to expect from countries that are not members of the cartel. Second, even if the members can agree about the ideal level of production, they have to allocate market shares among themselves. Huge sums of money are at stake in this zero-sum negotiation; not surprisingly, agreements often are hard to reach. Finally, even when OPEC members completely agree about total production and the allocation of production quotas, each has a short-term interest in cheating, because each producer can increase its own profit by exceeding its quota.

OPEC's difficulty managing oil supply varies depending on political and market conditions, if investment and production patterns or political events change the number of key players in the OPEC negotiations, the cartel management task will change, too. Agreements are simpler to reach and cheating is easier to detect and punish if fewer players are involved. Furthermore, cartels work better when the members are willing to sacrifice some of today's profits for the long-term benefits of a strong cartel, and the political and market conditions in the OPEC member states determine how much each country will sacrifice for future gains.

Each time the global oil supply-and-demand situation changes, OPEC members have to adjust their cartel agreement. Given that, before the disruption, the cartel was at least somewhat effective at increasing profits above the normal competitive level, most disruptions should hinder cartel cohesion. Each market disruption is an opportunity for intracartel conflict, hence an opportunity for the amount of oil flowing onto world markets to increase compared to the level that OPEC had preferred to offer in the past.

Like political risk, cartel behavior does not change the underlying importance of supply and demand in oil markets. Political risk and cartel behavior merely modify the expected responses across the oil industry to price changes and political shocks.

In the past decade, the authors of several widely read books and articles have raised alarms about the quantity of the world's remaining oil reserves. According to the peak oil hypothesis, the world recently passed an ominous milestone: half of the recoverable oil already has been consumed, and the rate of global oil production therefore has begun, or soon will begin, an irreversible decline. The implication, according to proponents of that hypothesis, is that, in the coming decades, oil prices will soar as supplies dwindle and demand grows. Some observers argue that the U.S. should use foreign policy tools to ensure access to the "American share" of oil supplies in that difficult environment; others ominously warn that it is exactly that sort of "mercantilism," which they view as an inevitable consequence of passing the oil supply peak, that will draw the U.S. into resource wars.

Plenty of Oil to Go Around

These pessimistic claims about peaking oil supplies should be treated with skepticism. For decades, analysts have contended that oil supplies were dwindling and that the peak rate of production soon would be reached. In fact, one prominent advocate of that argument once predicted that the global production peak would occur in 1989 but, since then, global crude oil production has grown by 23%, and oil supply (crude oil and other petroleum liquids) has risen by more than 28%. More telling, the world's ultimately recoverable resources (URR) have been growing over time, largely because many fields contain substantially more oil than originally was believed.

One reason URR is growing despite the world's continuing consumption of oil is that improved technology has allowed a far greater fraction of reserves to be extracted from oil fields. In 1980, 22% of the oil in the average field was recoverable but, with better technology, average recovery now is up to 35%, effectively increasing URR by more than 50%. The results of the growing URR and recovery rate are striking: in 1972 the "life-index" of global oil reserves (the length of time that known reserves could support the current rate of production) was 35 years; in 2003, after 31 more years of accelerating oil extraction, the life index stood at 40 years. In short, no one knows how much oil ultimately is recoverable from the Earth, and there is no compelling evidence that reserves are running out or that production is near its peak.

Although the simplest version of the peak oil hypothesis exaggerates the likelihood of impending oil shortages, there is a subtler cause for concern that has some merit—the world's remaining oil supplies increasingly are concentrated in politically volatile regions, particularly the Persian Gulf and Central Asia. Fears of instability in those regions could suppress investment in exploration and development of oil fields, which could raise prices. In addition, pessimists maintain, unstable oil production in the future could leave the U.S. vulnerable to sudden supply shocks.

Concern about the effect of peak oil on the geographic concentration of oil supplies has led foreign policy analysts to advocate costly initiatives to attempt to mitigate the instability in key oil-producing areas. One proposal is for the U.S. to do more to police the Persian Gulf and the oil-producing sections of Central Asia. More ambitious policies would aim at addressing the underlying political instability directly. Traditional realpolitik logic might suggest that the U.S. should support authoritarian leaders in oil-producing regions and even help them to quash unrest, although that option rarely is expressed openly. Alternatively, the U.S. could sacrifice the short-term stability provided by regional dictators in the hope that robust U.S. democracy-promotion efforts might enable peaceful democratic regimes to provide long-term stability. All three strategies are based on the view that the growing concentration of the world's oil reserves in unstable regions requires an enhanced U.S. effort to reduce that instability.

Those foreign policy prescriptions for responding to instability in oil-producing regions are unnecessary and unwise. If oil production becomes more and more concentrated in politically unstable regions, suppressing investment in the oil industry (raising prices) and increasing the frequency of supply disruptions (also raising prices), then possible policy responses should be evaluated on the basis of their ability to enhance supply and reduce price. Using that metric, investments in oil exploration and extraction technologies are far more attractive than foreign policies that support dictators or attempt to police or democratize violent regions.

Oil industry research and development has a solid track record for increasing oil supplies. Decades of investment in exploration technology have made it easier to find deposits, and improvements in extraction technologies have made it possible (and economically feasible) to recover oil from locations that once were inaccessible, such as under deep water. Improved extraction technology also has increased the fraction of the oil that can be recovered from fields. As a result, the average finding and development cost of a barrel of oil (adjusted for inflation) plummeted from $21 in 1979–81 to six dollars in 1997–99. The steady stream of technological innovation in the oil industry explains why URR has grown over the past half-century.

In contrast, past efforts to increase stability in oil-producing areas by supporting dictators, policing violent regions, or spreading democracy have a dubious track record. . . .

China's Huge Appetite for Oil

China's soaring demand for oil is one of the biggest changes to affect energy markets in recent times. China's growing thirst for oil, part of the broader global surge in energy consumption, will drive up prices, imposing costs on the U.S. economy. Some analysts see an even graver threat ahead stemming from Beijing's energy policy: China is negotiating preferential long-term purchase agreements that could deny Americans even the opportunity to bid for some oil. Those analysts fear that competition for oil supplies will lead the U.S. and China into a struggle they describe as "the geopolitics of oil." They implicitly recommend that the U.S. shift its foreign policy to work against the Chinese strategy—in essence, creating its own preferential agreements to guarantee U.S. access to oil and perhaps exclude China.

Fears about the implications of China's energy policy are blown way out of proportion. First, on the demand side, China's efforts to reach long-term oil purchase agreements will not affect aggregate global demand for oil; the prepurchase agreements merely will change the patterns of global oil trade—which specific barrels of oil China consumes—but not the overall level of consumption. The long-term agreements, therefore, will not affect oil prices significantly. Second, on the supply side, China's leap into the oil exploration and extraction business either will be economically neutral for the U.S. or, if Chinese investments increase aggregate global supplies, possibly advantageous to the American economy.

Until the mid 1990s, China produced more oil than it consumed; since then, China's consumption greatly has outpaced domestic production. China's economic growth creates a voracious appetite for oil, especially because much of the manufacturing investment that fuels the Chinese expansion is energy intensive, and Chinese consumers view personal cars as a symbol of their middle-class status. Each unit of Chinese gross domestic product increase therefore bumps up global energy consumption more than a comparable GDP increase in many other countries. Many oil analysts believe that Chinese demand accounts for a substantial part of the oil price jumps since 2000.

Meanwhile, as the appeal of communist ideology has faded, Chinese leaders have staked their political future on the country's economic performance and the ongoing rise in living standards. As a result, they have used price controls to insulate domestic consumers and industries from price increases for petroleum products. Protected from rising prices, Chinese consumers and industries unabatedly increase their consumption. The traditional geopolitics of oil argument goes like this: without a fundamental shift in Chinese political strategy, Chinese demand for oil may threaten the energy security of other consuming countries, notably the U.S. Because the Chinese recognize their sustained need for oil, the government encourages companies to sign long-term contracts to buy large quantities of oil from producers around the world, allegedly establishing "preferential relationships." They also have bought access to overseas fields by investing in established foreign oil companies and obtaining concessions to develop oil fields and rights to explore for new fields. Those acquisitions give the Chinese decision-making control over future oil supplies.

Meanwhile, Chinese diplomats cultivate relationships with the governments of countries with large oil reserves. Some analysts allege that such statecraft especially is helpful in the oil industry, because government-owned oil companies control the fields in many countries and, perhaps those governments will be persuaded to sell to the Chinese at below-market prices, particularly during an oil shock. Finally, the Chinese government and oil companies are negotiating overland pipeline deals to bring oil to China from Russia, the Caspian basin, and even the Middle East. Other analysts and a number of American politicians are worried that all of those moves reflect a coherent Chinese national energy policy, one that might lock up sources of oil supply, leaving less oil on the word market for relatively laissez-faire countries like the U.S.

The economic arguments against those fears are compelling. Whether or not China arranges its oil purchases years in advance, it will consume the same amount of oil. If China buys concessions from foreign governments to pump oil from their wells or to prospect for new fields on their territory and then chooses to ship the crude to Chinese customers rather than to sell it on the open market, the Chinese actions simply will free up oil pumped by other companies so that they then can sell to non-Chinese consumers. In other words, the Chinese arrangements may lock up supply, but they also sate a substantial portion of world demand. Even the Department of Energy study mandated by Congress—a study prompted by an overwhelming congressional vote to "protect" American energy security—found that the consequences of the Chinese oil strategy are "economically neutral."

Defenders of the geopolitics of oil argument attack those rebuttals by questioning a key assumption of the economic view. They ask, what if the Chinese government were willing to sacrifice profits to keep oil for the Chinese market—that is, what if China imported all of the oil from its foreign concessions, holding down oil prices on the Chinese domestic market, and refused to resell its oil, even if world market prices soared above the Chinese domestic price? That would reduce the supply of oil available to non-Chinese consumers, dramatically driving up oil prices outside China. Current Chinese price controls on petroleum products, after all, demonstrate the Chinese government's willingness to sacrifice economic efficiency for noneconomic goals, such as the political stability that they think cheap oil enhances.

What the pessimistic analyses overlook, however, is that a Chinese decision not to resell the oil China pumps (whether from foreign concessions or domestic production), despite the opportunity to make big profits, would be the same thing as China deciding to pay more for oil than other consumers. Remember, China's hypothetical decision not to sell oil to Americans even if world prices rose dramatically—during a supply disruption, for instance—would cost the Chinese the same amount of money that they could use to outbid Americans in a "free" oil market in which China had not made long-term deals with suppliers. The point is that China's current activities, whether or not they are characterized as mercantilist efforts to lock up oil supplies, make no difference to Americans' long-run ability to buy oil in the market. What might hurt American consumers is China's growing demand for oil, because that demand drives up prices.

In the end, though, Chinese ownership of oil, like political risk and cartel behavior, does not matter much when it comes to the U.S. being able to acquire and use the oil it needs to keep its economy and society humming.

Tom Whipple

NO

Peak Oil

What may be one of the most important debates of the 21st century has been taking place just below the threshold of public perception. This debate concerns how much longer it will be before world oil production starts to fall, forcing a transition to other sources of energy and probably a simpler lifestyle.

From time to time, a spike in gasoline prices or an election forces aspects of this debate to the public's attention, and various solutions—more drilling, more wind turbines, more biofuels—are proposed. Most of these solutions fail to grapple with the real issue—that world oil production, according to the International Energy Agency (IEA), has been essentially flat for several years and will soon steadily and irreversibly decline. Few want to hear this message for it will mean the rapid and dramatic decline in the world's standard of living. In many of the underdeveloped megacities in Africa and Asia supplies of kerosene and propane for cooking already are in short supply or are becoming unaffordable. Sputtering hydroelectric stations affected by droughts and unaffordable $100-per-barrel oil are resulting in electricity shortages and failing water and sanitation systems. For many, the peaking of the world's oil supply will raise issues not just of living standards but of survival.

For the last 40 years, the United States, which has been largely insulated from the early tremors of oil depletion, has imported all the oil its population has needed with large trade deficits and the sale of U.S. Treasury bills to foreign countries. Real trouble, however, is not far away, according to peak oil theory. Sales of U.S.-made automobiles are in free fall due to $4-per-gallon gasoline, and the U.S. airline industry is contracting rapidly.

The declining availability of oil products will mean steadily increasing prices for nearly everything else. As prices rise, the use of gasoline-powered automobiles, boats, all-terrain vehicles, and RVs will wane. Gradually, more and more people could be priced out of using their personal vehicles as freely as they do now.

Before we get too far ahead of ourselves, let's go back in history to understand where the idea of such a potentially dire future arose.

If you don't believe in the theory of evolution, then you are likely to believe that oil is abiotic and is constantly being formed and pushed up for our use from somewhere deep in the earth. For the rest of us, however, fossil fuels are understood to be composed of organic matter from organisms that lived hundreds of millions of years ago, which was buried and pushed deep underground by geological forces. Over the eons, that material was converted

From *Bulletin of the Atomic Scientists*, November/December 2008, pp. 34–37, 41.

into oil, coal, and natural gas by high temperature and pressure. Peak oil starts with not only the question of how much fossil fuels were formed but also how much can realistically be extracted from beneath Earth's surface.

Hydrocarbons within the earth come in many forms, ranging from the desirable light, low-sulfur crude oils to heavy oils, tar sands, kerogen (sort of a "pre-oil" substance found in prodigious quantities inside shale), and finally natural gas and coal. Fossil fuels are also found in a wide variety of locations. Some are found on land, some deep beneath the sea, and some below polar ice. Much of the debate over the future of world oil production stems from a failure to appreciate that all oil is not created equal. Unfortunately, most of the high-quality, cheap, and easy-to-extract oil has already been consumed.

From the very beginning of the oil age, some 150 years ago, experts have predicted that it was about to run out. Until 50 years ago, such predictions were idle speculation, as most of the world had not been fully prospected and no one had a clear idea about how much oil existed. The first real insight was gained in 1956, when Shell Oil geologist M. King Hubbert calculated that U.S. oil production in the lower 48 states would peak in 1970. Hubbert, of course, had the benefit of studying a region that, by that time, had been fully explored so that major new oil finds were unlikely.

The heart of Hubbert's theory is the observation that the rate of oil production from a well or a field generally follows a bell-shaped curve and reaches peak production after about half of the oil has been extracted.

When U.S. oil production actually did peak when Hubbert predicted and the country's demand for oil continued to rise, the United States could no longer control world oil prices as it had done for many previous years. Power shifted to OPEC, the Mideast oil cartel, which had substantial oil, little domestic consumption, and the ability to control production and thereby prices.

In 1973, the world learned a new lesson about oil, which haunts us to this day, when OPEC exercised its power over the global oil market to enforce an embargo on Israel's supporters in the Yom Kippur War. U.S. imports of Arab oil dropped within two weeks from 1.2 million barrels per day to 19,000. The United States experienced nationwide gasoline shortages and the quadrupling of gasoline prices. Shortages also developed during the early days of the Iran-Iraq War, when supplies of oil were reduced.

Later, however, in the 1980s and 1990s oil experienced its golden years. New discoveries in Alaska, the North Sea, Mexico, and off the coasts of Africa and Brazil led to a surfeit of oil. Crude prices stayed around $20 a barrel and supplies were plentiful. World oil consumption grew rapidly from 58 million barrels per day in 1980 to 76 million in 2000. While the Soviet Union may have been done in by cheap oil in the late 1980s (oil was selling so cheaply that it was below the Soviets' cost of production), in the United States, plans for oil conservation and efficiency that were formulated during the oil shocks of the 1970s went by the wayside, and the country went on an oil binge that lasted for the next 25 years. Consumption increased and reliance on imported oil (and the countries that exert control over supply and thus price) increased as well. Today, nearly 70 percent of oil consumed daily in the United States is imported.

Warning bells began to sound again when an article by two respected European geologists, Colin Campbell and Jean Laherrère, appeared in the March 1998 issue of *Scientific American* indicating that the world was rapidly consuming its known oil reserves. Crude oil was being pumped out of the ground much faster than new oil was being discovered. The authors predicted oil peaking globally within 10 years.

The article's publication marks the beginning of the peak oil debates that continue to this day. Most criticism of peak oil has centered on the idea that increasing oil prices would bring forth new supplies from unconventional sources, such as the Alberta tar sands, long considered too expensive to extract and process into usable oil. Many have pointed to developing technologies such as horizontal drilling, 3-D seismology, and oil field flooding, which peak oil critics believe will bring so much new oil to market that prices will fall, providing plentiful supply for at least another 30–40 years. The U.S. National Geodetic Survey contributed to this belief by releasing a report that estimated another trillion barrels of conventional oil were waiting to be discovered and produced.

But an important part of this debate, often left out by such projections, is that oil these days is not always the black stuff that comes up from a well. Although traditional petroleum constitutes the bulk of what is called oil, there are also liquids that are extracted from natural gas, heavy oil that is extracted from tar sands, and biofuels that are produced by distilling sugars from plants and processing plant oils. While conventional crude oil and the liquids that condense out of the natural gas that comes out of the ground with the crude currently amount to about 74 million barrels per day, total oil production, which is also known as "all liquids," amounts to 86 million.

For six years after Campbell and Laherrère's article not much happened. Oil production (of all liquids) climbed to about 80 million barrels per day in 2000 and then slipped back a couple of million barrels due to an economic downturn. Some thought they were witnessing the peak of world oil production in 2000, but this was not the case. After 2003, when oil production surged, peak oil doubters ridiculed those who had called the peak in 2000.

With the end of the recession in 2003, world oil production and prices began to climb rapidly. The inflation-adjusted price of oil had been steady at $20–$25 per barrel since the mid-1980s and had reached a low of $15 per barrel in 1998. From 2002, when oil was $26 per barrel, oil began a steady upward surge in both production and price. Crude went from an average of $30 per barrel in 2002 to a high of $147 in July 2007. World oil production (all liquids) climbed from 76 million barrels per day in 2002 to about 86 million in 2008.

By 2004, surging prices and consumption brought the issue of peak oil to wider attention. Blogs and websites on the subject began to proliferate (many blogs focusing on peak oil date from that year). New books foretold a very bleak future for the world without oil. Some were so apocalyptic in their pronouncements that their authors and peak oil adherents were labeled "doomers."

This development moved the peak oil debate in a slightly different direction as the so-called doomers argued with the "techno-fixers," those who believe that there are many alternate sources of energy, including nuclear, solar, wind, wave, and biofuel. For the doubters, peak oil is simply another item to be added to the list of possible, but improbable, dangers—cataclysmic comet strikes, thermonuclear war, and a super virus—that people should worry about, but not too much. Somewhere in all this uproar, which mostly took place online, the real issue of peak oil got lost. The question asked by peak oil is simply whether the world can produce enough new oil each year to keep up with falling production at older fields.

Peak oil is the last thing anyone really wants to hear or think about because it illustrates how humanity has blown through the planet's fossil fuel stores in a couple of centuries and that without these resources, life on Earth will become much rougher. Consider what happened to President Jimmy Carter some 30 years ago, when he told Americans to put on sweaters to deal with high fuel prices and was ridiculed for it. Few officials anywhere in the world will publicly say that worldwide oil production is nearing a peak, if that is truly the case, and that preparations should be made for this eventuality. Instead, U.S. politicians talk in circumlocutions, decrying the country's "addiction to oil" or touting a need for "energy independence." The press is also culpable, for as yet few stories connect all the dots in the peak oil story. Most reporters and editors content themselves with focusing on high gasoline prices. Few in the Western media have noticed that many places in the underdeveloped world are already suffering power and cooking fuel shortages due to unaffordable oil.

So what is likely to happen? Has oil really peaked? Production of crude oil has basically been flat since 2005, according to the U.S. Energy Department and the IEA, which both track worldwide oil production. Occasionally a single month will set a slightly higher production record, but annual production figures are no longer climbing. Although the "other liquids" number is climbing a bit as the world increases its production of natural gas–derived liquids and biofuels, it is important to realize that many of these alternate sources of liquid fuel contain substantially less energy than conventional oil does, so adding other liquids to world crude oil production overstates the amount of energy available.

In the last year, high prices have led to reduced consumption in parts of the world. U.S. oil consumption is down nearly 5 percent according to Energy, and in the rest of the Organization for Economic Co-operation and Development (OECD) countries consumption is flat. Chinese and Indian consumption, however, continues to grow rapidly.

What started out many years ago as a geological theory has become a much simpler issue that is readily observable—can new oil production keep up with the depletion of existing reserves? In recent years, "megaprojects analysis," a new way of approaching this issue, has attempted to answer this question.

Megaprojects analysis starts with an assessment of just how fast existing world oil fields are running dry. Ascertaining this information should be easy, but in many oil-producing countries depletion rates are a closely guarded state secret. Although there are many variables that determine the rate of oil depletion, such as oil field management and the use of water or gas flooding to force oil to the surface, the conservative consensus among oil watchers is that the rate is about 5 percent annually. This means that if no new projects begin producing oil in a given year, by the end of that year about 4 million fewer barrels per day will be produced. In short, the world's oil industry has to find 4 million barrels of new production each day, each year just to stay even, much less keep up with growing demand.

Yet most new production comes from giant, expensive projects that when fully operational produce no more than 50,000 barrels per day. Today such projects cost billions and take six to seven years to complete, hence the term "megaprojects."

Megaprojects are too big and expensive for a country to hide, so periodic updates on their progress are available from public announcements and in the trade press. Even with all this information, predicting how much new oil production will come online per year is still more art than science. Multibillion-dollar project deadlines have a tendency to slip, sometimes by many years, both as to when they will start production and when they will reach their peak production goal. Due to such variables, just how much new oil production will come online in the next few years is open to debate.

Despite all the unknowns, a consensus seems to be emerging among those tracking megaprojects, most notably Chris Skrebowski, editor of the London-based *Petroleum Review*, and a megaprojects team connected to the peak oil blog The Oil Drum, that at sometime between 2010 and 2012, new oil production will not be sufficient to keep up with depletion and that world oil production will begin dropping.

The reduction in oil production will stretch over decades or more likely centuries. Once major conventional oil fields go into decline, it is unlikely that new fossil fuel extraction technologies will come along to reverse that trend. Trillions of barrels of hydrocarbons certainly are deposited in Alberta's tar sands, in Colorado's shale, in Venezuela's Orinoco heavy oil fields, and under the polar ice caps. While such "oil" can be exploited, the cost, time, and energy required to extract it will not keep up with the coming decline in conventional sources.

A decline in world oil production may also come from a lack of demand. The price of oil has been dropping for the last two months because the markets collectively believe that a major worldwide recession is coming. If the current economic difficulties are serious and long-lasting enough, the reduction in demand could mask the geological peaking of oil for many years. With the recent rapid growth of China and India, it is also possible that future Asian demand will be high enough to offset any drop in demand from OECD and underdeveloped countries.

A last important factor in the peak oil equation is that the major oil exporters are using more and more of their own oil domestically, so there is less available to sell. This is particularly true of the Middle East and Russia. Statistics from the IEA indicate that total world oil exports are beginning to decline; in 2007 they were down by 2.3 percent, or 1 million barrels per day. Mexico's exports are dropping rapidly due to a very high depletion rate at its largest oil field, so fast that Mexico may become a net oil importer within three or four years.

This year has, without a doubt, been an eventful one in the U.S. economy. Crude prices soared from below $100 per barrel to $147 in July and then fell back to below $100 a barrel in September. The ongoing U.S. mortgage crisis has turned into an economy-wide liquidity crisis and now a worldwide economic crisis. In September, a pair of hurricanes slowed U.S. oil production and refining, which resulted in gasoline shortages across the southeastern part of the country.

U.S. voter concern about high gasoline prices has transformed this year's presidential election into a search for painless energy solutions with proposals to increase drilling and the production of biofuels, increase taxes on oil companies, punish speculators, and build more nuclear plants.

Yet no one is seriously talking about the world's flattening oil production, the likelihood of declining oil imports, the near certainty of much higher gasoline prices, or how we adjust to a world without cheap, abundant oil.

Alternative, renewable energy sources sources—solar, wind, tides, and biofuels—as yet provide only a tiny fraction of our energy needs. As the peaking of world oil production becomes more and more evident, our society will have to get by with considerably less energy, particularly from oil, in the next few decades, until new energy technologies can be brought into production and energy efficient forms of existing technologies become more widely adopted.

EXPLORING THE ISSUE

Can the United States Continue to Rely on Oil as a Major Source of Energy?

Critical Thinking and Reflection

1. When oil production peaks and begins to decline, analysts expect the price of fuel to rise. In what ways will this affect your life?
2. Do you expect to see "peak oil" in your lifetime? Why or why not?
3. Rising fuel prices will make life difficult for commuters (among others). In what ways might commuters change their lifestyle to cope?
4. Name two ways in which "peak oil" will increase use of alternative energy sources such as wind and sun.

Is There Common Ground?

There is no disagreement that the world is utterly dependent on ample supplies of energy, much of which is currently supplied by fossil fuels. The debate is mainly over how long we can continue to rely on fossil fuels.

1. Oil is only one of the fossil fuels available to us. We also use natural gas and coal, but no one seems concerned about "peak gas" or "peak coal." Why not?
2. How long are supplies of natural gas projected to last?
3. How long are supplies of coal projected to last?
4. Do we have time to develop alternatives to fossil fuels?

ISSUE 2

Is It Realistic for the United States to Move Toward Greater Energy Independence?

YES: Richard N. Haass, from testimony on "Geopolitical Implications of Rising Oil Dependence and Global Warming" before the House Select Committee on Energy Independence and Global Warming (April 18, 2007)

NO: Robert Bryce, from *Gusher of Lies: The Dangerous Delusions of "Energy Independence"* (Public Affairs, 2008)

Learning Outcomes

After studying this issue, students will be able to:

- Define "energy independence."
- Explain why our dependence on foreign energy sources worries many people.
- Explain why we cannot eliminate our dependence on foreign energy sources immediately, or perhaps at all.
- Explain how imported oil is associated with terrorism.

ISSUE SUMMARY

YES: Richard Haass argues that energy independence cannot be achieved if it means being able to do completely without imports of oil and gas. We can, however, move toward energy independence by raising gasoline taxes, making cars more fuel-efficient, and developing alternative energy sources.

NO: Robert Bryce argues that the American obsession with the idea of energy independence prevents honest, effective discussion of genuine energy challenges. We need to recognize and accept the difference between rhetoric and reality.

The United States gets most of its energy from coal, not oil. But oil does fuel the transportation sector—meaning cars, trucks, and airplanes—and the nation imports more than 60 percent of the oil it uses. That comes to almost 400 million barrels of oil per month, at a price in December 2010 of more than $90 per barrel. The nation sends huge amounts of money overseas and is at the mercy of international politics, which can cut off the flow at will, and terrorism, which can destroy pipelines and oil tankers. It lacks what is often called "energy security," the ability to ensure a steady and affordable flow of energy to consumers. To remedy this lack, politicians and others have long urged that the United States achieve "energy independence." In the extreme, this means producing all energy used within the United States within U.S. borders. More practically, it means reducing dependence on vulnerable foreign supplies—by stockpiling oil in the Strategic Petroleum Reserve and by developing stronger national sources—to the point where a disruption in supply of foreign oil will not cripple the country.

In February 2006, Barack Obama, speaking to the Governor's Ethanol Coalition, said that "for all of our military might and economic dominance, the Achilles heel of the most powerful country on Earth is the oil we cannot live without." The Bush Administration recognized that the problem exists but did not make "a serious commitment to energy independence. The solutions are too timid—the reforms too small. America's dependence on oil is a major threat to our national security, and the American people deserve a bold commitment that has the full force of their government behind it." When Bush signed the Energy Independence and Security Act of 2007, he noted that the Act would increase use of biofuels, increase automobile fuel economy standards, and call for more efficient lighting and appliances ("Remarks on Signing the Energy Independence and Security Act of 2007," December 19, 2007). Obama's suggested solutions began with more fuel-efficient cars and biofuels and move on to "a national commitment to energy security" exemplified by the appointment of "a Director of Energy Security to oversee all of our efforts [and] coordinate America's energy policy across all levels of government."

Would such an approach work? Philip J. Deutch, "Energy Independence," *Foreign Policy* (November/December 2005), does not believe the United States can stop relying on imported oil any time soon. Nor can it burn less coal. Nuclear power is needed. Energy conservation may help but it cannot get the energy monkey off our backs. "Energy independence may be hopeless in the next 20 years, but there is no doubt that emerging technologies will eventually bear the brunt of our energy burden." Steve Stein, "Energy Independence Isn't Very Green," *Policy Review* (April/May 2008), argues that replacing dependence on oil with dependence on coal (which adds to global warming) does not improve our security; nuclear power is the better answer.

Not everyone agrees that high oil prices or increasing oil imports threaten U.S. national security. Philip E. Auerswald, "The Myth of Energy Insecurity," *Issues in Science & Technology* (Summer 2006), argues that high oil prices may be hard to bear, but "in an open society with a market economy, only prices have the brute power to effect change on the scale required." We will shift to nonoil energy sources when high prices drive sufficient investment in alternatives. He does not discuss the vulnerability of oil supplies to politics or terrorism. Jane C. S. Long, "A Blind Man's Guide to Energy Policy," *Issues in Science & Technology* (Winter 2008), discusses the role of these threats, as well as that of climate change, and argues that it is time for a more deliberate approach. "We must become the makers of our energy, climate, economic, and security fate." Waiting for markets to solve the problem is an example of the kind of thinking (focused on economics) that got us into this mess. Gregory D. Miller, "The Security Costs of Energy Independence," *Washington Quarterly* (Spring 2010), finds that energy independence, because it threatens the revenues of oil-producing countries, may provoke international conflict; it would also have serious impact on international trade.

The high oil prices of 2008 drove many discussions of alternative approaches to meeting U.S. energy needs. Ken Zweibel, James Mason, and Vasilis Fthenakis, "Solar Grand Plan," *Scientific American* (January 2008), offer one of the most grandiose—and perhaps promising—schemes in their proposal to spend $420 billion from 2011 to 2050 to construct a vast array of photovoltaic cell and solar concentrator power plants in the American Southwest. By 2050, this array would be able to supply two-thirds of the nation's electricity and a third of its total energy requirements. By 2100, it would supply 100 percent of U.S. electricity and 90 percent of total energy. Transportation needs would be met by using electricity to generate hydrogen. Given that some very promising work on improved batteries and related devices is now under way, hydrogen may not be needed. For more on solar concentrators, see Susan Moran and J. Thomas McKinnon, "Hot Times for Solar Energy," *World Watch* (March/April 2008).

Jon Birger, "Oil from a Stone," *Fortune* (November 12, 2007), reminds us of oil shale, which three decades ago was being examined as a possible oil source. The United States has large amounts of this hydrocarbon-rich sedimentary rock, and new technology may allow large amounts of oil to reach the market for centuries to come. Marianne Lavelle, "Power Revolution," *U.S. News & World Report* (November 5, 2007), discusses progress in solar, wind, and geothermal energy. Joe Castaldo et al., "The Power of Being Green," *Canadian Business* (March 17, 2008), discuss several innovative approaches including that of Magenn Power, which plans to mount wind turbines on blimps floating at high altitude.

Paul Roberts argues in "The Seven Myths of Energy Independence," *Mother Jones* (May/June 2008), that many of the alternatives to oil come at a substantial environmental and political cost, and that even if we had good alternatives ready to deploy, it would take decades—as well as energy from present sources—to replace all the cars, pipelines, refineries, and other existing infrastructure. "Paradoxically, to build the energy economy that we want, we're going to lean heavily on the energy economy that we have." Some people argue that we can make the necessary infrastructure changes in remarkably little time; they recall how quickly the United States shifted its industry to war production in World War II. Robert Miller and Miles Imwalle, "Energy Independence Achievable with New Environmental Regulatory Approach," *Trends*

(November/December 2009), call for changing the way environmental laws are administered to speed implementation of alternative energy. Will we make such changes? That answer must come from politicians, and it will be shaped by how the costs of oil compare with those of deploying alternatives.

In the YES selection, Richard Haass, president of the Council on Foreign Relations, argues that energy independence cannot be achieved if it means being able to do completely without imports of oil and gas. We can, however, move toward energy independence by raising gasoline taxes, making cars more fuel-efficient, and developing alternative energy sources. In the NO selection, energy writer Robert Bryce argues that "energy independence" is a prized piece of political rhetoric that ultimately makes no sense. There are just no substitutes for oil in the near future. "Energy independence" has become an obsession that prevents honest, effective discussion of genuine energy challenges. We need to recognize and accept the difference between rhetoric and reality.

Richard N. Haass

Geopolitical Implications of Rising Oil Dependence and Global Warming

. . . The geopolitical implications of rising oil dependence and global climate change for the United States are great and likely to become even greater with time.

Let me address each of the questions you have posed to me and my fellow witnesses.

I will begin with how ever-increasing dependence on imported oil affects U.S. national security. The short answer is that it does, in many and important ways. Four stand out:

First, American and global dependence on the Middle East for oil artificially increases the importance of this part of the world. This is not to say it would not be important even if there were no oil in the region or if the United States and the world were not dependent upon the region's oil. The United States would still have important, even vital concerns relating to terrorism, nonproliferation, conflict resolution, Israel, and so on. But there is no denying that energy makes this part of the world far more vital than it would otherwise be and reduces American willingness and ability to tolerate developments that were they to occur in other regions would provoke less of a response. And just to be clear, let me stress that this concern for oil and gas is not tied to protecting the interests of the large oil companies but rather to maintaining adequate access on acceptable terms to a vital raw material.

Second, the fact that the United States imports roughly 60% of the oil it consumes leaves the U.S. economy vulnerable to supply interruptions that even in small amounts can cause price increases and in larger amounts cause not only price increases but economic disruption. The United States would be vulnerable economically to supply interruptions (and price spikes) even if it imported far less oil given the extent to which others are vulnerable and the degree to which U.S. economic fortunes are tied to those of others.

Third, the need to pay for oil imports exacerbates the already considerable current account deficit, which in turn further weakens the dollar and makes the United States more dependent on (and vulnerable to) the decisions of other governments. Approximately one-third of the annual current account deficit, or some $250 billion, is attributable to oil imports.

From House Select Committee on Energy Independence and Global Warming by Richard N. Haass, (April 18, 2007).

Fourth, American demand for oil contributes to upward pressures on prices and provides massive revenues to producers. One of the top five oil exporters to the United States is Venezuela, whose foreign policy is anti-American in large measure. The top two exporters of oil in the world, Saudi Arabia and Russia, carry out policies at home and abroad that at times run counter to American values and interests. Iran, the world's fourth largest exporter of oil, is in large part able to conduct the problematic foreign policy it does because of high oil revenues. In addition, massive inflows of oil revenues can be as much a liability as a windfall in another way in that they often work against efforts to promote market economies and the rule of law.

The second question posed asks whether it is urgent that the United States do something about this state of affairs. It is. It is also a national failure, a bi-partisan failure, that this country is consuming and importing as much oil as it is today, more than three decades after the first oil shock that accompanied the October 1973 Middle East conflict. It is a matter of some debate as to whether U.S. energy security has actually deteriorated despite that and subsequent crises: the United States is more dependent than ever on imports, but U.S. energy intensity is down and international markets seem better able to weather disruptions. But whatever the relative judgment on energy security, it is not what it needs to be in absolute terms. That said, it has taken us decades to get to where we are today, and will take decades for the situation to change fundamentally. There is, however, no reason to delay. Every day we as a country wait to act only increases the price we pay for the current state of affairs and makes it that much more difficult and costly for us to change them.

Should climate change be treated as a national security matter? The short and clear answer is "yes." Countries are unlikely to go to war over levels of greenhouse gas emissions, but they may well go to war over the results of climate change, including water shortages and large-scale human migration. Climate change, by contributing to disease, extreme weather, challenges from insects that attack both food production and people, water shortages, and the loss of arable land, will also contribute to state failure, which in turn provides opportunities for activities such as terrorism, illegal drugs, and slavery that exploit "sovereignty deficits." Development, democracy, and life itself will not thrive amidst such conditions.

The last two questions can best be answered together, as they ask for recommendations for reducing oil dependence and greenhouse gas emissions and addressing both climate change and energy security.

Energy security is not easy to define. It is a relative concept, in the sense that it is impossible to achieve total energy security—just as it is impossible to achieve full security (or complete invulnerability) in any realm. A traditional definition of energy security would be one that emphasized minimizing U.S. vulnerability to supply interruptions and price increases. This "reliability and affordability" approach to energy security is inadequate, as it does not capture the additional rationales for reducing consumption of oil (imported or otherwise) in order to curtail the flow of resources to unfriendly governments and to reduce the adverse impact on the world's climate. As a result, we need to adopt a broader definition of the concept. Energy security is directly related to the ability to manage the form and amount of energy produced, consumed, and imported so that the United States reduces its vulnerability to supply and price fluctuations, the flows of resources to unfriendly producer countries, and the adverse impact on the global climate.

A range of prescriptions, some familiar, some not, flows from this broader approach to energy security. One is the desirability of diversifying sources of oil and other energy supplies. Such diversification reduces the impact of losing for whatever reason access to the output of any single producer. The United States has done this in the oil realm, as only Canada provides the United States in the range of 20% of its total oil imports. 90% of U.S. crude oil imports are distributed to more than ten countries.

The United States can also help reduce its vulnerability to supply interruptions through contingency planning, including the maintenance of the strategic petroleum reserve (SPR) and various stand-by international sharing arrangements. Congress would be well-advised to assess both the adequacy and guidelines for use of the SPR. Also in need of overhaul is the International Energy Agency, which needs to be amended (or complemented by the International Energy Forum) so that major countries such as India and China are fully included in global planning.

The entire energy infrastructure—production areas, pipelines, pumps, refineries, terminals, power plants, and so on—needs to be made more robust and made more resilient. This involves better intelligence and law-enforcement cooperation, enhanced protection of critical sites, and provision for the redundancy of critical components. There is also no substitute for the ability to protect and clear critical transitways.

Supply diversification and related measures have their limits, however. The price of oil reflects global supply and demand, so the price of oil will rise if more than a negligible amount of oil is taken off the market. In addition, the United States is in principle more vulnerable to supply interruptions given the rise in terrorism and the increased role of national oil companies, who are more likely to reflect government policy when it comes to making decisions about production and sales.

Another way to increase diversification of supply is to increase domestic production, which is now below 7 million barrels a day. Expressed differently, the United States imports some 2/3 of the oil it consumes. It is doubtful new drilling (even with new technologies that increase recovery rates) could appreciably affect this number given the falling output of many mature wells and fields and the growing domestic demand for oil. Still, the United States ought to increase the amount of exploration and development that it allows, especially in coastal areas. Again, though, no combination of diversification of external oil supplies and increased domestic production can satisfy the demands of a comprehensive energy security posture.

Alternative forms of energy, including coal, natural gas, nuclear, solar, wind, geothermal, and biofuels, are also central to any discussion of energy security. One reality to contend with though is the fact that most of the oil produced and imported is used in the transport sector—and that most of what fuels the transport sector is oil. Massive substitution is not a near-term option.

In the medium and long-term, fuel-efficient "pluggable" hybrids that use electrical power appear promising. So as well does cellulosic biomass, which can substitute in significant quantities for gasoline without disrupting food supplies or requiring anything near the amount of energy to produce corn-based ethanol. One short-term step that should be taken is the removal of the tariff on ethanol imports.

Coal is and will remain the principal fuel for electricity generation. It generates half the electricity in the United States. Coal is readily available in the United States as well as in both China and India. It is also relatively inexpensive. China is building large coal-fueled plants at the rate of two per week; India is building them at a rate closer to two per month. The problem is that coal is a major contributor to greenhouse gas emissions. As the recently-released MIT study *The Future of Coal* makes readily apparent, the climate change problem will continue to worsen unless something can be done about coal. The reality, though, is that there is no realistic alternative to coal; the principal question is whether technology can be developed, proven and introduced with sufficient speed and on a sufficient scale to capture and then sequester the massive amounts of carbon dioxide existing and planned plants will produce. Governments ought to work with industry in creating an investment and regulatory environment that accelerates the emergence, testing, and fielding of such technology in the United States and around the world. In the meantime, the government should only authorize the construction of coal plants that use the most advanced, efficient and clean technologies and that are designed to incorporate emerging technologies designed to capture carbon.

Nuclear power is the ideal form of fuel for electricity production given that it adds hardly at all to climate change. Nuclear power stations now provide some 20% of U.S. electricity. There will be hurdles to maintaining, much less increasing, this percentage. Politics is one problem. The last reactor to be completed was ordered nearly four decades ago and became operational in 1996. There are currently 103 reactors operating. Even with 20 year extensions of their planned lifespan all existing reactors will be decommissioned by the middle of this century. Just replacing them will require building two reactors a year for the next fifty years. It is not clear this rate of construction in the United States (coupled with ambitious building programs elsewhere) is sustainable. Indeed, a forthcoming study (*Nuclear Energy: Balancing Benefits and Risks*) written by Charles D. Ferguson of the Council on Foreign Relations concludes that "Nuclear energy is not a major part of the solution to further countering global warming or energy insecurity. Expanding nuclear energy use to make a relatively modest contribution to combating climate change would require constructing nuclear power plants at a rate so rapid as to create shortages in building materials, trained personnel, and safety controls." Other analysts are more bullish about the prospects for nuclear power, although even if they are correct it will not prove transformational for decades if then. In addition, a greater emphasis on nuclear power will raise security challenges as well as demands for safe storage of spent fuel.

In short, developing alternatives will over time make a difference. But no energy security policy can be considered comprehensive without a significant emphasis on reducing the consumption of oil and oil products. The United States daily consumes some 21 million barrels of oil and oil products. The policy question is how best to slow or better yet reverse this growth.

Increasing the tax on gasoline would have the most immediate impact. U.S. taxes (18.4 cents per gallon at the federal level) are low by world standards. If politics required, an increase in the federal fuel tax could be offset by reductions or rebates in other taxes or in designating revenues for energy-related investments.

Tightening fuel efficiency standards is a good mid-term approach given the time it will take for more efficient cars and trucks to be built and to replace the existing fleet. One area deserving exploration is what might be done to accelerate the replacement of low-mileage vehicles with hybrids and relatively fuel-efficient cars and trucks.

All of the above would affect climate change. Climate change policy, however, is something different. Congress and the administration should start developing guidelines for the post-Kyoto Protocol, post-2012 world. They should work with state governments, business, and academic experts. It is essential that the United States be a full participant in any negotiations and in any resulting regime—and that it approach such negotiations with a national policy in place. Developing countries need to be a central (although not necessarily equal) participant in a post-Kyoto framework. Some sort of carbon tax or cap and trade system will likely work best. Factored into any plan should be a positive credit for forested areas that absorb carbon dioxide. Even before then, U.S. aid policy should be adjusted to provide financial incentives to discourage deforestation and encourage reforestation.

We will also need to consider whether and how future trade negotiations and the WTO process itself address climate change. Many of the innovations that will reduce emissions (such as nuclear power stations and cleaner coal plants and capture and sequestration technology) are costly. Questions such as how to treat subsidies and the role (if any) of tariffs to deal with producers who give short shrift to climate concerns require study.

I want to close with a few thoughts on this subject. Despite the formal name of this select committee, "energy independence" is beyond reach if by independence is meant an ability to do without imports of oil and gas. A recent Task Force (*National Security Consequences of U.S. Oil Dependency*) sponsored by the Council on Foreign Relations concluded "During the next twenty years (and quite probably beyond) it is infeasible to eliminate the nation's dependence on foreign energy sources." A more useful and realistic task is how to manage energy dependence or, better yet, how best to promote energy security.

Similarly, energy security cannot be promoted through any single policy or breakthrough. Rather, what is required is a family of policies. The U.S. government will need to adjust to help bring this about. The creation of this select committee is a step in the right direction; so, too, would be a directorate in the National Security council staff devoted to energy security and the inclusion of the secretary of energy more regularly and centrally in national security meetings. Energy security properly defined is now too intimately a part of overall security to be left out of the most important deliberations of our country.

Robert Bryce

Gusher of Lies: The Dangerous Delusions of "Energy Independence"

Introduction

The Persistent Delusion

Americans love independence.

Whether it's financial independence, political independence, the Declaration of Independence, or grilling hotdogs on Independence Day, America's self-image is inextricably bound to the concepts of freedom and autonomy. The promises laid out by the Declaration—life, liberty, and the pursuit of happiness—are the shared faith and birthright of all Americans.

Alas, the Founding Fathers didn't write much about gasoline.

Nevertheless, over the past 30 years or so—and particularly over the past 3 or 4 years—American politicians have been talking as though Thomas Jefferson himself warned about the dangers of imported crude oil. Every U.S. president since Richard Nixon has extolled the need for energy independence. In 1974, Nixon promised it could be achieved within 6 years. In 1975, Gerald Ford promised it in 10. In 1977, Jimmy Carter warned Americans that the world's supply of oil would begin running out within a decade or so and that the energy crisis that was then facing America was "the moral equivalent of war."

The phrase "energy independence" has become a prized bit of meaningful-sounding rhetoric that can be tossed out by candidates and political operatives eager to appeal to the broadest cross section of voters. When the U.S. achieves energy independence, goes the reasoning, America will be a self-sufficient Valhalla, with lots of good-paying manufacturing jobs that will come from producing new energy technologies. Farmers will grow fat, rich, and happy by growing acre upon acre of corn and other plants that can be turned into billions of gallons of oil-replacing ethanol. When America arrives at the promised land of milk, honey, and supercheap motor fuel, then U.S. soldiers will never again need visit the Persian Gulf, except, perhaps, on vacation. With energy independence, America can finally dictate terms to those rascally Arab sheikhs from troublesome countries. Energy independence will mean a thriving economy, a positive balance of trade, and a stronger, better America.

The appeal of this vision of energy autarky has grown dramatically since the terrorist attacks of September 11. That can be seen through an analysis of news stories that contain the phrase "energy independence." In 2000, the Factiva news database had just 449 stories containing that phrase. In 2001, there were 1,118 stories. By 2006, that number had soared to 8,069.

The surging interest in energy independence can be explained, at least in part, by the fact that in the post–September 11 world, many Americans have been hypnotized by the conflation of two issues: oil and terrorism. America was attacked, goes this line of reasoning, because it has too high a profile in the parts of the world where oil and Islamic extremism are abundant. And buying oil from the countries of the Persian Gulf stuffs petrodollars straight into the pockets of terrorists like Mohammad Atta and the 18 other hijackers who committed mass murder on September 11.

Americans have, it appears, swallowed the notion that all foreign oil—and thus, presumably, all foreign energy—is bad. Foreign energy is a danger to the economy, a danger to America's national security, a major source of funding for terrorism, and, well, just not very patriotic. Given these many assumptions, the common wisdom is to seek the balm of energy independence. And that balm is being peddled by the Right, the Left, the Greens, Big Agriculture, Big Labor, Republicans, Democrats, senators, members of the House, George W. Bush, the opinion page of the *New York Times*, and the neoconservatives. About the only faction that dismisses the concept is Big Oil. But then few people are listening to Big Oil these days.

Environmental groups like Greenpeace and Worldwatch Institute continually tout energy independence. The idea has long been a main talking point of Amory Lovins, the high priest of the energy-efficiency movement and the CEO of the Rocky Mountain Institute. One group, the Apollo Alliance, which represents labor unions, environmentalists, and other left-leaning groups, says that one of its primary goals is "to achieve sustainable American energy independence within a decade."

Al Gore's 2006 documentary about global warming, *An Inconvenient Truth*, implies that America's dependence on foreign oil is a factor in global warming. The film, which won two Academy Awards (for best documentary feature and best original song), contends that foreign oil should be replaced with domestically produced ethanol and that this replacement will reduce greenhouse gases. (In October 2007, Gore was awarded the Nobel Peace Prize.)

The leading Democratic candidates for the White House in 2008 have made energy independence a prominent element of their stump speeches. Illinois senator Barack Obama has declared that "now is the time for serious leadership to get us started down the path of energy independence." In January 2007, in the video that she posted on her Web site that kicked off her presidential campaign, New York senator Hillary Clinton said she wants to make America "energy independent and free of foreign oil." Former North Carolina senator John Edwards believes the U.S. needs "energy independence from unstable and hostile areas of the world."

The Republicans are on board, too. In January 2007, shortly before Bush's State of the Union speech, one White House adviser declared that the president would soon deliver "headlines above the fold that will knock your socks off in terms of our commitment to energy independence." In February 2007, Arizona senator and presidential candidate John McCain told voters in Iowa, "We need energy independence. We need it for a whole variety of reasons." In March 2007, former New York mayor Rudolph Giuliani insisted that the federal government "must treat energy independence as a matter of national security." He went on, saying that "we've been talking about energy independence for over 30 years and it's been, well, really, too much talk and virtually no action. . . . I'm impatient and I'm single-minded about my goals, and we will achieve energy independence."

On April 26, 2007, another Republican candidate, Mitt Romney, the former governor of Massachusetts, used the *Jerusalem Post's* e-mail list to conflate the issues of oil, terrorism, Israel, and energy independence in a fund-raising appeal for his presidential campaign. The e-mail message, which showed a large picture of Iranian president Mahmoud Ahmadinejad, asked several questions, including "Do you believe that those who support terrorism against America and against the state of Israel should be held accountable?" The next question: "Do you agree that we must become energy independent and stop sending $1 billion a day to nations like Iran and Syria who use that money against us?" (Syria exports modest amounts of crude oil.)

The Democratic Party, which won control of the House and Senate in the November 2006 elections, has made energy independence a key talking point. About the time of the elections, Nancy Pelosi, the congresswoman from San Francisco who became Speaker of the House, issued the Democrats' "New Direction" agenda. The third point on that list—right after raising the minimum wage and repealing certain tax incentives—is "invest in research and development to promote energy independence." It says the Democrats will achieve energy independence "within ten years. We should be sending our energy dollars to the Midwest, not the Middle East. America's farmers will fuel America's energy independence."

A Democratic think tank, the Center for American Progress, which was created by a group of politicos from the Clinton administration, has launched a campaign called "Kick the Oil Habit," an effort that seems to imply America can quit using oil with the same ease that a smoker might give up cigarettes. In May 2006, the group's lead spokesman, actor Robert Redford, appeared on TV talk shows and wrote opinion pieces in which he said the U.S. should quit using oil altogether so that it can get away from "dictators and despots." The solutions proposed by Redford and the Democrats: more ethanol, biofuels, and hybrid vehicles. During an appearance on CNN's *Larry King Live*, Redford said that he supported corn ethanol production because "it's cheaper. It's cleaner. It's renewable. And you know what? It's American because we grow it."

In January 2007, Andy Grove, the former chairman of giant computer-chip maker Intel Corp., penned an opinion piece for the *Wall Street Journal* in which he decried the lack of progress toward energy independence: "Even though the importance of the energy independence issue has been recognized and emphasized by every president since 1974, our vital national objective is vanishing like a mirage in the distance." Grove went on to claim that our use of foreign energy "gives great power to other nations over our destiny."

In September 2007, S. David Freeman, a longtime advocate of renewable energy who once chaired the Tennessee Valley Authority and has headed several other electric utilities, released a book called *Winning Our Energy Independence: An Energy Insider Shows How*. Freeman's book calls for a multidecade effort to close America's older coal and nuclear power plants while focusing on more efficient plug-in hybrid cars. A press release publicizing the book says that "Freeman charges that the reason we aren't already using more renewable energy is that the oil companies and electrical utilities have waged a slick campaign to deceive Americans."

In October 2007, a book with a similar theme—*Freedom from Oil: How the Next President Can End the United States' Oil Addiction*—rose to number 8 on the *Washington Post's* bestseller list. The book, by David Sandalow, a senior fellow at the Brookings Institution and a former official in the Clinton administration, touts the potential of plug-in hybrid cars, biofuels, and fuel efficiency to cut America's oil consumption. The front cover of the book has a blurb from Al Gore which says that when Sandalow "writes about energy and the environment, we should all pay close attention."

Polls show that an overwhelming majority of Americans are worried about foreign oil. A March 2007 survey by Yale University's Center for Environmental Law and Policy found that 93 percent of respondents said imported oil is a serious problem and 70 percent said it was "very" serious. That finding was confirmed by an April 2007 poll by Zogby International, which found that 74 percent of Americans believe that cutting oil imports should be a high priority for the federal government. And a majority of those surveyed said that they support expanding the domestic production of alternative fuels.

The energy independence rhetoric has become so extreme that some politicians are even claiming that lightbulbs will help achieve the goal. In early 2007, U.S. Representative Jane Harman, a California Democrat, introduced a bill that would essentially outlaw incandescent bulbs by requiring all bulbs in the U.S. to be as efficient as compact fluorescent bulbs. Writing about her proposal in the *Huffington Post*, Harman declared that such bulbs could "help transform America into an energy efficient and energy independent nation."

While Harman may not be the brightest bulb in the chandelier, there's no question that the concept of energy independence resonates with American voters and explains why a large percentage of the American populace believes that energy independence is not only doable but desirable.

But here's the problem. . . . It's not and it isn't.

Energy independence is hogwash. From nearly any standpoint—economic, military, political, or environmental—energy independence makes no sense. Worse yet, the inane obsession with the idea of energy independence is preventing the U.S. from having an honest and effective discussion about the energy challenges it now faces.

[We] need to acknowledge, and deal with, the difference between rhetoric and reality. The reality is that the world—and the energy business in particular—is becoming ever more interdependent. And this interdependence will likely only accelerate in the years to come as new supplies of fossil fuel become more difficult to find and more expensive to produce. While alternative and renewable forms of energy will make minor contributions to America's overall energy mix, they cannot provide enough new supplies to supplant the new global energy paradigm, one in which every type of fossil fuel—crude oil, natural gas, diesel fuel, gasoline, coal, and uranium—gets traded and shipped in an ever more sophisticated global market.

Regardless of the ongoing fears about oil shortages, global warming, conflict in the Persian Gulf, and terrorism, the plain, unavoidable truth is that the U.S., along with nearly every other country on the planet, is married to fossil fuels. And that fact will not change in the foreseeable future, meaning the next 30 to 50 years. That means that the U.S. and the other countries of the world will continue to need oil and gas from the Persian Gulf and other regions. Given those facts, the U.S. needs to accept the reality of *energy interdependence*.

The integration and interdependence of the $5-trillion-per-year global energy business can be seen by looking at Saudi Arabia, the biggest oil producer on the planet. In 2005, the Saudis *imported* 83,000 barrels of gasoline and other refined oil products per day. It can also be seen by looking at Iran, which imports 40 percent of its gasoline needs. Iran also imports large quantities of natural gas from Turkmenistan. If the Saudis, with their 260 billion barrels of oil reserves, and the Iranians, with their 132 billion barrels of oil and 970 trillion cubic feet of natural gas reserves, can't be energy independent, why should the U.S. even try?

An October 2006 report by the Council on Foreign Relations put it succinctly: "The voices that espouse 'energy independence' are doing the nation a disservice by focusing on a goal that is unachievable over the foreseeable future and that encourages the adoption of inefficient and counterproductive policies."

America's future when it comes to energy—as well as its future in politics, trade, and the environment—lies in accepting the reality of an increasingly interdependent world. Obtaining the energy that the U.S. will need in future decades requires American politicians, diplomats, and businesspeople to be actively engaged with the energy-producing countries of the world, particularly the Arab and Islamic producers. Obtaining the country's future energy supplies means that the U.S. must embrace the global market while acknowledging the practical limits on the ability of wind power and solar power to displace large amounts of the electricity that's now generated by fossil fuels and nuclear reactors.

The rhetoric about the need for energy independence continues largely because the American public is woefully ignorant about the fundamentals of energy and the energy business. It appears that voters respond to the phrase, in part, because it has become a type of code that stands for foreign policy isolationism—the idea being that if only the U.S. didn't buy oil from the Arab and Islamic countries, then all would be better. The rhetoric of energy independence provides political cover for protectionist trade policies, which have inevitably led to ever larger subsidies for politically connected domestic energy producers, the corn ethanol industry being the most obvious example.

But going it alone with regard to energy will not provide energy security or any other type of security. Energy independence, at its root, means protectionism and isolationism, both of which are in direct opposition to America's long-term interests in the Persian Gulf and globally.

Once you move past the hype and the overblown rhetoric, there's little or no justification for the push to make America energy independent. And that's [my] purpose . . . to debunk the concept of energy independence and show that none of the alternative or renewable energy sources now being hyped—corn ethanol, cellulosic ethanol, wind power, solar power, coal-to-liquids, and so on—will free America from imported fuels. America's appetite is simply too large and the global market is too sophisticated and too integrated for the U.S. to simply secede.

Indeed, America is getting much of the energy it needs because it can rely on the strength of an ever-more-resilient global energy market. In 2005, the U.S. bought crude oil from 41 different countries, jet fuel from 26 countries, and gasoline from 46. In 2006, it imported coal from 11 different countries and natural gas from 6 others. American consumers in some border states rely on electricity imported from Mexico and Canada. Tens of millions of Americans get electricity from nuclear power reactors that are fueled by foreign uranium. In 2006, the U.S. imported the radioactive element from 8 different countries.

Yes, America does import a lot of energy. But here's an undeniable truth: It's going to continue doing so for decades to come. Iowa farmers can turn all of their corn into ethanol, Texas and the Dakotas can cover themselves in windmills, and Montana can try to convert all of its coal into motor fuel, but none of those efforts will be enough. America needs energy, and lots of it. And the only way to get that energy is by relying on the vibrant global trade in energy commodities so that each player in that market can provide the goods and services that it is best capable of producing. . . .

EXPLORING THE ISSUE

Is It Realistic for the United States to Move Toward Greater Energy Independence?

Critical Thinking and Reflection

1. In what sense is (or is not) imported oil linked to terrorism?
2. If we choose to develop alternative energy technologies, it will still take many years to reduce dependence on oil. Why?
3. What is "energy interdependence"?
4. In what sense does arguing for "energy independence" amount to arguing for foreign policy isolationism?

Is There Common Ground?

Both sides in this debate agree that the United States is dependent on foreign energy (oil) sources. One side insists that we can reasonably speak of moving toward "energy independence." The other insists that the very term "energy independence" is nonsense, for the world's nations are inescapably interdependent in energy matters.

1. Imagine that we build a vast network of solar and wind power facilities. Neither sun nor wind are "on" all the time, but if the network is spread over enough of the Earth's surface, transmission lines can move energy from wherever it is being produced to where it is needed. In what sense does this illustrate "energy interdependence"?
2. In Europe, the Desertec proposal to install vast arrays of solar collectors in the Sahara desert is moving forward (see http://www.guardian.co.uk/business/2009/nov/01/solar-power-sahara-europe-desertec). Does this provide Europe with "energy independence"?
3. In Australia, a solar updraft tower is being developed (see http://www.enviromission.com.au/EVM/content/home.html). Might this technology provide a major step toward "energy independence"?

ISSUE 3

Is America Ready for the Electric Car?

YES: Michael Horn, from "Roadmap to the Electric Car Economy," *The Futurist* (April 2010)

NO: Rick Newman, from "A Stuttering Start for Electric Cars," *U.S. News & World Report* (April 2010)

Learning Outcomes

After studying this issue, students will be able to explain:

- The advantages of electric vehicles over fossil fuel–powered vehicles.
- The shortcomings of electric vehicles.
- Why even perfect electric vehicles will not replace fossil fuel–powered vehicles quickly.
- Why electric-vehicle charging systems should be standardized.

ISSUE SUMMARY

YES: Michael Horn argues that the technology already exists to replace gasoline-burning cars with electric cars and thereby save money, reduce dependence on foreign oil sources, and reduce pollution. All we need is organization and determination.

NO: Rick Newman argues that because electric car technology is still new, expensive, and unreliable, it will be at least a decade before consumers are willing to shift from gas burners to electric cars.

The 1973 "oil crisis" heightened awareness that the world—even if it was not yet running out of oil—was extraordinarily dependent on that fossil fuel (and therefore on supplier nations) for transportation, home heating, and electricity generation. Since the supply of oil and other fossil fuels is clearly finite, some people have worried that there would come a time when demand could not be satisfied, and our dependence would leave us helpless. At the same time, we have become acutely aware of the many unfortunate side-effects of fossil fuels, including air pollution, strip mines, oil spills, global warming, and more.

The 1970s saw the modern environmental movement gain momentum. The first Earth Day was in 1970. Numerous steps were taken by governments to deal with air pollution, water pollution, and other environmental problems. In response to the oil crisis, a great deal of public money went into developing alternative energy supplies. The emphasis was on "renewable" energy, meaning conservation, wind, solar, geothermal, and tidal energy. However, when the crisis passed and oil supplies were once more ample (albeit it did cost more to fill a gasoline tank), most public funding for alternative-energy research and demonstration projects vanished. What work continued was at the hands of a few enthusiasts and those corporations that saw future opportunities.

The electric car is not a new idea. Such cars existed in the early years of the twentieth century, although they soon gave way to gasoline-burning cars, which could run longer on a charge (or fill-up). The idea has been revived in recent decades, however. Its most common form is the gas-electric hybrid (e.g., the Toyota Prius), but all-electric cars have been built. GM's EV never hit the market (see the documentary *Who Killed the Electric Car?* available on DVD), but the Chevy Volt (www.chevrolet.com/pages/open/default/fuel/electric.do) is already rolling off assembly lines. Unfortunately, perhaps, the Chevy Volt is not a pure electric car; it is a "plug-in hybrid," meaning that its battery can be charged from a wall socket, but it also uses a gas-powered engine to charge its battery while on the road. Several all-electric cars will be at dealers soon, but they will not be cheap. Improvements in price and performance are frequently promised, but they depend mostly on finding a way to store an amount of electrical energy equivalent to the energy content of a tankful of gasoline in a compact package that does not cost a huge amount to replace. A great deal of research effort is currently devoted to improving batteries and developing alternative electricity-storage devices such as supercapacitors. Entrepreneurs are also working on battery-leasing plans and other methods of making electric cars more affordable.

The prospective benefits of electric cars are well established. In August 2009, the Center for Entrepreneurship & Technology at the University of California, Berkeley, published Thomas A. Becker, Ikhlaq Sidhu, and Burghardt Tenderich, "Electric Vehicles in the United States: A New Model with Forecasts to 2030" (http://cet.berkeley.edu/dl/CET_Technical%20Brief_EconomicModel2030_f.pdf). Among their premises were that electric cars will have batteries that are not charged in situ but are swapped for charged batteries (Shai Agassi is working to establish such a system in Israel at the moment; see Steve Hamm, "The Electric Car Acid Test," *Businessweek* [February 4, 2008]) and that charging networks will be financed by pay-per-mile contracts. In their baseline forecast, by 2030 electric cars will account for 64 percent of U.S. light vehicle sales. Of light vehicles on the road, electrics will be 24 percent. Oil imports will be almost 40 percent lower than if all we do is improve the efficiency of internal combustion engines. The annual trade deficit will be down by as much as a quarter trillion dollars. Employment will be boosted by as much as 350,000 jobs. If the electric cars are charged using nonpolluting sources of electricity (wind, solar, etc.), there will be $100–200 billion in savings in health care costs due to reduced pollution. And carbon dioxide emissions will be greatly reduced. Unfortunately, the extent of those emissions reductions depends on where you live and how your electricity is generated; see Michael Moyer, "The Dirty Truth about Plug-In Hybrids," *Scientific American* (July 2010). For an overview of what is now available, see "Should You Plug In?" *Consumer Reports* (October 2010).

In 2010, the National Research Council's Committee on Assessment of Resource Needs for Fuel Cell and Hydrogen Technologies published "Transitions to Alternative Transportation Technologies—Plug-in Hybrid Electric Vehicles" (www.nap.edu/catalog.php?record_id=12826). The more cautious focus—on plug-in hybrids instead of all-electric cars—makes the results not quite comparable to the earlier report. In addition their conclusions are less optimistic. For instance, they forecast that the maximum practical market penetration would result in 2030 in an on-the-road fleet that is no more than 15 percent plug-in hybrids. There will still be marked reductions in gasoline use, pollution, and carbon emissions.

According to Julian Edgar, "Electric Cars Now!" *Autospeed* (http://autospeed.com/cms/A_111205/printArticle.html), recent improvements in battery and charging technology mean that "The arguments against electric cars are now so weak they're effectively gone." Indeed, it is hard to find anyone who seriously argues that electric cars are a bad idea. The real question is whether people will buy them, and costs of vehicles and batteries will play a large part in shaping the answer to this question. Reliability of the technology may be demonstrated very soon. The Chevy Volt—a plug-in hybrid—will be on the market in 2011. Also in 2011, Nissan's all-electric Leaf is scheduled to hit the U.S. market; see http://nissanusa.com/leaf-electric-car. More electric cars are due soon from Ford, Hyundai, Volvo, Toyota, and Audi. The next few years will tell the tale.

In the YES selection, retired aerospace scientist Michael Horn argues that the technology already exists to replace gasoline-burning cars with electric cars and thereby save money, reduce dependence on foreign oil sources, and reduce pollution. All we need is organization and determination. In the NO selection, Journalist Rick Newman argues that because electric car technology is still new, expensive, and unreliable, it will be at least a decade before consumers are willing to shift from gas burners to electric cars. It will be longer than that before they replace more than a small fraction of the gas burner fleet.

YES 

Michael Horn

Roadmap to the Electric Car Economy

By the middle of this century, the United States may have completely transitioned from gasoline to electric vehicles, or EVs. Its economy will then enjoy an EV-energy bonus, somewhat like the peace bonus at the end of the Cold War, but this one will result from saving half of the money that U.S. consumers previously spent on oil imports to make gasoline for all their cars.

By then, the bankruptcies of GM and Chrysler will be a long-forgotten anomaly in the history of the auto industry, because once the auto companies replace the 254.4 million gasoline-powered cars in the United States with electric ones by mid-century, they will create a manufacturing boom that completely wipes out the losses that they sustained during the 2008 recession.

These electric cars will have come a long way from the twentieth-century electric prototypes that required drivers to stop frequently to recharge the batteries, making highway driving nearly impossible. In the coming decades, highways might be outfitted with guardrails that emit harmless radio-frequency charging waves. Special coils inside each car will receive the waves and harness them to recharge the battery continuously while the car is driving. The human passengers will have the luxury of never having to stop for a refueling.

The chore of pumping gas will seem akin to shoveling coal, and the ignominy of driving out of a gas station with the smell of gasoline on your hands from the pump handle will be memories only older folks might recall. The electric car, emitting no fumes and being virtually silent in operation, will make the gasoline-powered cars of the past seem utterly primordial by comparison.

Late in the century, historians may even debate why it took so long to make the transition from oil to electricity, when oil was so inefficient, involved so much peril, and all the elements necessary for transition to electric were well in hand by the end of the twentieth century.

Gasoline's Problems

But back to the present, where for the last 100 years, the U.S. economy has continued to depend on the gasoline-fueled, internal-combustion engine for powering cars and trucks. This engine is only a small technological step above the steam engine. The fact that it's primitive is not the problem; it's that it is so inefficient. For every $1 that we spend on gas, 85¢ is wasted in heating the engine block and the surrounding air (the reason it needs both an oil and water pump to cool it); only 15¢ goes into moving the car down the road.

Obtaining the gasoline is fraught with serious environmental and political perils. We are facing many questions concerning global warming, not the least of which is how much automobile exhaust contributes. Then there's the question of what happens to our frail economy if our oil supplies are disrupted. But our societal attitude is curiously sanguine concerning how imperiled we make ourselves when we rely so heavily upon the gasoline-powered engine. We seriously need to reconsider our options.

Continued Dependence On Oil

Over the last 30 years, each American president has vowed to reduce the nation's consumption and dependence on oil (i.e., gasoline) by increasing gas mileage of all cars and light trucks through a program called CAFE (Corporate Average Fuel Economy). After 30 years, however, the United States is still locked into an oil economy, and CAFE, which has done little to alleviate the situation, continues to be its only option for reducing oil imports.

President Obama has continued to follow the course of previous presidents and recently raised the CAFE standards from 27 mpg to 35 mpg. He also enacted a $3 billion cash-forclunkers program in an effort to get gas-guzzlers off the road. At the same time, billions are being spent on alternative-fuel cars, with research money still being spent on getting fuel out of corn. America is desperately trying to wean itself off oil imports from unstable sources around the world, but has yet to come up with a winning formula for doing so. When the CAFE was introduced in the 1970s, the average fuel economy mandated was 18 mpg. Now more than three decades later the fuel standards have gotten to 35 miles per gallon.

At that rate of development, it would take about 350 years for the internal-combustion engine cars to reach the more than 200 mpg efficiency already realized by battery-powered cars today! So why not put all available resources into making the electric car the successful replacement to the gasoline-powered car? It's been tried before, and we see the result all around us: Electric cars are far from becoming mainstream.

The problems were not so much with the cars, but with the environment into which the cars were cast. Back in 1990, the California Air Resources Board (CARB) introduced new regulations mandating that progressively larger shares of auto manufacturers' fleets of cars must produce zero emissions: By 1998, according to the CARB mandate, 2% of all the vehicles on California's roadways would have to be zero-emissions vehicles (ZEVs); by 2003, 10%. The only ZEVs circa 1990 were battery-powered cars. So, somewhat reluctantly, the auto industry began projects to develop battery-powered cars. GM and Toyota produced very impressive electric cars (based on relatively primitive battery technology compared to today) that eventually attained ranges of over 100 miles.

After huge investments and low sales, however, General Motors and Chrysler filed a lawsuit against CARB in 2002 to reverse the ZEV regulations. The auto companies got their way, and CARB was forced to repeal its mandates for 2003 and 2004. The program resumed in 2005, but with vastly lowered expectations. Today, of the more than 28 million cars registered in California, only 30,400 of them—less than a tenth of one percent—run on batteries.

Regardless of the reasons, the electric cars were doomed from the start because there were few places to recharge them, recharging them was time consuming, and the vehicles have limited range.

The electric car still faces the obstacle of having places to recharge, and now the Obama administration is putting money into electric car technology research without, once again, providing for an infrastructure to support it. It's like planning an aquarium and buying the fish before you build the tanks for the fish to swim in. Right now the electric car is like a fish out of water.

The other side of this issue is the auto industry, which has been burned by the hundreds of millions of dollars lost in the last round of electrics that failed to capture the imagination of all but a small sector of the car-buying public. Auto manufacturers realize the issues and limitations of the electric car and don't want to invest millions or maybe billions more in a car that might not sell. So the bottom line becomes this: Above and beyond getting society away from dependence on oil, what makes the electric car worth all the trouble?

Benefits of the Electric Car

The thing that makes the electric car worth all the trouble is its efficiency. Money and energy are interchangeable. When we waste energy, we waste money, and we waste a tremendous amount of money on oil. Half of all the oil that the United States imports goes into making gasoline. Switching to electric cars would eliminate about half of U.S. oil consumption and dependency.

According to the Energy Information Administration, America used about 137.8 billion gallons of gasoline in 2008. With an average price of $3.55 per gallon that year, the nation spent nearly half a trillion dollars on gasoline! But the worst part is that, since the gasoline engine is only 15% efficient, 85% of the money spent went up in smoke. That's $400 billion that was simply wasted. Any economy that has to contend with that degree of wasted resources is going to be in trouble, sooner or later.

In EVs' favor, most of the extra electricity needed to charge up the new swarm of electric cars wouldn't be coming from burning oil, because it's not the fuel of choice for the electric power industry. Moreover, most of the cost to charge up a battery-powered electric car actually goes into moving the car and is not wasted. That's what I mean by efficiency. Even the charging process is pretty efficient, at better than 85%, according to a study by Sandia Labs.

Along with all the efficiency there's also the immediate and complete cessation of direct air pollution. Despite what detractors of the electric car say, there is no pollution coming out of an electric car itself. There may be pollution created at the point where fossil fuels are burned to make electricity, but at least there are not engines pumping out fumes on every street corner of your city. And besides, there are ways to cap pollution from power plants and keep it from escaping into the atmosphere.

Making the EV Future Viable

Despite public perception of electric vehicles as slow, ugly "golf carts," they can be made to accelerate as fast as any gasoline-powered cars. In fact, an electric motor can accelerate a car faster than any gas engine, and it has done so many times—just check out the YouTube video, "unveiling of the Tesla." The electric car is also virtually silent and very clean, with no oil drips on the driveway or fumes filling up the garage while you warm up the engine in the winter—their motors don't even need to be warmed up!

Electric cars only expend energy when they're moving, so if you're stuck in a traffic jam, at least you can take comfort in knowing that your electric isn't using any energy at all. Also, it's nearly as efficient going 50 mph as it is going 5 mph. Electric cars can be made large or small, and can carry as many people and look as cool or sophisticated as any gas-powered car. They can be made in two-wheel or four-wheel drive. Electric motors are small enough to be the basis of more-flexible interior cabin designs.

The electric motor will also last much longer than its gasoline-powered counterpart because it contains only one moving part. The electric car will be another electrical appliance that runs for years with virtually no maintenance. Just think, when was the last time you needed to have your clothes dryer taken in for an oil change or new spark plugs? They both have similar electric motors. A gasoline car needs something done to it starting around the first three months and then every 3,000 miles—from oil changes to timing-belt replacements, ad nauseam. That in itself is part of the problem in breaking the auto industry away from the gas-engine cars: The industry's business model is based as much on service and parts as it is on sales. Look how much more space at the dealership is devoted to the service and parts department compared to the sales showroom. Clearly, the auto industry will need a better business model than the planned obsolescence that got it so much mileage in the 1950s, '60s, and '70s.

Being so reliable, the electric car may spawn some other novel approaches that would change the way we look at cars. For example, as styling fads change over a few years, instead of buying whole new cars, vehicle owners might just have the recyclable, outer body panels swapped to keep up with the latest trends or just for the sake of change. The auto industry would realize a way to augment money lost on service and parts by marketing newly styled outer panels over the durable electric-drive machinery. GM proposed such a swappable body in its "Skateboard" design, built around a fuel-cell electric vehicle.

Prospects for an "Electric Car Economy"

The electric car economy is one that has eliminated its dependence on oil and transitioned to a vastly more efficient means of powering its cars and trucks. It has removed the addiction to oil that binds consumers to perilous sources of energy. It has removed the issues that distract researchers from developing alternative sources of energy. And it's an economy that provides everyone, businesses and individuals alike, with hundreds of billions of dollars to spend on other things besides the oil they formerly imported to make gasoline. It's an economy that's pumped up by its newly found efficiency. But to create the electric car economy, we first have to make the electric car a real and complete replacement for the gas-powered car.

The first thing required to make the electric car a viable replacement for the gasoline-powered car is to create the infrastructure that supports the electric car. This infrastructure will be, by its very far-flung, distributed nature, enormous. According to the U.S. Census Bureau, there are 117,000 gas stations in the United States—or as they prefer to put it, one gas station for every 2,500 people. That's the reason that the gasoline-powered car is so successful—it has a vast gasoline distribution infrastructure.

To be as successful, the electric car needs a comparable infrastructure, but we won't need to start from scratch. Electricity is much simpler than liquid or gaseous fuels, which need to be piped or trucked and then stored in huge tanks at the point of sales. Electricity is available everywhere through power lines. Gas stations already have plenty of electric power to run anywhere from four to 12 big pumps and dozens of high-wattage light fixtures. Every gas station will need to conform to a government regulation to have at least one electric car-battery charger to begin with, by a reasonable deadline. Eventually, they'll need more than one charger per station, as people catch on that the electric car can now go further than the city limits. The planned Smart Grid will help accommodate the gradually rising load.

In order for the plan to work, these battery chargers will have to charge a car battery quickly, unlike the home plug-in power sources that take all night. Fortunately, the rapid-charge technology is a reality right now. In fact, one such system, called PosiCharge, was able to charge an electric car battery made by Altairnano in less than 10 minutes.

Standardization is another issue that must be addressed. The batteries and charging-control modules on electric cars have to be standardized so that any vehicle can be recharged anywhere. The plan won't work if you need to go to special recharging stations for a charger that works for your car. At present, different automakers are planning to use different chemistries and internal structures in their batteries. Consequently, not every battery would fit a given service station's charging system. This would put an electric car in the same position a gas-powered car would be in if there was no gas station in the community. How far would a gas-powered car go if there were no gas stations? The government and the auto industry will need to implement standards so that there is a uniform infrastructure of chargers and a uniform code, and that cars everywhere comply with the uniform code. That way, drivers everywhere will be able to quickly charge up their cars without damaging them.

My idea is really quite simple, but it will require some government regulations: Put rapid electric chargers in every gas station, because that's where people already go for gas. Drivers can pull into their local gas station, just as they always had in the past, and quickly charge up their shiny new electrics. Nothing much changes for the driving public except that everyone will be saving that huge amount that was once wasted on gasoline.

The advantages of promoting an all-electric-car economy are manifold, including the enormous savings in money and the substantial reduction in need for foreign oil supplies, which threaten further economic distress. Air quality would improve significantly. The electric-car economy represents a monumental plan that doesn't actually cost that much—an estimated $3.5 billion. With the right organization and determination, a future of clean, efficient, oil-free electric cars is attainable.

Rick Newman

A Stuttering Start for Electric Cars

YOUNG TECHNOLOGY MEANS CONSUMERS MAY SIT ON THE SIDELINES

You might get the impression, over the next year or so, that driving as you know it will never be the same again. The electric vehicle, you see, is about to arrive.

The much-heralded Chevrolet Volt, able to travel up to 40 miles on a battery charged through a household outlet, is set to go on sale late this year. So is the Nissan Leaf, with a range of up to 100 miles. A plug-in version of the Toyota Prius will be right behind them. Once the numbers are official, fuel economy for these newfangled machines, when converted to conventional measures, could easily exceed the mystical mark of 100 miles per gallon.

Such eye-popping mileage, plus a dramatic cutback in tailpipe emissions, represents the kind of automotive breakthrough that clean energy advocates have been seeking for decades. And now, they've got powerful allies in government and industry. The Obama administration has offered generous subsidies to manufacturers that build electric vehicles and consumers who buy them. In response, many automakers are ramping up their EV plans, with venture capitalists starting to gamble on electric cars, too. And a number of cities are developing plans to build charging stations and an EV-friendly infrastructure.

The only question is whether drivers will go along. They may just sit on the sidelines as electric vehicles start to roll off the assembly lines. Despite loads of hype and a government jump-start, price premiums will still be high, even with government subsidies. Practical limitations may force Americans to drive differently. And automakers themselves will develop other innovations that buyers may prefer. "Consumers are happy with hybrids and gas-powered cars that get 30 miles per gallon," says Mike Omotoso, a powertrain analyst with J. D. Power & Associates. "As long as gas stays below $3 per gallon, it will probably stay that way."

On paper, the appeal of EVs is, well, electrifying. Unlike hybrids, which are powered by a gas engine and a battery-powered electric motor working more or less in tandem, electrics will run purely on battery power, at least part of the time. That will allow them to take advantage of electricity rates that, mile for mile, are significantly cheaper than gas. General Motors has designed the Volt to travel up to 40 miles on an electrical charge and be chargeable from a 120-volt household outlet, the kind you use to plug in your washer or dryer. Once the battery runs low, a gas-powered engine kicks in and powers a generator that can move the car for about 300 miles more.

The Prius Plug-In will go 10 to 15 miles on battery power before reverting to the same hybrid operation as in the conventional Prius. On the Volt and the Prius, the gas-powered engine spares drivers "range anxiety": the worry that your car will run out of juice before you reach your destination. The Leaf, by contrast, would be a "pure electric" vehicle able to travel 100 miles or so on a charge—but without a second power source to prevent drivers from being stranded. "Don't worry, there will be lots of charging reminders," says Mark Perry, a top product planner at Nissan.

Saving Money

The potential savings in fuel costs are dramatic. The government hasn't yet developed official mileage ratings for electric vehicles, but General Motors drew headlines last year when it said that the Volt could get the equivalent of 230 mpg in city driving. Nissan scoffed at GM's methodology, saying that by the same metrics the Leaf would get 300 mpg. Tony Posawatz of GM's Volt program acknowledges that the Volt's official mileage number could be lower than 230 mpg, because it will have to account for time spent tooling around on the less efficient gas engine. But overall mileage, he says, "will be better than any conventional car out there." That means the Volt would handily beat the current Prius, which has a combined 50 mpg and is the reigning gas-mileage champ among ordinary production cars.

Comparing EVs with conventional cars is bound to be confusing, with a learning curve that regulators and consumers will have to scale together. The fuel economy for the $109,000 Tesla Roadster, for example—the only full-speed electric vehicle on U.S. roads today—is 110 watt-hours per kilometer, allowing it to travel 244 miles on a charge that takes about 3½ hours. Enthusiasts estimate that the mpg-equivalent ranges from 135 mpg to 400 mpg, depending on how the equivalency between gas and electricity is computed. (The Roadster also rockets from zero to 60 in 3.9 seconds, highlighting one feature of electric powertrains that drivers will love: instantly available torque, which translates into zesty acceleration.)

Drivers may ultimately compare fuel efficiency through another measure: cost per mile. If gas costs $3 per gallon and typical mileage is 25 mpg, it will take 4 gallons of gas to go 100 miles, at a total cost of $12. The per-mile cost is one hundredth of that, or 12 cents per mile. For the Prius, at 50 mpg, fueling costs are half that: 6 cents per mile.

GM says that 100 miles of battery-powered travel in the Volt will require about $2.75 worth of electricity. That adds up to less than 3 cents per mile based on average electricity rates of about 11 cents per kilowatt-hour. If drivers charge their Volt overnight, when off-peak electricity rates are much lower, the cost could be as little as 1 cent per mile, according to Posawatz. At those rates, the annual cost savings on fuel could easily be more than $1,000 in typical driving.

Potential Problems

That's on paper. In the real world, lots of unanticipated things can go wrong. Toyota's recent recalls of millions of vehicles, for instance, involved gas pedals and brakes, hardly exotic components. Some experts think electronics may have played a role in the "sudden acceleration" reported by some Toyota owners, which highlights the risks that come even with the gradual introduction of new technology.

Electric vehicles represent an entirely new set of hardware and software, and it could take years to work out the bugs. Lithium-ion batteries, for instance, have been used for years in cellphones and laptops, but scaling them up for use in cars is a technological leap that's only in the early stages. Poor driving techniques or other unforeseen factors could produce driving range that's lower than manufacturers expect. And any bad press could dent public confidence in the precocious cars.

Then there's the cost of the vehicles themselves. Consulting firm CSM Worldwide estimates that new EV technology generates a price premium of about $20,000 per vehicle, mostly on account of the costly batteries. With gas at $3, it would take nearly 15 years, longer than the life span of most cars, to earn that back through savings on fuel. That's why manufacturers will eat some of the cost, and a $7,500 federal tax credit will bring prices down further. But EVs will still be expensive compared with regular cars. Manufacturers haven't announced prices yet, but the Volt is likely to cost about $40,000 before the tax credit. The Leaf could be $30,000 or so. Consumer reluctance to pay extra for unproven technology is one reason the take rate for EVs is expected to be very low. J. D. Power estimates that by 2015, plug-ins and pure electrics will account for just 0.3 percent of all car sales, or approximately 50,000 vehicles. The Prius alone tallies about 150,000 in annual sales today.

EVs could still change the automotive landscape, however, and the history of hybrids helps explain how. Toyota and Honda began developing hybrids in the early 1990s, with help from the Japanese government. In the United States, the Honda Insight appeared in 1999, the Prius in 2000. While praised for their fuel-sipping engines, as automobiles, they were widely panned. The Insight two-seater was odd and impractical. The pod-shaped Prius was slow and unsure on the highway. Still, some environmentalists flocked to the hybrids, and the unexpected endorsement of a few celebrities pushed Prius sales above expectations.

A Bigger Impact

Toyota lost money on the Prius for at least a decade, and only now does it say its hybrids are profitable. And the Prius has grown from a niche vehicle into Toyota's No. 4 seller. The Prius also paved the way for hybrid versions of the Camry, Highlander, and several Lexuses. Competitors like GM, which once dismissed hybrids as a glorified science experiment, now embrace them. Overall, hybrids represent nearly 3 percent of the market today, and J. D. Power predicts they'll account for nearly 9 percent of sales in 2015.

EVs could make a bigger impact sooner. GM actually developed an electric vehicle in the 1990s—the EV-1—which it abandoned because of low gas prices, weak consumer interest, and the lack of a backup engine. The technology is back because of tough new fuel efficiency and pollution requirements, adopted in both the Bush and Obama administrations, that will force automakers to find new ways to boost fuel economy 4 percent per year, on average, through 2015—an aggressive target that can't be met simply by tweaking existing engines.

But the American market may be the last to see a widespread rollout of EVs, largely because of low gas prices. In Europe and Japan, high taxes keep gas prices well above $6 per gallon, which cuts in half the "payback period" required to recoup the added cost of new technology. With gas prices at $6, for example, an EV would pay for itself in fuel savings after 6.2 years, according to CSM; with $8 gas, that drops to 4.5 years. And greater sales always bring down the cost per unit. If EVs catch on, the added cost could fall by 15 percent a year or so, which is similar to the cost decrease for hybrids and other new technology.

CSM and other forecasting firms believe electric vehicles will start to penetrate the mainstream market by 2020 or so, as costs come down, the technology becomes more reliable, and consumers make the mental shift into a new driving paradigm. In the meantime, collateral benefits of electrification will filter into the mainstream fleet in smaller ways. More cars, for example, will start to feature "start stop" systems, standard on hybrids, which shut down the engine when the car is idling and can boost fuel economy by 5 percent or more. Virtually every automaker plans to roll out more hybrids. There will also be more "clean diesels" and perhaps ethanol-powered vehicles.

The most striking thing about the auto fleet in 2015 will be the shrinking proportion of conventional gas-powered cars, which J. D. Power expects to be just 73 percent, down from 89 percent today. That in itself could foretell an automotive revolution. But what will fuel it is unknown.

EXPLORING THE ISSUE

Is America Ready for the Electric Car?

Critical Thinking and Reflection

1. What are the advantages of electric cars?
2. What are the disadvantages of electric cars?
3. How will standardization benefit the electric car economy?
4. What can government do to encourage adoption of electric cars?
5. Would electric vehicles be more acceptable in a less spread-out society (such as, for instance, Holland)?

Is There Common Ground?

There is little disagreement on the basic question of whether electric vehicles are a "good idea." However, optimists see the technology as ready to be adopted, with the big question being why society is not rushing to adopt the technology. Pessimists do not think the technology is quite ready yet.

1. When the gasoline-powered car was new, there was very little infrastructure to support it. How much paved road was there? Where were gas pumps located, and how common were they?
2. Electric vehicles will also need infrastructure. Roads exist, but what about charging stations? How many will be needed? Where will they be? And how long will it take to provide enough?
3. Do electric vehicles seem more acceptable if you consider only urban driving? Why?

ISSUE 4

Is Shale Gas the Solution to Our Energy Woes?

YES: Diane Katz, from "Shale Gas: A Reliable and Affordable Alternative to Costly 'Green' Schemes," *Fraser Forum* (July/August 2010)

NO: Deborah Weisberg, from "Fracking Our Rivers," *Fly Fisherman* (April/May 2010)

Learning Outcomes

After studying this issue, students will be able to:

- Explain how shale gas threatens water supplies and even human health.
- Discuss the relative merits of putting energy-related decision making in the hands of private investors or government.
- Explain how an influx of cheap fossil fuel will affect the development of alternative (e.g., wind and solar) energy systems.
- Discuss whether increasing prices through regulation of an industry, in order to protect public health, is justifiable.

ISSUE SUMMARY

YES: Diane Katz argues that new technology has made it possible to release vast amounts of natural gas from shale far underground. As a result, we should stop spending massive sums of public money to develop renewable energy sources. The "knowledge and wisdom of private investors" are more likely to solve energy problems than government policymakers.

NO: Deborah Weisberg argues that the huge amounts of water and chemicals involved in "fracking"—hydraulic fracturing of shale beds to release natural gas—pose tremendous risks to both ground and surface water, and hence to public health. There is a need for stronger regulation of the industry.

Fossil fuels have undeniable advantages. They are compact and easy to transport. In the form of petroleum and natural gas and their derivatives, they are well suited to powering automobiles, trucks, and airplanes. They are also abundant and relatively inexpensive, although the end of the era of oil abundance is in sight, and prices are rising. However, fossil fuels also have disadvantages, for their use puts carbon dioxide in the air, which threatens us with global warming. Oil is associated with disastrous oil spills such as the one that resulted from the failure of a British Petroleum drilling rig in the Gulf of Mexico in 2010. Coal mining leaves enormous scars on the landscape, and coal burning emits pollutants that must be controlled. Natural gas alone seems relatively benign, for although it emits carbon dioxide when burned, it emits less than oil or coal. It produces fewer air pollutants, it cannot be spilled (if released, it can cause explosions and fires, but outdoors it mixes with air and blows away), and obtaining it has not meant huge damage to the environment. Much of the United States' demand for natural gas is met by domestic production, but demand is rising and imports—now at about 15 percent of demand—will have to rise to keep up, just as they have with oil. Lacking new sources of natural gas or a shift to coal, nuclear power, or alternatives such as wind and solar power, the nation must inevitably become more dependent on foreign energy suppliers.

It has long been known that large amounts of "unconventional" natural gas reside in deep layers of sedimentary rock such as shale. However, this gas could not be extracted with existing technology, at least not at a price that would permit a profit once the gas was sold. In recent years, this has changed, for drilling technology now allows drillers to bend drill holes horizontal to follow rock layers. Injecting millions of gallons of water and chemicals at extraordinarily high pressure can fracture (or "frack") the rock surrounding a drill hole and permit trapped gas to escape. See Richard A. Kerr, "Natural Gas from Shale Bursts onto the Scene," *Science* (June 25, 2010). Mark Fischetti, "The Drillers Are Coming," *Scientific American* (July 2010), notes that the Marcellus shale formation, which stretches from upstate New York through Pennsylvania to Tennessee, may contain enough gas to meet U.S. needs for 40 years. There are other shale formations in the United States, Canada, and Europe. The total U.S. supply may be enough to meet needs for a century; see Steve Levine, "Kaboom!" *New Republic* (May 13, 2010). See also Paul Stevens, "Cheap Gas Coming?" *World Today* (August–September 2010).

Not surprisingly, many people are concerned about the environmental impacts of "fracking" and disposing of used water, chemicals, and drilling wastes. Richard A. Kerr describes threats to groundwater in "Not Under My Backyard, Thank You," *Science* (June 25, 2010). But the industry insists that it will deal responsibly with its wastes and hastens to reassure people living near drilling sites. Alex Halperin, "Drill, Maybe Drill?" *American Prospect* (May 2010), describes the debate over shale-gas drilling in upstate New York. The area has suffered large job losses, something the shale-gas industry may remedy. Many landowners—including farmers—see the potential for huge boosts to their income. But the industry has reportedly persuaded people to lease drilling rights on their property by making promises that cannot be kept.

Environmental impacts are a huge concern. So far, regulations are slowing development of the industry in New York—a moratorium on drilling permits has been proposed; see Theresa Keegan, "Controversy Rages in Hydro-Fracking Debate," *Hudson Valley Business Journal* (June 28, 2010). In neighboring Pennsylvania, the drillers are already producing large quantities of natural gas and spilled fracking fluid has caused problems. The potential problems are discussed in Brian Colleran, "The Drill's About to Drop," *E Magazine* (March/April 2010). James C. Morriss, III, and Christopher D. Smith, "The Shales and Shale-Nots: Environmental Regulation of Natural Gas Development," *Energy Litigation Journal* (Summer 2010), contend that if companies in the industry act to prevent problems before regulators require such action, this both demands a better understanding of the technology and prevents future litigation.

According to Marc Levy and Mary Esch, "EPA Takes New Look at Gas Drilling, Water Issues," *AP* (www.google.com/hostednews/ap/article/ALeqM5jrnCodmMZhlWjyiXJP_JYQUd77BgD9H33C703) (July 20, 2010), in 2004 an Environmental Protection Agency (EPA) study said that fracking was little or no threat to drinking water and Congress exempted fracking from federal regulation. Now, however, the EPA is holding hearings and will conduct a $1.9 million study to reevaluate fracking technology; see Tom Zeller, Jr., "E.P.A. Considers Risks of Gas Extraction," *The New York Times* (July 23, 2010). Preliminary results of the study are expected in 2012. The EPA's Web page on fracking and this study is at www.epa.gov/ogwdw000/uic/wells_hydrofrac.html.

Concern over such problems has prompted the New York State Department of Environmental Conservation to revisit its 1992 Generic Environmental Impact Statement (GEIS), noting that "After a comprehensive review of all the potential environmental impacts of oil and gas drilling and production in New York, the Department found in the 1992 GEIS that issuance of a standard, individual oil or gas well drilling permit anywhere in the state, when no other permits are involved, does not have a significant environmental impact. A separate finding was made that issuance of an oil- and gas-drilling permit for a surface location above an aquifer is also a nonsignificant action. . . ." However, the new fracking technology warrants further review based on "required water volumes in excess of GEIS descriptions, possible drilling in the New York City Watershed, in or near the Catskill Park, and near the federally designated Upper Delaware Scenic and Recreational River, and longer duration of disturbance at multi-well drilling sites." The Department's "Draft Supplemental Generic Environmental Impact Statement on the Oil, Gas and Solution Mining Regulatory Program" is available at www.dec.ny.gov/energy/58440.html.

In the YES selection, Diane Katz argues that the new "fracking" technology has made it possible to release vast amounts of natural gas from deep shale deposits. As a result, we should stop spending massive sums of public money to develop renewable energy sources. The "knowledge and wisdom of private investors" are more likely to solve energy problems than government policymakers. In the NO selection, Deborah Weisberg argues that the huge amounts of water and chemicals involved in "fracking" pose tremendous risks to both ground and surface water, and hence to public health. There is a need for stronger regulation of the industry.

Diane Katz

Shale Gas: A Reliable and Affordable Alternative to Costly "Green" Schemes

Governments at every level across North America are collectively showering billions of tax dollars on "green energy" schemes in an effort to avert global warming and end our "dependence on foreign oil." But in the political arena, there is precious little attention being paid to a far more affordable alternative energy source with great potential to reduce both fossil fuel emissions and imports of Middle Eastern oil.

In contrast to government tax breaks, preferential loans, grants, and other forms of subsidies to wind and solar projects, private investors are moving capital into the production of "shale gas." Trapped within dense sedimentary rock, this "unconventional" natural gas was for decades considered too costly to retrieve. But advances in drilling technologies, along with the rising cost of conventional natural gas, have transformed the economics of shale gas extraction. Consequently, the vast stores of shale gas buried a thousand meters or more below the surface of North America (and beyond) have the potential to dramatically alter both environmental politics and geopolitics.

The actual volume of recoverable shale gas remains imprecise as supplies are still being mapped and evaluated. The National Energy Board estimates Canada's volume to be 1,000 trillion cubic feet, with similar reserves in the United States. Europe also may be home to nearly 200 trillion cubic feet of shale gas.

In Canada, there are major shale gas "plays" in the Horn River Basin and the Montney Formation, both in British Columbia. Major exploration for shale gas is also occurring in the Colorado Group in Alberta and Saskatchewan, the Utica Shale in Quebec, and the Horton Bluff Shale in New Brunswick and Nova Scotia.

When burned, shale gas emits just half the carbon dioxide of coal. Unlike wind and solar power, which produce power intermittently, natural gas is continuously available to produce the steam that powers turbines in the production of electricity. In addition, distribution networks for natural gas already exist, meaning that there is less need to build costly infrastructure. These and other advantages of shale gas call into question the massive public outlays for more problematic "renewable" power sources.

According to energy analyst Amy Myers Jaffe, shale gas "is likely to upend the economics of renewable energy. It may be a lot harder to persuade people to adopt green power that needs heavy subsidies when there's a cheap, plentiful fuel out there that's a lot cleaner than coal, even if [natural] gas isn't as politically popular as wind or solar."

That very dynamic stymied energy mogul T. Boone Pickens in his plan to build the world's largest wind farm in the Texas Panhandle. The plan called for the construction of a wind farm with 687 turbines, driving the production of 1,000 megawatts of electricity—the equivalent of a nuclear power plant.

Shortly after the debut of the project in 2008, natural gas prices declined, making wind energy not competitive enough to attract the $2 billion needed in financing. As Pickens told the *Dallas Morning News,* "You had them standing in line to finance you when natural gas was $9 [per million Btu] . . . Natural gas at $4 [per million Btu] doesn't have many people trying to finance you." The lack of a transmission line to move the wind power to urban centers also contributed to his decision to kill the project, Pickens said.

But governments across Canada have virtually unlimited financing at their disposal in the form of tax revenues, and thus are forcing taxpayers to subsidize costly "renewable" energy projects and transmission build-outs, even though more efficient alternatives exist. The government of Ontario, for example, is forcing utilities (read consumers) to buy "green" power at more than double the market rate for conventional electricity.

In the past, the fine grain of shale rock made tapping the natural gas within particularly difficult. The National Energy Board describes shale as "denser than concrete" and thus virtually impermeable. But from the tenacity of a lone Texan, a productive method to set the gas flowing has emerged. As the *Sunday Times* reports:

> It all began in 1981 when Mitchell Energy & Development, a Texas gas producer, was, quite literally, running out of gas. [George] Mitchell, who founded the firm, ordered his engineers to look into tapping shale, which drillers usually passed through to get to the oil and gas fields below them. . . . For years, [the shale] had been ignored, but Mitchell had a hunch about their potential. "I thought there had to be a way to get at it," he said. "My engineers were always adamant. They would say, 'Mitchell, you're wasting your money.' And I said, 'Let me.'" It took 12 years, more than 30 experimental wells and millions of dollars before he came up with the technical solution.

That technical solution is known as "hydraulic fracturing" (or "fracking"), which involves injecting at high pressure a mixture of water, sand, and chemicals into the shale to fracture the rock and allow the release of the natural gas therein. In conjunction with fracking, horizontal drilling is used to maximize the surface area of the borehole through which the gas is collected.

Some environmentalists complain that the chemical compounds used in fracking threaten to pollute soil and groundwater, and they decry the volumes of water used in the production process. In addition, some global warming alarmists oppose the development of new stores of fossil fuel. But in many instances, fracking is conducted thousands of feet below aquifers, and the strata are separated by millions of tons of impermeable rock. Moreover, ever larger quantities of the water used in fracking are recycled. The industry also maintains that stringent regulatory standards are in place to protect the environment. And, as detailed in another article in this edition of *Fraser Forum,* all sources of energy—"renewables" included—involve environmental trade-offs.

Initially, fracking and horizontal drilling were too costly for widespread adoption. But as oil prices rose, these techniques became more cost-effective. Since then, economies of scale and technological innovations have "halved the production costs of shale gas, making it cheaper even than some conventional sources."

Energy analysts expect further cost reductions in shale gas production as major oil and gas companies invest in new technologies. For example, production costs have fallen to $3 per million Btu at the Haynesville Formation, which encompasses much of the US Gulf Coast, down from $5 or more at the Barnett Shale in the 1990s.

The turnabout in shale gas fortunes is all the more remarkable given predictions in the past decade that Canada and the United States were running low on natural gas. US Federal Reserve Chairman Alan Greenspan, for example, declared in 2003 that the United States would have to import liquid natural gas to meet demand.

Doing so would have increased reliance on supplies from Russia and Iran, hardly an appealing prospect for anyone intent on "energy independence." Before the shale gas boom, both countries were thought to control more than half of the known conventional gas reserves in the world. Now, however, Canada and the United States have access to huge domestic stores.

This could cause dramatic shifts in global petro-politics. As energy analyst Amy Myers Jaffe notes, "Consuming nations throughout Europe and Asia will be able to turn to major US oil companies and their own shale rock for cheap natural gas, and tell the Chavezes and Putins of the world where to stick their supplies—back in the ground."

The new accessibility to shale gas will also moderate the influence of OPEC and any potential natural gas cartel by providing affordable and reliable alternative sources of energy. Indeed, US production of natural gas in March hit an historical monthly high of 2.31 trillion cubic feet, topping Russia to become the largest producer in the world. Consequently, natural gas exports once headed to North America are instead heading to Europe, thereby forcing Russia to lower prices for its once-captive customers.

Illustrating the new political tectonics is the recent agreement between Chevron and Poland for natural gas development and production. According to Dr. Daniel Fine of the Mining and Minerals Resources Institute at MIT, "When Chevron announces that they have gas [in Poland], then Russia is shut out" from having a monopoly in Eastern Europe.

Canada will also feel the effects of the energy market shifts. For example, the expansion of US supplies means that Canada will need to find new export opportunities for its natural gas. However, this should not cause problems, analysts say, because supplies of conventional natural gas are declining elsewhere while fuel demands for transportation and electricity are growing.

The private sector is adept at adjusting to shifting trends. For example, a shipping terminal for natural gas imports to be built by Kitimat LNG Inc. was redesigned for exports to the Pacific Rim due to "increases in supply throughout North America—including in the US, Canada's traditional export market."

Unfortunately, federal and provincial governments remain wedded to energy policies that lack the knowledge and wisdom of private investors and fail to account for the dynamic nature of the market. Vast infusions of subsidies obscure the true costs of various energy sources, while disparate regulations and mandates inhibit the unfettered competition that would otherwise determine the most efficient and beneficial fuels. Policy makers and politicians could dramatically improve energy policy by releasing their ham-fisted grip on the energy market.

Deborah Weisberg

Fracking Our Rivers

On Christmas Day 2007, George Watson returned home from a family dinner to find one of his prized Black Angus cows dead alongside Hargus Creek, a stream that runs through his southwestern Pennsylvania farm.

Over the next three months, Watson lost 16 more cattle—all of which had been bred—making it, as he said, "a double loss." Up to three in one day were found lying near the water. A series of calves died soon after birth.

"I've been raising cattle for 22 years and never had anything like that," said Watson, a Vietnam veteran, who also was having problems with discolored, sludgy well water. A local vet tested the dead cows, but failed to find anything abnormal. Looking back now, Watson wishes he'd had someone test the water in the creek.

Although natural gas wells were being developed all around him, rumors of illegal wastewater dumping in local streams, and a 43-mile fish kill on Dunkard Creek in the same Monongahela River watershed two years later, fueled his darkest fears.

"After my cows died, I suspected it was from brine and waste being dumped, although I can't prove it now," said Watson, who later leased the mineral rights on his farm to Range Resources for $3,000 an acre plus 15 percent production royalties. Drilling hadn't begun as of late last year.

Range is one of 40 companies driving the boom in hydraulic fracturing for natural gas in Pennsylvania, where 53,000 wells are turning pastures and woods into industrial sites. Although hundreds of thousands more have changed the landscape in at least 31 states, Pennsylvania and New York have an abundance of Marcellus Shale wells and, unlike out West, they are close to end users. While vertical drilling and "hydrofracking" for gas has existed for decades, new technologies enable extractors to go more than a mile deep and a mile horizontally to fracture the Marcellus—and release embedded gas—using millions of gallons of sandy, chemical-laden water.

Dunkard Fish Kill

CONSOL Energy's Morris Run borehole and other sources in the Dunkard watershed are under investigation by several federal and state agencies, including the Pennsylvania Attorney General's Office, over possible illegal discharges of hydraulic fracturing fluid, since the level of total dissolved solids, including chlorides, in Dunkard Creek was higher than anything previously associated with coal bed methane wastewater, the only discharge permitted at Morris Run.

"There's pretty strong evidence there was more than coal bed methane water going down that borehole," said Charlie Brethauer of Pennsylvania DEP's water management section. "As far as allegations of illegal activity, I think there's something to it, although to what extent, we don't have any idea yet. We haven't ruled out 'fracking' fluid."

Ed Pressley and his wife Verna live along Dunkard Creek in Brave, Pennsylvania, and watched in horror as fish began going belly up in September 2009 in what would become a massive loss of wildlife that continued for a month. The shells of rare mussels popped open, said Verna, and muskellunge and smallmouth bass bled to death from their gills.

"Kids were putting fish into buckets trying to save them—the tears were running down their cheeks—but there was nowhere to take the fish to," said Verna, a retired science teacher. "We counted 600 dead fish—the stench was overwhelming—just below our dam. It was one of the most devastating emotional experiences of my life."

What made it especially heartbreaking for the Pressleys is that their dream was to turn their property into a living classroom, where children could study kingfishers, blue herons, mudpuppies, turtles, and other forms of wildlife sustained by the water. They were negotiating a conservation easement agreement with the US Department of Agriculture that would protect their land against development for generations to come, and with American Rivers to have a relic industrial dam removed from their section of the stream.

"The folks at Agriculture and American Rivers say they're going to stick with it," Verna said as she stood along Dunkard and peered into the eerily empty water last fall. "But it's going to be years before you'll see fish in here again. I know it's not going to happen in my lifetime."

An EPA interim report about Dunkard's demise cites the presence of golden algae, a toxic organism indigenous to southern U.S. coastal waters, but never before documented in Pennsylvania. Whether it got to Dunkard on migratory birds' feet, drilling equipment that originated in Texas, or by some other means may never be known, but the EPA confirmed that excessive levels of total dissolved solids turned Dunkard so salty the algae were able to thrive.

Golden algae was later found on Whitely Creek, a stocked trout fishery in the same watershed, said Brethauer, who indicated it is likely to spread to other streams.

While the gas drilling industry touts hydraulic fracturing as America's path to energy independence—the Natural Gas Supply Association claims there are enough reserves to meet the nation's needs for a century—some watchdogs say weak regulations and poor enforcement are fueling an environmental nightmare.

The 2005 Energy Policy Act exempts injection of hydraulic fracturing fluids from a key provision in the Safe Drinking Water Act, and federal regulations governing wastewater disposal are limited, according to Deborah Goldberg of Earthjustice, a nonprofit environmental law firm. "Gas wastewater treatment is mostly left to states to regulate and monitor, and most states are way behind the curve."

Ron Bishop, a biochemistry lecturer at SUNY College at Oneonta and a nationally certified chemical hazards management expert, put it this way: "You have to go through more permitting hoops to put a new garage on your property than to drill for gas."

Pennsylvania is in the process of tightening limits on total dissolved solids that can be discharged in rivers and streams, and New York is considering new permitting requirements—generating a de facto moratorium on drilling—although many environmental stakeholders, including New York City, say they aren't strong enough to protect watersheds such as the Delaware River, which provides drinking water to 17 million people.

"Government has to ramp up its regulations tremendously," said Jeff Zimmerman, an attorney for Damascus Citizens for Sustainability and Friends of the Upper Delaware River, groups which formally have protested the New York proposal. "Until an environmentally infallible extraction system can be assured without qualification, the gas drilling industry should not be allowed to operate. It must be failsafe. A single mistake or uncontrolled accident can wipe out, for years and years, important resources, such as those of Dunkard Creek."

Federal lawmakers are also considering legislation—the Fracturing Responsibility and Awareness of Chemicals (FRAC) Act—that would reverse the Clean Water Act exemption and force industry to disclose the names of all of the hundreds of chemicals used in the hydrofracking process, including those traditionally guarded as proprietary information. Pennsylvania makes the names of chemicals available, but not the proportions.

"Some of them are really nasty, like toluene and benzene, which are known to cause cancer," said Bishop. "Others are harmful to wildlife. DB-NPA is a biocide commonly added to fracking water to kill bacteria and algae. Even in amounts too tiny to show up on chemical tests, it's lethal to bay oysters, water fleas, and brown trout."

The staggering volume of fracking fluid used in each horizontal well—up to 6 million gallons of water and 50,000 pounds of chemicals—means environmental impacts can occur on a massive scale, Bishop said. Spills at drill sites and well casing failures—the two most common problems associated with hydrofracking—can cause escaping fluids to contaminate ground and surface water, and gas to migrate underground.

Violations

PADEP cited drillers for more than 450 violations last year. Cabot Oil Co. was charged with a series of spills that polluted a wetlands and killed fish in Stevens Creek, a Susquehanna River tributary in northeast Pennsylvania.

In a separate matter, Cabot is being sued by 15 Dimock residents who claim drilling operations contaminated their drinking water and caused them to suffer neurological and gastrointestinal ills. They are seeking a halt to drilling plus establishment of a trust fund to cover their medical care.

[On Jan. 9, 2010, PADEP also announced it had fined Atlas Resources $85,000 for violations at 13 different well sites in Greene, Fayette, and Washington counties. The violations included failure to restore well sites after drilling, failure to prevent discharges of silt-laden runoff, and for discharging industrial waste including production fluids onto the ground at 7 of the 13 sites. THE EDITOR.]

Among the many environmental threats or impacts associated with hydrofracking—including huge withdrawals of water from lakes and streams and erosion and sedimentation from truck traffic on rural roads—one of the more concerning is disposal of wastewater, since about half the liquid used in fracking flows back with additional toxins, including brine six times saltier than ocean water, Bishop said. "This hazardous, industrial waste must be disposed of, but there's no good answer as to how or where. Texas and Oklahoma allow deep well injection, but it doesn't work in Pennsylvania and New York because our rock 10,000 feet down isn't porous enough to absorb the waste."

Before water can be discharged into streams it must be strained, desalinated and restored to an acceptable pH level, but few sewage treatment plants are equipped to deal with the volume and chemical composition of fracking water, and many streams have reached their capacity for assimilating more total dissolved solids, Bishop said. "It's a gigantic problem."

Chris Tucker of Energy In-Depth, a coalition of trade groups managed by the Independent Petroleum Association of America, agrees wastewater disposal is one of the industry's biggest bugaboos.

"Everyone knows we have to get on top of it," he said. "Producers are taking a lot of the wastewater from Pennsylvania into Ohio for deep-well injection. The industry is also looking at mobile recycling facilities, but they're getting quoted one cent a gallon. Consider what that would cost when you're dealing with 3 or 4 million gallons of water."

Although Range Resources' CEO John Pinkerton insists that his company's wastewater poses no threat to freshwater streams, Range has turned to recycling in Pennsylvania, where one-acre impoundments and miles of aboveground pipes circulate frack wastewater among several wells. Range also is exploring additional technologies, including crystallization and evaporation—essentially boiling wastewater and skimming off the salt which could be sold for road de-icing.

"We don't know how much Marcellus play there will be but wastewater disposal will keep pace. If it doesn't, the drilling will cease," Pinkerton said. "We have millions of dollars invested in each well. We have to know where every gallon coming out of the ground will go. It's in our best interest to do it right. To do otherwise would be business suicide."

As a fly fisher, Pinkerton considers himself an environmentalist, and he said natural gas extraction is the only practical alternative to foreign oil and coal. "The idea that we can go to 100 percent renewables before you and I pass away is ludicrous. We need a portfolio of energy solutions—a balanced energy policy—so if oil goes to $300 a barrel, we're not stuck. If we don't figure this out, we're dead meat."

He said every industry has risks and impacts—"you've got to cut down trees to print your magazine," he said—"but temporary inconveniences are necessary for tremendous, long-term gain, unless we all want to walk or ride horses to work."

Both Pinkerton and Tucker decry direct EPA permitting, which the FRAC Act would require. "It wouldn't just slow us down, it would bring us to a stop for four or five years," said Tucker, who points to the job growth he claims his industry has spawned. "We put 48,000 people to work in Pennsylvania and zero in New York because of the de facto moratorium. Where I come from, Wilkes-Barre/Scranton, gas is a godsend for folks who are economically depressed."

Fly fishing guide Glenn McConnell said he felt better about leasing the mineral rights to his land in the Pennsylvania Wilds after Range agreed to address Trout Unlimited.

"The drillers are just as concerned about the environment as you and me," McConnell said. "They don't want to make a bad name for themselves. If something isn't right, they'll correct it immediately."

PA Council Trout Unlimited environmental chair Greg Grabowicz is more focused on problem prevention. "We want assurances that operations will be fail-safe. Our immediate concern is whether DEP can enforce even existing regulations, with such a small staff and so many wells," said Grabowicz, a professional forester. "There's no doubt Pennsylvania's watersheds will change dramatically over the next 30 years from new roads and pipelines, but only time will tell if drillers run into problems that cause catastrophes."

Others, though, already have seen impacts to their favorite coldwater fisheries, including Sam Harper, the DEP water management program chief monitoring Dunkard, who has a camp in the Allegheny National Forest. "There's been a dramatic change in the South Branch of Tionesta Creek, where I fish," he said. "We're seeing a lot fewer brook trout and a lot more roads leading to wells."

And there are likely to be more impacts to woodland streams as ozone from diesel-powered trucks and drilling equipment cause leaf burn and deforestation, according to Al Appleton, a former New York City Department of Environmental Protection commissioner, who serves as technical advisor to Damascus Citizens.

Appleton said too little is also made of the millions of gallons of water sucked from lakes and streams for each hydrofracking operation.

"They may not impact flow during certain times of the year, but drilling isn't a seasonal business," he said. "These companies are withdrawing significant amounts of water constantly."

While PADEP raised drilling permit fees last year to help pay for more site inspections, it also streamlined the permit approval process to 28 days with completion of a basic application—even though the agency admits the need to put more teeth into existing regulations. "Environmentalists focus on wastewater, but the biggest issue for us is what happens at the site," said PADEP spokesman Tom Rathbun. "Is the well 'cased' properly? Are the water pipes built properly?

What about how trucks are crossing streams? That's where our focus needs to be."

In the meantime, lawmakers expect to hold hearings on hydrofracking and to request an EPA study on its effects on the environment, according to Kristopher Eisenla, an aide to FRAC Act co-sponsor Congresswoman Diana DeGette (D-Colorado). "The industry has had a free ride for so long, if greater oversight costs it a few more bucks, in the interest of public health, it's worth it."

[In the Sept. 2009 issue, John Randolph in his page 2 article "The Threats Posed by Marcellus Drilling" identified "the single largest threat to Pennsylvania (also new York, West Virginia, and Ohio) wild-trout streams since the coal/steel era of the Industrial Revolution." After that issue went to subscribers, a "total" fish kill on 43 miles of Dunkard Creek in Pennsylvania raised the question again: "Is the Commonwealth of Pennsylvania protecting its waterways?" THE EDITOR.]

EXPLORING THE ISSUE

Is Shale Gas the Solution to Our Energy Woes?

Critical Thinking and Reflection

1. Do we need energy so badly that we should ignore risks to water supply and human health?
2. In what sense is the knowledge and wisdom of private investors preferable to that of government policymakers when it comes to deciding what to do about energy?
3. How will ample supplies of cheap natural gas affect development of renewable energy supplies such as wind and solar power?
4. Is it true that if greater oversight of an industry (such as the shale gas industry) costs a few more bucks, in the interest of public health, it is worth it?

Is There Common Ground?

Even if the proponents of unrestrained exploitation of shale gas by fracking are right when they say it solves our energy problems, the supply of shale gas will not last forever. The public—and its health—will remain, as will concern over carbon emissions and the need for ample amounts of energy to run our civilization.

1. Should we, as suggested by Diane Katz, stop investing public money in developing alternative energy sources? If not, why not?
2. Is government regulation essential to protect public health? Visit the Public Health Service at http://www.usphs.gov/aboutus/mission.aspx to explore one agency's approach.
3. Another kind of fossil fuel we have not yet tapped in any major way is shale oil. (See the Bureau of Land Management's oil shale information at http://ostseis.anl.gov/guide/oilshale/.) Discuss the potential benefits (and environmental costs) of exploiting this resource.

ISSUE 5

Should We Drill for Offshore Oil?

YES: Stephen L. Baird, from "Offshore Oil Drilling: Buying Energy Independence or Buying Time?" *The Technology Teacher* (November 2008)

NO: Mary Annette Rose, from "The Environmental Impacts of Offshore Oil Drilling," *The Technology Teacher* (February 2009)

Learning Outcomes

After studying this issue, students will be able to:

- Explain the case against offshore drilling.
- Explain the case for offshore drilling.
- Describe how oil and gas development affect human and environmental health.
- Describe what oil spilled during offshore drilling does to marine organisms.

ISSUE SUMMARY

YES: Stephen Baird argues that the demand for oil will continue even as we develop alternative energy sources. Drilling for offshore oil will not give the United States energy independence, but the nation cannot afford to ignore energy sources essential to maintaining its economy and standard of living.

NO: Mary Annette Rose argues that the environmental impacts of exploiting offshore oil—including toxic pollution, ocean acidification, and global warming—are so complex and far-reaching that any decision to expand U.S. oil drilling must be based on more than public opinion driven by consumer demands for cheap energy, economic trade imbalances, and politics.

Petroleum was once known as "black gold" for the wealth it delivered to those who found rich deposits. Initially those deposits were located on land, in places such as Pennsylvania, Texas, Oklahoma, California, and Saudi Arabia.

As demand for oil rose, so did the search for more deposits, and it was not long before they were being found under the waters of the North Sea and the Gulf of Mexico, and even off the beaches of California.

In 1969, a drilling rig off Santa Barbara, California, suffered a blowout, releasing more than three million gallons of oil and fouling 35 miles of the coast with tarry goo. John Bratland, "Externalities, Conflict, and Offshore Lands," *Independent Review* (Spring 2004), calls this incident the origin of the modern conflict over offshore drilling for oil. He notes that since then most accidental oil releases have been related to transportation (as when an oil tanker runs into rocks; the *Exxon Valdez* spill was a striking example; see John Terry, "Oil on the Rocks—the 1989 Alaskan Oil Spill," *Journal of Biological Education* (Winter 1991)). Underwater oil releases have largely been prevented by the development of blowout-prevention technology. But Santa Barbara has not forgotten, and residents do not trust blow-out prevention technology. See William M. Welch, "Calif.'s Memories of 1969 Oil Disaster Far from Faded," *USA Today* (July 14, 2008).

Some people do not seem disturbed by the prospect of oil blowouts or spills. Ted Falgout, director of the port at Port Fourchon, Louisiana, looks at the forest of oil rigs in the Gulf of Mexico and "sees green: the color of money that comes from the nation's busiest haven of offshore drilling. 'It's OK to have an ugly spot in your backyard,' Falgout says, 'if that spot has oil coming out of it.'" See Rick Jervis, William M. Welch, and Richard Wolf, "Worth the Risk? Debate on Offshore Drilling Heats Up," *USA Today* (July 13, 2008).

In 2008, oil and gasoline prices reached record highs. Many people were concerned that prices would continue to rise, with the result being rapid investment in alternative energy sources such as wind. At the same time, those who favored increased drilling, both on and offshore, began to call for the government to open up more land for exploration (see Fred Barnes, "Let's Drill," *The Weekly Standard,* May 26, 2008). In its last few months in office, the Bush Administration issued leases for lands near national parks and monuments in Utah and lifted an executive order banning offshore drilling. Both measures were considered justified because they would reduce dependence on foreign sources of oil, ease a growing balance of payments problem, and ensure a continuing supply of oil. Critics pointed out that any oil from new wells, on land or at sea, would not reach the market for a decade or more. They also stressed the risks to the environment and called for more attention to alternative energy technologies.

In the YES selection, Stephen Baird argues that the demand for oil will continue even as we develop alternative energy sources. Drilling for offshore oil will not give the United States energy independence, but the nation cannot afford to ignore energy sources essential to maintaining its economy and its standard of living. He claims the environmental objections just do not add up. In a direct response to Baird's essay, Mary Annette Rose argues in the NO selection that the environmental impacts of exploiting offshore oil—including toxic pollution, ocean acidification, and global warming—are so complex and far-reaching that any decision to expand U.S. oil drilling must be based on more than public opinion driven by consumer demands for cheap energy, economic trade imbalances, and politics.

Stephen L. Baird

Offshore Oil Drilling: Buying Energy Independence or Buying Time?

Skyrocketing fuel prices, unprecedented home foreclosures, rising unemployment, escalating food prices, increasing climate disasters, and the continued war on two fronts have prompted greater public support for renewed offshore drilling for oil. A Gallup poll conducted in May of 2008 found that 57 percent of respondents favored such drilling, while 41 percent were opposed. [. . .] The political landscape is also being changed in favor of offshore drilling, with the results of a Zogby poll (Zogby International has been tracking public opinion since 1984) showing that three in four likely voters—74 percent—support offshore drilling for oil in U.S. coastal waters, and more than half (59 percent) also favor drilling for oil in the Alaska National Wildlife Refuge. [. . .] The tide is turning in favor of offshore drilling, with environmental concerns given less thought because of the increasing financial strain being realized by a majority of the American public. The debate on offshore drilling has captured headlines in newspapers, stirred debate on talk radio, and has been at the forefront on the nightly news.

The rising tide for support of offshore drilling recently gathered momentum when, on July 14, 2008, President George W. Bush lifted a 1990 executive order by the first President Bush banning offshore drilling, while at the same time calling for drilling in the Arctic National Wildlife Refuge. As of August 2008, however, a 1982 congressional ban is still in place, making Bush's action a symbolic gesture, and now the congressional ban is being debated in terms of both environmental issues and U.S. energy independence. In an almost complete reversal of policy, on July 30, 2008, the U.S. Department of the Interior released a news report saying that the nation's energy situation has dramatically changed in the past year. Secretary of the Interior, Dirk Kempthorne, said, "Areas that were considered too expensive to develop a year ago are no longer necessarily out of reach based on improvements to technology and safety." Kempthorne went on to say that, "The American people and the President want action, and a new initiative (the development of a new oil and natural gas leasing program for the U.S. Outer Continental Shelf) can accelerate an offshore exploration and development program that would increase production from additional domestic energy resources." President Bush is urging Congress to enact legislation that would allow states to have a say regarding operations off their shores and to share in the resulting revenues. [. . .] Shortly after the Interior Department released plans for jumpstarting new offshore oil exploration, on August 16, 2008 the Speaker of The House, Nancy Pelosi, dropped her opposition to a vote on coastal oil exploration and expanded offshore drilling (with appropriate safeguards and without taxpayer subsidies to big oil) as part of broad energy legislation to be addressed when Congress returned in September. [. . .] Today, with the high price of oil and a widening gap between U.S. energy consumption and supply, the ban on offshore oil drilling is being rethought by the general public, politicians, and the oil industry.

The energy stalemate between environmentalists and industry that has inhibited U.S. offshore oil production since the late 1960s is being broken, environmental arguments no longer add up, and working Americans are now taking energy policy inaction personally. According to a Pew Research Center poll conducted in July 2008, 60 percent of respondents considered energy supplies more important than environmental protection, and a majority of young Americans, 18–29, now consider energy exploration more important than conservation. [. . .]

Addressing Environmental and Safety Concerns

Though offshore drilling conjures up fears of catastrophic spills, (such as the 80,000 barrels that spilled six miles off Santa Barbara, California, inundating beaches and aquatic life in January 1969), the petroleum industry rightly argues that safety measures have improved considerably in recent years. According to the U.S. Minerals Management Service, since 1975, 101,997 barrels spilled from among the 11.855 billion barrels of American oil extracted offshore. This is a 0.001 percent pollution rate. That equates to 99.999 percent clean—compare that with Mother Nature herself, as 620,500 barrels of oil ooze organically from North America's ocean floors each year. [. . .]

The United States has been a leader in the creation of the modern offshore oil industry and has pioneered many new safety technologies, ranging from blowout preventers to computer-controlled well data designed to help oil companies' efforts to prevent disasters. Sensors and other instruments now help platform workers monitor and handle the temperatures and pressures of subsea oil, even as drilling is occurring. Hurricanes have become manageable, with oil lines now being capped at or beneath the ocean floor. Even if oil platforms snap loose and blow away, industrial seals restrain potentially destructive petroleum leaks from hundreds or even thousands of feet below the ocean's surface. In August and September of 2005, the 3,050 offshore oil structures endured the wrath of Hurricanes Katrina and Rita without damaging petroleum spills. While 168 platforms and 55 rigs were destroyed or seriously damaged, the oil they pumped remained safely encased, thanks to heavy underwater machinery. The U.S. Minerals Management Service concluded, "Due to the prompt evacuation and shut-in preparations made by operating and service personnel, there was no loss of life and no major oil spills attributed to either storm." [. . .] If it can be done in an environmentally friendly fashion—and with oil companies themselves footing the bill—increasing opportunities for new offshore drilling might be worthwhile.

Offshore territories and public lands like the Alaska National Wildlife Refuge (ANWR) that don't allow drilling have been estimated to contain up to 86 billion barrels of oil according to the U.S. government's Energy Information Administration. Although analysts say that amount of oil will not greatly affect the price of oil, and that renewed offshore drilling would have little impact on gas prices anytime soon, in the short term, oil prices could go down slightly if Congress lifts its moratorium on new offshore drilling because the market would factor in the prospect of additional oil supplies later on. A spokeswoman for the American Petroleum Institute said that, "If we had new territory, we could hypothetically make a big find." [. . .] Offshore drilling might not be the end-all solution to our oil dependence, but any serious energy proposal has to be comprehensive and should include more oil supply and production from the outer continental shelf.

How Dependent Are We on Foreign Oil?

Although the United States is the third largest oil producer (the U.S. produces 10 percent of the world's oil and consumes 24 percent), most of the oil we use is imported. The U.S. imported about 60 percent of the oil consumed in 2006. [. . .] About half of the oil we import comes from the Western Hemisphere (North, South, Central America, and the Caribbean including U.S. territories). [. . .] We imported only 16 percent of our crude oil and petroleum products from the Persian Gulf countries of Bahrain, Iraq, Kuwait, Qatar, Saudi Arabia, and the United Arab Emirates. During 2006, our five biggest suppliers of crude oil were: Canada (17.2%), Mexico (12.4%), Saudi Arabia (10.7%), Venezuela (10.4%), and Nigeria (8.1%). It is usually impossible to tell whether the petroleum products that you use came from domestic or imported sources of oil once they are refined. [. . .] According to the United States Energy Information Administration, the United States spends more than $20 billion, on average, per month to purchase oil, gasoline, and diesel fuel from abroad.

The negative aspects of this dependency are fairly obvious, and they have been well documented. First, oil imports contribute heavily to the United States' trade deficit, which is at record levels. Second, the United States is forced to make political decisions that it might not make otherwise (invading Iraq, cooperating with hostile governments such as Venezuela and Nigeria, looking the other way at Saudi Arabia's reactionary regime, etc.) because it needs their oil. Third, up to now the availability of oil at a fairly reasonable price has left the United States to continue down a path of using more and more energy. [. . .] From Nixon to now, every sitting President has promised to make sure that we wouldn't have a future energy problem . . . though we certainly do now.

> Richard Nixon, 1974: "We will lay the foundation for our future capacity to meet America's energy needs from America's own resources."

> Gerald Ford, 1975: "I am proposing a program, which will begin to restore our country's surplus capacity in total energy. In this way, we will be able to assure ourselves reliable and adequate energy and help foster a new world energy stability for other major consuming nations."
>
> Jimmy Carter, 1980: "We must take whatever actions are necessary to reduce our dependence on foreign oil—and at the same time reduce inflation."
>
> Ronald Reagan, 1982: "We will ensure that our people and our economy are never again held hostage by the whim of any country or oil cartel."
>
> George H. W. Bush, 1990: "The Congress should, this month, enact measures to increase domestic energy production and energy conservation in order to reduce dependence on foreign oil."
>
> George W. Bush, 2008: "And here we have a serious problem, America is addicted to oil." [. . .]

It is somewhat misleading when politicians talk about "America's addiction to oil" because there are some mitigating factors that make this situation a lot less dire than it might seem. The two largest foreign suppliers of oil to the United States are friendly to us: Canada and Mexico. These countries have increased their exports to the United States for the past decade and are well-positioned to continue doing so. Thus, fears that the United States will be dependent on "enemy" regimes are overblown, and the price of oil is a world-market price, so the United States is not being gouged. The United States can buy oil from anywhere (except where it imposes sanctions, like Iran), and it doesn't really matter if the oil comes from internal sources or imports. In fact, the United States exports some oil from Alaska, because it is more efficient to send that oil to Japan than it is to send it down to refineries in California. [. . .] The world oil markets are very competitive, and the locating, capturing, refining, and selling of oil is a very complex system. Saying that the United States is too dependent on foreign oil and that this will spell disaster in the near future is not an accurate statement. A more accurate statement would be to say that we have many wasteful energy habits, and that we need to focus on how to reduce our energy use and to expand alternative energy sources without drastically affecting our lifestyles and our economy. But oil is essential to our country's normal functioning, and therefore more American oil must be part of an American energy solution.

Why Drill Offshore for New Oil?

Is more drilling for American oil an essential part of lowering energy costs and freeing us from dependence on foreign sources of energy? Opening up new areas for exploration in the Outer Continental Shelf and the Alaska National Wildlife Refuge in the United States, even if new supplies won't actually reach our gas tanks for several years, would immediately impact the amount of upward speculation on long-term commodity investment in oil. Oil speculators would see a greater supply ahead and that the future of oil would be less constrained on the supply side. Also, fears of Middle Eastern turmoil or South American unrest that could disrupt supply shipments would be much less of a reason to drive up the price of crude if a stable United States could supply additional millions of barrels of oil.

Today, oil drilling is prohibited in all offshore regions along the North Atlantic coast, most of the Pacific coast, parts of the Alaska coast, and most of the eastern Gulf of Mexico. The central and western portions of the Gulf of Mexico therefore account for almost all current domestic offshore oil production, providing 27 percent of the United States' domestic oil production. The areas under the congressional ban contain an estimated additional 18 billion barrels of oil. This estimate is considered conservative since little exploration has been conducted in most of those areas during the past quarter of a century due to the congressional ban. Estimates tend to increase dramatically as technology improves and exploration activities occur. [. . .] Major advances in seismic technology and deep-water drilling techniques have already led the Interior Department's Minerals Management Service to increase its original estimate of untapped Gulf of Mexico oil from 9 billion barrels to 45 billion barrels. In short, there could be much more oil under the sea than previously thought.

The Interior Department has already taken steps for new offshore oil exploration, announcing plans for a lease program that could open up new areas off the coasts of Florida, Georgia, Texas, North Carolina, Virginia, and other coastal states to drilling if Congress lifts the ban. Randall Luthi, director of the department's Minerals Management Service, which handles offshore oilfield regulations and leases, said, "The technology has improved . . . the safety systems we now require have greatly improved . . . and the industry has a good record." According to Luthi, new tools such as high-tech computers that make exploration easier and tougher building materials on platforms are making offshore drilling safer. Seismic technology and directional drilling techniques let oil companies drill 100 exploratory wells from a single offshore platform, reducing the number of derricks and therefore the potential for problems. Automatic shut-off valves underneath the seabed can cut the flow of oil immediately if there's a problem or a storm coming. Blowout prevention equipment can automatically seal off pipes leading to the surface in the case of an unexpected pressure buildup, and undersea pipelines and wellheads can be monitored with special equipment such as unmanned, camera-equipped, and sensor-laden underwater vehicles. [. . .] New drilling technologies and the industry's track record in the Gulf of Mexico show that offshore drilling for oil is safer than it ever has been, proving that you can drill and still be environmentally friendly.

Conclusion

The demand for energy is going up, not down, and for a long time, even as alternative sources of energy are developed, more oil will be needed. The strongest argument against drilling is that it could distract the country from the pursuit of alternative sources of energy. The United States cannot drill its way to energy independence. But with the developing economies of China and India steadily increasing their oil needs in their latter-day industrial revolutions, the United States can no longer afford to turn its back on finding all the sources of fuel necessary to maintain its economy and standard of living. What is required is a long-term, comprehensive plan that includes wind, solar, geothermal, biofuels, and nuclear—and that acknowledges that oil and gas will be instrumental to the United States' well-being for many years to come.

Mary Annette Rose

The Environmental Impacts of Offshore Oil Drilling

Stephen L. Baird's article in the November 2008 issue of *The Technology Teacher* describes a contemporary debate about opening more U.S. land and coastal regions to oil and gas exploration and production (E&P). While Baird's thesis—"informed and rational decisions can be reached through the understanding of how complex technological systems can impact the environment, our economy, our politics, and ultimately our culture" [. . .] epitomizes the goal of a technologically literate citizen, his article is a stark contradiction to this call for understanding. His one-sided argument is built upon public opinion driven by consumer demand for cheap energy, economic trade imbalance, and politics. Baird fails to connect the offshore oil and gas E&P to the toxins and greenhouse gases these technological processes release to the marine environment and the atmosphere. Decades of empirical evidence indicates that all stages of offshore oil and gas activity have consequences for the health and survivability of marine plants and animals, humans, and our planet.

In the following, I counter Baird's proposition that "environmental arguments no longer add up" [. . .] by identifying a few of the impacts of offshore oil E&P on the environment. My hope is that this analysis will better prepare teachers to foster the development of *critical-thinking* skills in their students. These skills are prerequisite to assessing the impacts of technology upon the environment and society [. . .] and essential for making environmentally sustainable choices.

Technology Assessment

When we ask our students to assess the impacts of technology, we ask them to engage in a process of inquiry, a cognitive journey of questioning assumptions, hypothesizing, gathering and reviewing evidence and trends, and testing their hypotheses against the body of evidence. This process, known as Technology Assessment (TA), refers to an examination of the potential or existing risks and consequences of developing, adopting, or using a technology. TA begins by bounding the study to identify time horizons, impact zones, stakeholders, and a host of relevant technical and environmental information. Tools, such as cross-impact analysis, mathematical models, and regression analysis, are used to analyze this data and to predict outcomes and risks associated with possible decisions. TA results in a list of "if/then" statements, options, trade-offs, or alternative future scenarios, which decision makers use to inform policies, make investments, and plan for the future.

Nature of Petroleum (Hydrocarbons)

To examine environmental impacts, we should begin by looking at the nature of crude oil (petroleum). Petroleum is a fossil fuel that forms from the remains of prehistoric vegetation and animals as a result of millions of years of heat and pressure. It is a complex mixture of hydrocarbons, several minor constituents (e.g., sulfur), and trace metals (e.g., chromium). The chemical composition of crude oil varies by the age of the geologic formation from which it came. When crude oil is released into the environment, biological, physical, and chemical processes (referred to as weathering) alter the oil's original characteristics. [. . .] Lighter oils tend to be volatile, reactive, and highly flammable, while heavier crudes tend to be tarry and waxy and contain cancer-causing polycyclic aromatic hydrocarbons (PAH) and other toxic substances.

Offshore Oil Exploration and Production

One challenge for offshore (waters beyond three miles from the shoreline) E&P operators is to control the dynamic changes in temperature and pressures when drilling into rock formations located deep beneath the ocean. As depths increase, the pressure of drilling fluids (muds) is used to counter the deep-sea pressures related to depth and the pockets of high-pressure and high-temperature (HPHT) gases. If not contained, these HPHT gases result in dangerous oil-well blowouts that emit a buoyant plume of oil, produced water, and pressurized natural gas (methane).

The 1969 blowout off the Santa Barbara, California coast that spilled 80–100,000 gallons of crude oil and inundated local beaches was an ecological disaster, killing thousands of birds, fish, and marine mammals. [. . .] However, this pales in comparison to the 1979 blowout at the Ixtoc 1 offshore oil rig in the Gulf of Mexico, which spewed an estimated 140 million gallons of crude oil until that well was capped nine months later. [. . .]

Modern technology and the research and monitoring systems of the Department of Interior's Minerals Management Service (MMS), which manages E&P in the outer continental shelf (OCS), have reduced the frequency of these ecological catastrophes. However, weather, tectonic events, equipment failure, transportation accidents, human error, and deliberate unethical choices continue to make oil spills a reality.

Today, there are nearly 4,000 active platforms in the OCS. [. . .] MMS [. . .] indicates that 115 platforms were destroyed and 600 offshore pipelines were damaged by Hurricanes Katrina and Rita in 2005. For these hurricanes, "124 spills were reported with a total volume of roughly 17,700 barrels of total petroleum products, of which about 13,200 barrels were crude oil and condensate from platforms, rigs, and pipelines, and 4,500 barrels were refined products from platforms and rigs." [. . .] In 2008, 60 platforms were destroyed during Hurricanes Gustav and Ike [. . .]; data on oil spills and damage to pipelines has not yet been released.

Offshore E&P also requires transportation vessels and terrestrial storage. As a direct result of Hurricane Katrina, an above-ground storage tank in St. Bernard Parish, Louisiana, spilled over 25,110 barrels of mixed crude oil. [. . .] The oil inundated about 1,700 homes and several canals. Testing conducted in 2006 confirmed that contaminants, including PAHs, arsenic, and other toxics were above acceptable risk standards. [. . .]

Wastes from Offshore E&P

The less dramatic, yet more disturbing impacts of offshore oil E&P relate to the volume and type of wastes these processes generate. Wastes include produced water, drilling fluids (muds), cuttings (crushed rock), diesel emissions, and chemicals associated with operating mechanical, hydraulic, and electrical equipment, such as biocides, solvents, and corrosion inhibitors.

Produced Water. By volume, 98% of the waste from E&P is *produced water,* with estimates at 480,000 barrels per day in 1999. [. . .] Produced water is a water mixture consisting of hydrocarbons (e.g., PAH, organic acids, phenols, and volatiles), naturally occurring radioactive materials, dissolved solids, and chemical additives used during drilling. Glickman [. . .] concluded that "hydrocarbons are likely contributors to produced water toxicity, and their toxicities are additive, so that although individually the toxicities may be insignificant, when combined, aquatic toxicity can occur." [. . .] Furthermore, studies document that sediments become contaminated with these toxins and that the concentration has a direct correlation with produced water discharges. [. . .]

Citing "relief to coastal waters, which support spawning grounds, nurseries, and habitats for commercial and recreational fisheries: reducing documented aquatic 'dead zone' impacts; reduction of potential cancer risks to anglers from consuming seafood contaminated by produced water radionuclides; and reducing potential exposure of endangered species to toxic contaminants" the EPA [. . .] banned the release of produced waters to inland and coastal waters (extending to three miles from shore). However, offshore E&P operations in U.S. waters may legally discharge treated produced water directly into the ocean or inject it into underground wells.

Drilling Fluids. Drilling muds and cuttings are of environmental concern because of their potential toxicity and the large volume that are discharged during drilling. Three types occur, including oil-based (OBM; diesel or mineral oil serves as base fluid), water-based (WBM), and synthetic-based muds (SBM). The EPA [. . .] requires zero discharge of OBM—to dispose, OBM is shipped to onshore oil field waste sites or injected into disposal wells at sea. WBM and SBM typically contain arsenic, barium, cadmium, chromium, copper, iron, lead, mercury, and zinc. Barium and barium compounds are used in drilling muds because they act as lubricants and increase the density of mud, thereby sealing gases in the well.

Drilling in deep water (>1,000 ft) uses rotary bits that chip through thousands of feet of rock to access oil and gas deposits. Diesel-powered engines provide the power to operate the drilling rig and drive the drill. As paraphrased from Continental Shelf Associates, Inc. [. . .], the general sequence of events entails initial "open hole" drilling where a drill bit positioned within a drill pipe chips away at the ocean floor. As the drill bit spins, mud (WBM) and water are forced at high velocity around the drill bit to force rock chips (cuttings) up and out to the seabed. After a known distance, the drill bit and pipe are removed and a wellhead is installed. WBM is typically discharged and replaced with SBM. Additionally, a marine riser system is connected to the wellhead to return fluids, muds, and cuttings to the drill rig where they are separated. Cuttings are discharged in a plume from the platform, and mud is recycled back to the drill bit. [. . .]

Continental Shelf Associates, Inc. [. . .] examined the impact of synthetic-based drilling fluids (SBF: mixtures of organic isomers) by comparing indicators from near and far distances from four E&P drilling sites in the Gulf of Mexico. [. . .] Cuttings and SBM extended several hundred meters from the well site and up to 45 cm in thickness. Analyses indicated that near-field sediments were toxic to amphipods (crustaceans). Chemicals associated with both WBM and SBM waste solids in near-field sediments contributed to sediment toxicity. Significantly higher mercury and lead concentrations were found in near-field sediments than in far-field sediments for some sites. Red crabs had high concentrations of toxins, such as arsenic, barium, chromium, and mercury.

Ethical Climate

Offshore E&P is conducted by a small number of operators, most of which are multinational oil corporations with single-year income larger than the GDP of entire nations; e.g., ExxonMobil Corporation [. . .] reported $40.610 billion in net income for 2007. The isolated nature of offshore drilling operations fosters a climate conducive to environmentally irresponsible behaviors. For instance, as reported by the EPA [. . .], two large electrical transformers located on Platform Hondo, Exxon's Santa Ynez Unit, leaked nearly 400 gallons of fluids contaminated with PCBs into the Pacific Ocean. In one instance, a transformer leaked for almost two years before repairs were made. Cleanup workers were not provided protective equipment to protect themselves against direct contact with and inhalation of PCBs. Exxon agreed to a settlement of $2.64 M in violation of the federal *Toxic Substances Control Act.*

Environmental Impacts

There are known detrimental impacts upon the marine environment for all phases of offshore E&P. [. . .] While natural seepages contribute more hydrocarbons to the marine environment by volume, the quick influx and concentration of oil during a spill makes them especially harmful to localized marine organisms and communities. Plants and animals that become coated in oil perish from mechanical smothering, birds die from hypothermia as their feathers lose their waterproofing, turtles die after ingesting oil-coated food, and animals become disoriented and exhibit other behavior changes after breathing volatile organic compounds.

When emitted into the marine environment, oil, produced water, and drilling muds may adversely impact an entire population by disrupting its food chain and reproductive cycle. Marine estuaries are especially susceptible, as hydrocarbons and other toxins tend to persist in the sediments where eggs and young often begin life. However, the severity and effects of oil exposure vary by concentration, season, and life stage. The oil spill from the Ixtoc 1 blowout threatened a rare nesting site of the Kemp's Ridley sea turtle, an endangered species. Field and laboratory data on the nests of turtle eggs found a significant decrease in survival of hatchlings, and some hatchlings had developmental deformities. [. . .]

Marine organisms that live near an existing or sealed wellhead or an oil spill area experience persistent exposure to a complex web of hydrocarbons, petroleum-degrading microbes, and toxic substances associated with drilling muds and produced water. Abundance and diversity of marine life, especially those living near or in the seabed, decline. The growth and reproduction rates of entire populations that live in the water column may decline for months after a spill [. . .], natural defense mechanisms necessary to deal with disease (immune suppression) become compromised [. . .], and genetic mutations may occur. Many of these toxins (e.g., arsenic, chromium, mercury, and PAH) move up the food chain and biomagnify, i.e., increase in concentration.

One of the most disturbing trends is the evidence that common hydrocarbon contaminants (e.g., PAH) act as endocrine disrupters. Endocrine disrupters are chemicals that can act as hormones or anti-hormones in aquatic ecosystems, thus disrupting normal reproductive and developmental patterns. [. . .] Evidence also suggests that polychlorinated biphenyls (PCBs), a known carcinogen, also exhibit these endocrine effects. [. . .]

Climate Change and Ocean Acidification. Greenhouse gases (GHG) are generated directly by offshore oil E&P and indirectly by enabling future emissions of oil as it is refined, distributed, and consumed, primarily by the transportation sector. In the U.S., 2007 emissions of methane from petroleum E&P was estimated at 22 MMTCO_2e (million metric tons of carbon dioxide equivalents) in addition to total contributions of petroleum at 2,579.9 MMTCO_2e. [. . .] These GHGs are primary drivers that disrupt the carbon cycle involving the biosphere, atmosphere, sediments (including fossil fuels), and the ocean. The consequences of this disruption include climate change (global warming, melting of ice at the poles, and ocean acidification). The ocean acts as a carbon sink, absorbing CO_2. As CO_2 increases, the acidity of the ocean increases and carbonate becomes less available to marine organisms that need it to build shells and skeletal material. Corals, calcareous phytoplankton, and mussels are especially susceptible to acidosis, which "can lead to lowered immune response, metabolic depression, behavioral depression affecting physical activity and reproduction, and asphyxiation." [. . .]

Human Health Impacts

Workers, victims of oil spills, and rescue workers are exposed to a host of chemical hazards. When people come in dermal contact with drilling fluids, muds, and cuttings, they can experience dermatitis; as exposure increases, impacts can include hypokalemia, renal toxicity, and cardiovascular and neuromuscular effects. [. . .] Exposure to volatile aromatic hydrocarbons (e.g., benzene) results in respiratory distress and unconsciousness. Long-term exposure can cause anemia, leukemia, reproductive problems, and developmental disorders. [. . .] Exposure to fine particulate matter, nitrogen oxides, sulphur, and dozens of hydrocarbons (e.g., PAH) emitted from diesel and gasoline engines, is linked to a variety of health impacts, including asthma attacks, cancer, endocrine disruption, and cardiopulmonary ailments. Because toxins bioaccumulate in fish, people who eat fish and shellfish from affected waters may experience nervous system effects, such as impairment of peripheral vision and seizure. Children and fetuses are especially vulnerable; exposure to toxins impairs physical and cognitive development.

Conclusion

The environmental impacts of offshore oil exploration and production are complex and far-reaching. Petroleum, produced water, and the chemicals used to extract petroleum from under the ocean have mechanical, toxic, carcinogenic, and mutagenic impacts on marine life and the humans who eat its marvelous bounty. But more profoundly, the combustion of petroleum is a major contributor of carbon dioxide to the atmosphere which, in turn, drives global warming and ocean acidification.

Therefore, the decision to open additional U.S. lands and oceanic territories to oil exploration and production should be based on more than public opinion and some unquestioned assumption that the U.S. has a right to consume about 25% of the world's energy and contribute 21% of the worlds' CO_2 emissions. Engaging students in a process of technology assessment is a viable pedagogy to help students develop dispositions and critical-thinking skills. These skills will enable students to not only recognize narrow, one-sided perspectives, but be better prepared to seek and apply valid data and analytical strategies to the critically important decisions that could impact life on this planet. In an age of complexity, these skills are essential for making environmentally sustainable choices. [. . .]

EXPLORING THE ISSUE

Should We Drill for Offshore Oil?

Critical Thinking and Reflection

1. List three strong arguments against offshore drilling.
2. List three strong arguments for offshore drilling.
3. In what ways does oil and gas development (onshore or offshore) affect human well-being?
4. How does nature handle oil spills, including both natural seeps and human-caused spills?

Is There Common Ground?

One cannot help but wonder whether, in the wake of the 2010 BP Deepwater Horizon spill in the Gulf of Mexico, Stephen L. Baird has changed his mind about whether environmental objections to offshore drilling hold up. Read Joel K. Bourne, Jr., "The Gulf of Oil: The Deep Dilemma," *National Geographic* (October 2010), and Michael Grunwald, "BP Oil Spill: Has the Damage Been Exaggerated?" *Time* (July 29, 2010), and answer the following questions:

1. How bad has the environmental impact of the spill been?
2. Is the environmental impact as bad as people initially expected?
3. If your answer to the previous question is "No," describe at least three factors that help to explain the smaller impact.
4. Should we continue to worry about the environmental impacts of offshore drilling?

ISSUE 6

Should Utilities Burn More Coal?

YES: Steven F. Leer, from "Role of Coal in Future Energy Policy," testimony at the hearing on "The Role of Coal in the New Energy Age" before the House Select Committee on Energy Independence and Global Warming (April 14, 2010)

NO: Susan Moran, from "Coal Rush!" *World Watch* (January/February 2007)

Learning Outcomes
After studying this issue, students will be able to: • Describe the environmental impacts of using coal for energy. • Explain the need for carbon capture and storage (CCS) technology. • Explain why the use of coal will continue and even expand.

ISSUE SUMMARY

YES: Steven Leer argues that the world will continue to use coal, massively and in rapidly growing quantities. The question is not whether global coal use will continue and grow, but rather whether carbon emissions from coal will grow. That answer depends on whether we can make carbon capture and storage (CCS) technology both effective and affordable.

NO: Susan Moran argues that U.S. utilities are building and planning to build a great many coal-burning power plants, often hoping to get them in operation before legislation restricting carbon emissions forces them to find alternatives.

Coal is so plentiful in the United States that the nation has been called the Saudi Arabia of coal. If the world runs out of oil, the United States will still (for awhile) have fossil fuel in plenty for electricity generation and—because coal can be liquefied or turned into a liquid much like petroleum—for transportation. However, coal is infamous for its environmental impacts. In the past, burning coal has put clouds of smoke, soot, fly ash, and sulfur dioxide into the air. It has been responsible for killer smogs in Donora, Pennsylvania, and London, England, as well as elsewhere, which helped spur the passage of clean air legislation and the development of technologies to make coal burning much cleaner. Coal-smog remains a problem in some parts of the world; see D. Mira-Salama et al., "Source Attribution of Urban Smog Episodes Caused by Coal Combustion," *Atmospheric Research* (June 2008).

Burning coal also releases more carbon dioxide, the major greenhouse gas, than does burning any other fossil fuel. If society is serious about limiting or preventing global warming by controlling carbon emissions, burning coal calls for somehow capturing the carbon dioxide emitted and keeping it out of the air. However, people such as Mike Carey, president of the Ohio Coal Association, continue to argue that carbon emissions have nothing to do with global climate change and therefore emissions need not be controlled (see Carey's testimony at the April 14, 2010, hearing on "The Role of Coal in the New Energy Age" of the House Select Committee on Energy Independence and Global Warming).

But the environmental impacts do not lie only in the *use* of coal. As a solid, it cannot be pumped from beneath the surface of the Earth. It must be dug out, either through underground mines or through strip mines. Underground mines are infamous for killing miners when they collapse, polluting waterways, and draining aquifers on which people depend (see Brad Miller, "Draining the Life from the Land," *Earth Island Journal* (Autumn 2002)). When they catch fire, as one did in Centralia, Pennsylvania, they may burn for decades (see Jeff Tietz, "The Great Centralia Coal Fire," *Harper's Magazine* (February 2004)). Surface mining leaves great holes in the landscape; in Appalachia, it carves off mountaintops and fills in valleys; see Patrick C. McGinley, "From Pick and Shovel to Mountaintop Removal: Environmental Injustice in the Appalachian Coalfields," *Environmental Law* (Winter 2004), and Jim Motavalli, "Once There Was a Mountain," *E Magazine* (November/December 2007).

Despite the problems, coal remains an attractive fuel because it is plentiful and relatively cheap. In China, which like the United States has extensive coal deposits, it is used in great quantities to provide the power for economic development. Chinese cities are extraordinarily smoggy, but China feels impelled to use coal despite its environmental effects. It is also developing coal as a replacement for oil as feedstock for chemical industries and as a source of liquid fuels. However, China is also developing other energy sources, including nuclear, wind, and hydropower, and planning to slow the pace of economic development; see "A Large Black Cloud," *The Economist* (March 15, 2008).

A great deal of work has gone into finding ways to control the pollutants, including carbon dioxide, emitted by coal-burning power plants. Much of that work has been funded by the U.S. Department of Energy (DOE). However, in January 2008, the DOE announced that it was "restructuring" its FutureGen program, among the projects described in Eli Kintisch's "Making Dirty Coal Plants Cleaner," *Science* (July 13, 2007). According to *New Scientist* ("US Government Pulls the Plug on Flagship Clean Coal Project" (February 9, 2008)), the restructuring "decision was based on rising costs of the project and notes that money will continue to be contributed to CCS [carbon capture and storage] development. Some scientists believe the move has set back CCS technology by three to five years."

On April 15, 2008, the House Science and Technology Committee held a hearing on DOE's restructuring plans. According to Paul Thompson, chairman of the FutureGen Alliance, "The FutureGen program is a global public-private partnership formed to design, build, and operate the world's first near-zero emission coal-fueled power plant with 90 percent capture and storage of carbon dioxide (CO_2). It will determine the technical and economic feasibility of generating electricity from coal with near-zero emission technology. FutureGen has five years of progress behind it. More than fifty-million dollars have been obligated to the effort with the majority spent. It is positioned to advance integrated gasification combined cycle (IGCC) and carbon capture and storage (CCS) technology faster and further than any other program in the world."

Such developments raised questions about the commitment of the U.S. government to cleaning up dirty energy technologies. Jeff Goodell, *Big Coal: The Dirty Secret Behind America's Energy Future* (Houghton Mifflin, 2006), says that our reliance on coal is so great that it is difficult to fund work that will lead to change. However, the Obama Administration has revived the program as FutureGen 2.0 (www.fossil.energy.gov/news/techlines/2010/10033-Secretary_Chu_Announces_FutureGen_.html). Will it succeed in its aim of showing that coal can be burned cleanly? Christine MacDonald, "Pipe Dreams: The Question of Clean Coal," *E Magazine* (September/October 2009), argues that such success is decades away at best.

Meanwhile, construction of coal-burning power plants continues and environmentalists continue to try to stop them; see Glenn Unterberger, Brendan Collins, and Sabrina Mizrachi, "Litigation Challenging Coal Plants, One Permit at a Time," *Natural Resources & Environment* (Spring 2008). Yet, there is a need for energy, and some parts of the world are far more dependent on coal than the United States. Seyward Darby, "Clearing the Air," *Transitions Online* (April 14, 2008), notes that the Czech Republic generates 60 percent and Poland 90 percent of its electricity with coal, and coal emissions cause serious environmental problems in these countries. There is a real need for clean coal technologies. If the United States adds as many coal-burning power plants as Susan Moran says have been proposed (more than 150), U.S. need will be even greater than it is today. Perhaps DOE is betting that U.S. electricity generation will shift toward nuclear power, as discussed in Issue 12. There are other options as well; James Ridgeway, "Scrubbing King Coal," *Mother Jones* (May/June 2008), says they include nuclear power and wind, geothermal, and solar power; see also Unit 4 of this book. Richard A. Kerr, "How Much Coal Remains?" *Science* (March 13, 2009), notes that we are likely to stop using coal before we run out of it. James B. Meigs, "The Myth of Clean Coal," *Popular Mechanics* (February 2010), argues that we should not "allow clean-coal myths to divert us from real-world alternatives that work today."

In "Clean Coal," testimony before the Senate Finance Committee (April 26, 2007), Nina French argues that the continued use of coal is critical for sustainable, inexpensive, secure, and reliable power generation. Coal meets these needs, and the technology exists—and is being improved—to ensure that coal is "clean," meaning that it emits less sulfur, mercury, and carbon. In the YES selection, coal company executive Steven Leer argues that the world will continue to use coal, massively and in rapidly growing quantities. The question is not whether global coal use will continue and grow, but rather whether carbon emissions from coal will grow. That answer depends on whether we can make carbon capture and storage (CCS) technology both effective and affordable. In the NO selection, environmental journalist Susan Moran argues that U.S. utilities are building and planning to build a great many coal-burning power plants, often hoping to get them in operation before legislation restricting carbon emissions forces them to find alternatives. What is needed is a broad mix of energy sources.

YES 

Steven F. Leer

Role of Coal in Future Energy Policy

. . .

General Background Information on Coal and Coal Emissions

Coal is a major contributor to the energy mix in this country, and in the world. Coal contributed 23% of the energy consumed in the U.S. in 2008, and 27% of the energy consumed in the world in 2006 (USDOE/EIA AER-2008). In the U.S., the vast majority of coal consumption is for electric power generation. Coal was used to generate 48.5% of the electricity generated in the U.S. in 2008 (USDOE/EIA AER-2008), and 40% of the global production of electricity in 2005 (International Energy Agency, ETP-2008). The world relies on coal because it is abundant and inexpensive—typically a small fraction of the cost of oil, natural gas, or biomass. The fact that the U.S. relies so heavily on coal is a large part of the reason that, in 2006, U.S. consumers paid about 58% of what Japanese consumers paid for electricity. In France, the U.K., Germany, Italy, and Denmark the fractions were 72%, 56%, 72%, 46%, and 32% (USDOE/EIA website, data compared in US dollars per kilowatthour).

The benefits of coal use to the U.S. economy are substantial. A 2006 report by Dr. Adam Rose (*The Economic Impacts of Coal Utilization and Displacement in the Continental United States*, 2015) concluded that "in 2015, U.S. coal production, transportation, and consumption for electric power generation will contribute more than $1 trillion of gross output . . . to the economy of the lower-48 United States." Coal and coal-based electricity are clearly key components of our energy mix and our economy.

China is an important player in global coal markets. In 2006, while the U.S. consumed 22 Quadrillion Btu's of coal, 90% for power generation, China consumed 25 Quadrillion Btu's of coal for electricity, and another 24 Quadrillion Btu's in its industrial sector, and a final 3 Quadrillion Btu's in residential and commercial sectors. All told, China burned over twice as much coal as the U.S. in 2006 (USDOE/EIA IEO-2009). More current estimates are that Chinese coal use is now three times that of the U.S., and continues to grow at about 20% of the total U.S. coal consumption rate each year. With China leading the way, non-OECD nations constitute 63% of current world coal use, and are projected by DOE/EIA to contribute 94% of the growth in

United States Congress, April 14, 2010.

world coal use between 2006 and 2030 (USDOE/EIA AEO-2009). We have all heard the anecdotes about energy growth in Emerging Asia, and most of them are true. China is building approximately one new coal-based power station every week. China has been adding the equivalent of the entire power grid of the UK each and every year. Chinese car sales exceeded U.S. car sales for the first time ever in 2009.

Consider the prospect of China, which had a private car ownership rate of 4 cars per 1000 people in 1999, compared to about 700 vehicles per 1000 people in the U.S. or 400 vehicles per 1000 people in South Korea, growing in vehicle intensity to just the South Korean rate. That is over 500 million vehicles. If those vehicles are fueled with petroleum, the impact on global oil markets would be very large, not to mention the impact on CO_2 emissions. If the sector were dominated by electric vehicles, then the implications for power generation, coal combustion and climate are equally significant, in the absence of effective and affordable CCS technology.

And it's not just China. In fact, coal has been the fastest growing fuel source on the planet this past decade—with global coal consumption up a staggering 41% in just the past eight years. To put U.S. coal consumption into perspective, the U.S. accounts for just one seventh of global coal use today—and that fraction is shrinking rapidly as coal consumption around the world grows, while U.S. consumption remains roughly constant.

So, why is the world turning to coal? Above all, it is because coal is the fuel source that the world's fastest growing economies—China and India in particular, but Russia and Indonesia and most of the rest of the developing world as well—have in greatest abundance. None of their actions support—and I believe it would be naïve of us to think—that such countries will turn their backs on such a vast storehouse of reliable, secure and low-cost energy.

In fact, with competition for energy resources intensifying, such countries are even beginning to look beyond their own borders for fossil energy resources, including coal. During the past year, we have seen state-operated Chinese and Indian companies acquire coal reserves and mines in other countries, with the view of ensuring a sufficient source of energy for the decades to come. Private Chinese, Indian and Russian steel and energy companies are following this same strategy.

Coal's future is not resource-constrained. EIA estimates that the world has 929 billion tons of recoverable coal reserves, enough for about 137 years of production at current rates, and the U.S. has the greatest share of those reserves, about 28% of the global total. By providing affordable heat and power, coal has raised—and is continuing to raise—the standard of living and quality of life for literally billions of people.

Arch Coal takes pride in the fact that we produce about 16% of the coal mined each year in the U.S., providing fuel for about 8% of our national electricity generation. But we also recognize that coal is a major contributor to manmade emissions of greenhouse gases. Arch Coal is committed to playing a constructive role in helping advance federal legislation that both addresses climate concerns and preserves the tremendous economic and human benefits associated with low-cost and secure energy from coal.

Recommended Principles to Follow in Addressing Global Warming

Arch Coal supports legislation to reduce global greenhouse gas emissions. We believe that this can be done in a manner that maintains U.S. and global prosperity, and that the two most important keys to this are the timing of reductions and a collaborative government/industry effort to commercialize improved, lower cost emission mitigation technologies. I would offer that CCS technologies and their global deployment provide the only technologically feasible and politically achievable path for stabilizing CO_2 concentrations in the atmosphere within the next 40 years. We must recognize that even if we could eliminate 100% of the CO_2 emissions in the U.S., it would not stabilize CO_2 concentrations in the atmosphere. We must develop CCS technologies that can be shared and deployed around the globe.

Our recommended strategy to address global warming includes the following principles:

Provide reasonable targets and timetables. Elements of this principle include recognition that not all the technologies needed to achieve long term goals are available, and will require time to mature and penetrate markets. Also embodied in this principle is the need to establish national policy with a single federal climate program—thus avoiding duplicative or overlapping measures, or a patchwork of state and tort-based activities.

Maintain America's competitiveness in the global economy. This principle goes beyond the basic components of targets and timetables, and includes measures to assure effective cost-containment such as a compliance safety valve or ceiling price for carbon that is certain, reasonable, economically achievable, and consistent with the need to allow time for emerging mitigation technologies to achieve commercial viability before requiring broad deployment. In addition, any program should encourage the expedited development and use of domestic and international offset projects to ensure progress in reducing global emissions at minimal cost.

Foster development and deployment of emerging low-emission technologies. There are a number of these, but for coal and natural gas the key technology is CCS. We must forge public/private sector partnerships now to invest in carbon capture, transport, storage, and conversion to beneficial uses. We must address the regulatory framework under which such technologies would be structured, including rules for injection and long-term stewardship and liability issues. The timing of reductions and introduction of emission standards must, when considered together with financial incentives, serve to encourage and not frustrate early deployment of the technology.

Promoting Improved Technologies to Reduce Emissions from Coal

Technology Is the Solution

It has been stated repeatedly in recent years that there is no silver bullet in addressing the climate challenge. We disagree. We believe that there is in fact a singular solution—albeit a multi-faceted one—and that solution is technology. This concept is not original with Arch Coal; it was well presented in *Ending the Energy Stalemate*, a December 2004 report by the National Commission on Energy Policy. NCEP concluded, correctly we believe, that current technologies were not up to the task of providing the needed reductions in greenhouse gases, and recommended setting aside a portion of "cap and trade" allowances for technology development. Advanced coal use technologies and carbon capture and storage (CCS) were both explicitly identified by NCEP as important to this technology development concept. Since the NCEP report, most comprehensive climate change mitigation bills have included measures to recycle a portion of compliance revenues to reduce the cost of advanced coal-based technologies, although it should be noted that CCS is applicable to any fossil fuel, and not limited to power production.

Of course, the list of ways in which technology can and should be brought to bear in meeting today's energy challenges is a lengthy one. We need to harness technology to help consumers use power more prudently and cost-effectively; to store energy from intermittent renewable resources such as wind and solar; to boost thermal efficiencies at power plants; and to facilitate the electrification of the automotive fleet, to name just a few. But perhaps most importantly, we must commercialize CCS technology.

In the developed world, we often hear that carbon capture and storage technology will be necessary in order for coal to continue to be used. The background information presented above should make it abundantly clear that nothing could be further from reality. The facts are that the world will continue to use coal, period—massively and in rapidly growing volumes. The question is not whether global coal use will continue and grow, but rather, whether emissions from coal will grow. The answer to that question will hinge on our ability to make CCS technology effective, and just as important, affordable.

The rest of the world has reached the same conclusion. In a recent report entitled *Breaking the Climate Deadlock: A Global Deal for Our Low-Carbon Future*, Former UK Prime Minister Tony Blair summed it up as follows: "The vast majority of new power stations in China and India will be coal-fired. Not 'may be coal-fired'; will be. So developing carbon capture and storage technology is not optional, it is literally of the essence."

We believe that helping bring CCS technology to maturation would represent an enormous contribution to the global effort to stabilize GHG concentrations in the atmosphere. If we can move the technologies forward and drive down their costs, we can not only address our domestic greenhouse gas emissions more effectively, but also equip the developing world with the kinds of tools that will provide an improved standard of living in a climate-compatible manner.

Furthermore, there is already an excellent technological foundation in place upon which we can build. Virtually every aspect of carbon capture and storage has been proven to work, and many of the key pieces are currently being used at commercial scale. For instance, the U.S. is already injecting millions of tons of carbon dioxide into the ground each year to increase recovery in declining oil fields. Coal gasification—particularly for chemical production but increasingly for power generation—is widely deployed, and provides one avenue for isolating carbon dioxide from fossil fuel prior to combustion. American Electric Power—in what we view as a watershed event—recently began capturing carbon dioxide from a portion of the flue gas of its Mountaineer power plant in West Virginia and is currently injecting it underground for permanent storage. Through a project jointly funded by AEP and DOE, they have already begun planning on a commercial-scale version of the same technology on the same generating unit.

I must emphasize that CCS remains an emerging technology. There is still not a single commercial scale power plant in the world which captures its CO_2 and injects it into a geological formation for permanent storage. The planned project by AEP will inject 1.5 million tonnes per year of CO_2 deep underground. Success at Mountaineer will be a major accomplishment, but we must remember that individual coal-fired generating units in the U.S. often emit over 6 million tonnes per year of CO_2—four times the rate to be captured by the Mountaineer demonstration project. As noted above, coal-fired power plants in the U.S. cumulatively emit over 2000 million tonnes of CO_2 per year, so the scale of the challenge is large. Moreover, the current technology options are all quite costly, and there are aspects of the process which go beyond our current legal infrastructure, such as addressing long-term stewardship and liability for stored CO_2.

Benefits of CCS

CCS technology offers three distinct types of benefits. The first, and most obvious, is that for the types of aggressive emission reduction goals that are currently being projected, it can greatly reduce compliance costs. The second is a corollary benefit: that by reducing costs; CCS enables society to seek larger overall emission reductions. It should be noted that these first two benefits are associated with all fossil fuels used with large stationary sources, not just coal. The third type of benefit is that by making an abundant low cost energy resource compatible with environmental goals, CCS allows the world to continue to derive the economic and geopolitically stabilizing benefits associated with coal.

The International Energy Agency, in *Energy Technology Perspectives—2008,* a report prepared to support the G8 Plan of Action on climate change, stated **"CO_2 capture and storage for power generation and industry is the most important single new technology. . . ."** In IEA's modeling, CCS accounted for 19% of global CO_2 reductions. Perhaps just as important, IEA evaluated multiple scenarios, and the cost of climate mitigation in a world *without* CCS was 97% more expensive than a scenario in which CCS was assumed to be demonstrated and affordable.

CCS technology is important because the electric power sector is crucial to meeting climate change goals, and because coal dominates U.S. and global power generation. The DOE/EIA analysis of H.R. 2454 concluded that "The vast majority of reductions in energy-related emissions are expected to occur in the electric power sector." For the main scenarios evaluated by EIA, the power sector contributed 80–88% of such reductions. EPA's analyses project that U.S. electricity prices, which averaged about 10 cents per killowatthour for residential customers in 2007, will increase about 80% (in constant dollars) by 2050 under HR. 2454, and by nearly 100% if critical power technologies, including CCS, are unavailable. (EPA did not evaluate the impact on electricity prices of substantial reductions in the cost of CCS.) Hence, CCS technology is crucial to both achieving targeted emission limits, and for making the global warming mitigation program affordable.

In February 2009, BBC Research and Consulting conducted a study for four major labor unions and the American Coalition for Clean Coal Electricity. The study examined the micro-economic and employment effects of deploying CCS technology on just one-fourth 7 of the U.S. coal-based generation fleet. Effects evaluated included the economic activity associated with the construction and operation of the power plants with CCS, the suppliers and support services industry for those units, and "induced effects"—the jobs created by purchases by employees in the first two categories such as for homes, automobiles, and groceries. BBC concluded that this initial CCS deployment would result in 5.5–7.0 million job-years, and after the massive ($300–400 billion) construction period was concluded, 175,000 to 250,000 permanent jobs to operate the technology.

BBC did not include an analysis of the job impacts of further research to drive down CCS costs, making electricity less costly than otherwise, and thereby improving American competitiveness in global markets. Neither did it consider the beneficial impact on capital markets of being able to retrofit existing power plants with CCS, rather than replace them with new power plants using other technologies, such as nuclear energy. Of course, reduced demand for energy capital means more capital is available for other job-producing investments. Such "macro-economic" benefits could greatly exceed the direct employment impacts. One could argue that if CCS achieved its ultimate goal of both domestic and international deployment, then that American advantage would be lost. However, there is a more persuasive argument that such success would yield even greater benefits in resolving the global climate problem, and in providing greater prosperity for all through a global reduction in basic energy costs.

The National Coal Council, an official Federal Advisory Committee to the Secretary of Energy, took a longer view on the benefits of CCS in its December 2009 report, *Low–Carbon Coal: Meeting U.S. Energy, Employment and CO_2 Emission Goals with 21st Century Technologies*. The report concluded that "Extensive deployment [through 2050] of coal-based generation with CCS will have far-reaching socioeconomic benefits, yielding **over 28 million job-years** from new construction and revitalizing the industrial sector of the U.S. GDP will be increased by **$2.7 trillion**. Further, continuing operation and maintenance of the facilities would support over **800,000 permanent jobs.**" [emphasis in original]

One might believe that *any* of the emerging "green" technologies would reduce mitigation costs and create jobs in America. However, not all technologies are created equal, and the current existing generating base and energy infrastructure provide an inherent advantage to some technologies. Whereas wind turbines and solar systems can easily be imported from overseas, the U.S. clearly has the global lead in CCS technology. And the coal and most of the natural gas that fuels CCS-equipped power plants in the U.S. will be produced in the U.S.—with obvious security of supply implications—helping our economy and providing domestic employment.

A final benefit of CCS technology, generally missing from current analyses, is its ability to reduce emissions from the transportation sector. If you carefully examine EPA's analysis of H.R. 2454, you will find that petroleum use remains relatively constant through 2050. While we seek an 83% reduction in greenhouse gas emissions, in the face of a growing population and increasing prosperity, the best that improvements in mileage can provide is the ability to stay even with current emissions. Electric vehicles could disrupt this environmental stagnation, both domestically and globally. But the increased consumption of electricity to displace oil used in the transportation sector must come from additional generating capacity. Multiple sources will be needed, and coal with CCS is certainly capable of contributing in a significant way. Success in substituting electricity for oil conveys benefits that go beyond mitigation of global warming. It is easy to see that the rise of Asian economies will place ever increasing pressure on limited supplies of crude oil and transportation fuels, often produced by countries that do not like us. The potential for CCS to relieve some of that pressure should not be casually overlooked.

Realizing the Potential

So, how should we proceed? Based on the principles articulated above, Arch believes that near- and mid-term targets should be harmonized with technology availability, which suggests a more modest target in 2020. Moreover, we believe it is imperative that we put in place cost containment mechanisms that provide greater certainty. We applaud Chairman Waxman and Chairman Markey for allowing the expansive use of offsets in meeting compliance targets—although we would have preferred fewer restrictions and limitations on those offsets. But we remain very concerned that the projected offsets market would prove to be far less robust and liquid than currently envisioned. If that proves to be the case, the cost of compliance would likely increase dramatically, as the EPA analysis suggests.

Timing is crucial. One of the greatest dangers to affordable long-term solutions is that overly aggressive near-term targets would prompt power generators to look for a short-term fix by turning to more expensive but lower-carbon fuels such as natural gas. We have been down that path before—in the first years of this decade—and the result was higher power costs and lower reliability. The National Academies of Science just released a report on America's Energy Future in which they cautioned against just such thinking. In the end, to reach the goals in recent legislative proposals and supported by various studies, we will need to apply CCS to natural gas-based power systems as well as coal-based power systems. A sudden "dash to gas" would likely eliminate interest in longer-term application of CCS, and the technology would freeze in its current state of development. All this brings me back to the key point, which is that without robust CCS technologies we cannot stabilize CO_2 concentrations in the atmosphere within the next 40 years. . . .

With respect to CCS, Arch believes that this technology is essential to meeting climate goals, and to [en]suring an affordable solution. We would encourage Congress to take the steps necessary to work with industry to accelerate the timeline for widescale deployment of this crucial technology.

Arch supports the underpinning pro-technology philosophy of the 2004 NCEP report, and most comprehensive climate bills proposed since then, and recommends:

- A substantial government/private sector collaborative effort to construct a significant number of power plants using CCS with saline geologic storage to demonstrate a portfolio of CCS technologies, and
- Continuing R&D and further cost-sharing by government on a large initial deployment of CCS facilities, on both new and existing fossil fuel based power plants, and
- Creation of a legal framework for CCS that overcomes recognized *non-technology* barriers to the technology, including certainty in the environmental rules that apply to CCS and long-term liability.

Elements of the above concepts can be found in HR. 2454, but perhaps the most complete CCS legislative proposal to date is found in a draft bill released on March 22 by Senators Rockefeller and Voinovich. This discussion draft includes:

- Continued government support for the DOE CCS research and development program.
- A CCS "Pioneer" program to supplement the DOE CCS demonstration program, with an immediate effort to demonstrate 20 GW of CCS systems. Private sector costs for these units would be supplemented by a "wires charge" placed on sale of fossil-based electric power, by federal loan guarantees, and by tax incentives.
- An "Early Adopter" program to provide tax incentives to foster deployment of another 62 GW of CCS systems.
- A performance standard that would require all new coal-based power plants to use CCS technology, once the above programs have demonstrated that the technology is effective and reliable.
- A placeholder for future language addressing the long-term liability issue.

Arch is optimistic about the Rockefeller-Voinovich package because it could move forward immediately. We should not squander valuable time needed to advance this critical technology.

An alternative approach to greatly increase the number of CCS demonstration facilities in the short term and reduce the amount of CO_2 released per unit of energy consumed would be to expand current legislative proposals for a federal renewable electricity standard to include other clean electricity options: specifically, fossil fuel generation with CCS, advanced nuclear power generation, and improved efficiency at existing power generation facilities. This broadening of the RES was rejected by the Senate Energy Committee, but Arch continues to believe that it would be a pragmatic mechanism to establish a portfolio of improved low carbon options from which new generation markets could choose.

These recommendations constitute a pragmatic action plan to achieve aggressive environmental goals, both domestically and globally, without sacrificing economic prosperity. The proposals outlined above would allow the nation to begin building CCS systems at a pace that would otherwise be unachievable. Industry has repeatedly demonstrated its support for this technology-based solution to global warming, but we need to see shared determination and support from the federal government to get the job done.

Susan Moran

Coal Rush!

For all the talk about renewable energy, the extraction industry is alive and well in the United States. In Texas, utility company TXU Energy has proposed building 11 new coal-fired power plants. American Electric Power, of Columbus, Ohio, is also seeking approval for several new coal plants. In the interior West, Minnesota-based Xcel Energy is building one of the nation's largest coal plants in Pueblo, Colorado, while planning for others in the next few years.

These are but a sampling of the 150-plus new coal-fired plants being proposed throughout the United States, according to the Department of Energy's National Technology Laboratory. The vast majority of them will use conventional coal burning technology. (Elsewhere in the world, hundreds of new coal-fired plants are planned or under construction, some 550 of them in China alone. Of the 2004 U.S. total emissions of 5.9 billion metric tons of carbon dioxide (CO_2), electric power generation contributed 2.3 billion metric tons, or 39 percent, and coal-fired plants accounted for 82 percent of that. Because CO_2 is the dominant human-caused greenhouse gas, that unheralded feat makes power plants among the biggest culprits behind climate change. And while power plant emissions of sulfur dioxide, nitrogen oxides, and particulates have dipped since the early 1990s, thanks to federal legislation, CO_2 emissions continue unabated. In fact, coal is making a comeback.

For nearly three decades no new U.S. coal plants came on line. But suddenly that's changed. "No question about it, this is a coal rush," says Robert McIlvaine, a coal industry consultant in Northfield, Illinois. "This is on par with the biggest expansion coal has ever seen, assuming all the ones proposed are actually built," adds Travis Madsen, a policy analyst with U.S. Public Interest Research Group (PIRG).

Nationwide support for coal dates back several decades. Between 1950 and 1997 the coal industry received more than US$70 billion in federal subsidies, or nearly $1.5 billion a year, according to a 2006 report from PIRG. In the Energy Policy Act of 2005, Congress approved an additional $7.8 billion for coal, including several billion for a "clean coal" research and development program.

But what accounts for the current feverish pace? Utilities point to all of us, our insatiable appetite for the must-have gadgets of the modern world, such as laptops and iPods, as well as plasma televisions, refrigerators, and other household appliances. And as the population grows, that demand—absent any serious drive for higher efficiency—will only head upward. "In the last decade our customer base has gone up 20 percent, and the amount of electricity per household is up 10 percent," says Mark Stutz, a spokesman for Xcel, which generates 66 percent of its kilowatthours of electricity from coal. "That's a lot of demand."

Coal is especially attractive now because its prices are low and stable relative to those of natural gas, which fluctuate unpredictably. But some critics of coal suspect that the current rush has more to do with fear and greed. With the specter of climate change looming ever larger in the public consciousness, utilities are anticipating that the time will soon come when legislators will slap a limit on carbon emissions from electric power generators and perhaps other industrial sources, and that the more coal-fired capacity the producers build before that day of reckoning, the higher their share of the total cap will be.

Of course, utilities have choices. Some climate change is inevitable, and it will get worse before it gets better. But utilities can still do plenty to stabilize and even reverse the course of climate change by dramatically cutting back on carbon emissions and by advancing solar, wind, and other renewable energy sources. "We're at a really interesting fork in the road," says Jim White, a paleoclimatologist at the University of Colorado in Boulder. "Do we recognize that continuing to burn these fuels will cause a lot more problems than they solve, and therefore do we promote alternative energy supplies? Or are we going to go down the road of coal?"

Gambles

Many utilities are hedging their bets by investing to some degree in renewables and even nuclear power, and a few have even asked the federal government for uniform, nationwide carbon restrictions now. But the industry in general is choosing the coal road for the time being. Charlotte, North Carolina-based Duke Energy, which generates 54 percent of its electricity from coal and 46 percent from nuclear power, in September appealed to state utility regulators to construct two coal-fired plants totaling 1,600 megawatts of generation capacity. "Abandoning coal and moving to another source, such as natural gas, is not a viable economic option for our company or its customers," James Rogers, Duke Energy's chief executive officer, wrote in an article published in July for the American Air & Waste Management Association. He said the company's views on climate change "reflect the need to ensure a future role for coal in our nation's electricity generation portfolio."

Ohio-based American Electric Power is about as coal dependent as they come; its coal-fired plants account for 73 percent of its capacity. The company plans to build several new coal plants, including some with cleaner-burning coal (called integrated gasification combined cycle, or IGCC) technology that can be equipped to capture and sequester carbon. "Coal is a domestic source and there's a vast amount of it now," says Melissa Henry, a spokeswoman for American Electric, which owns units in coal-rich West Virginia, as well as in Tennessee, Indiana, Kentucky, and other states. She said the company's reliance on coal will likely continue at the same level in the future, but that investments in IGCC and other "clean" coal technologies will help mitigate the company's carbon footprint. "The reality is that we will have to continue to use coal to generate electricity to meet demand in the U.S.," she says.

Perhaps the most brazen example of the current rush to coal is TXU Energy, which boasts it is planning "a Texas-sized $10 billion investment" in 11 conventional coal-fired power plants across Texas. Collectively, they would emit 78 million metric tons of CO_2 per year, doubling the utility's emissions. Those 78 million tons are more than the total 2001 emissions of 21 different U.S. states or many countries, including Sweden, Denmark, and Portugal. While several utilities are introducing cleaner, advanced technologies, TXU's proposed power plants are to be pulverized coal-fired generators—old technology that is difficult and expensive to retrofit later to capture carbon.

(In late summer, the company hit a speed bump, however, when judges in Texas ruled that the proposed pollution controls of one of TXU's proposed plants—a 1,720-megawatt generator—weren't proven and that the plant's emissions could harm downwind cities. That ruling was heralded by environmentalists as it suggests that utilities may be under increasing scrutiny in their push to build more coal plants nationwide.)

TXU and other companies generally point to steep natural gas prices and customers' rabid demand for more electricity as the incentives for their coal plant buildup. But some analysts question whether this strategy even makes sense economically, much less politically and morally. "A lot of utilities are ignoring the whole issue. They're pretending there's not going to be a climate policy" says Bruce Biewald, president of Synapse Energy Economics, a research and policy organization in Cambridge, Massachusetts. "We're saying, 'get your heads out of the sand.' Your particular decision, say, to build out fossil-fuel plants in Texas while pretending there'll be no costs to carbon emissions is simply imprudent."

While there is no mandated nationwide emission limit for carbon dioxide, thanks largely to the fact that the Bush administration wouldn't ratify the Kyoto Protocol, several bills being mulled in Washington call for a cap and various forms of carbon trading programs to help reward cleaner emitters and punish the dirtier ones. In Europe such a market, called the European Trading System, already exists, and California and other individual states, as well as a group of northeastern states, are already setting themselves up for a cap-and-trade system. In essence, such a system could allocate to utilities and other emitters allowances based on their emissions levels, or the caps could be auctioned, as many economists prefer.

Synapse estimates that if U.S. companies had to comply now with the Kyoto Protocol the cost of emitting carbon would be $20 to $50 a ton of carbon dioxide. (TXU's proposed new coal plants' emissions, therefore, might cost the utility up to $3.9 billion every year if it had to pay outright for them.) Biewald says companies that try to "grandfather" in new coal plants so they can get free allowances are banking on a "highly risky strategy." The key precedent in the United States that Biewald and others point to is the successful sulfur dioxide cap-and-trade program spurred by the 1990 Clean Air Act. It allocated allowances to polluters based on their emissions levels from several years prior to the ruling. Another model is the California Climate Registry, which allows utilities and other companies to claim credit for acting early. The message is, do good now and you'll be rewarded once a cap-and-trade system is in place. . . .

Energy Costs

Whatever kind of emissions mandates the courts or politicians may impose, many utilities (backed by Department of Energy incentives and research projects) are developing "cleaner" coal burning technologies. The most conservative form applies to conventional pulverized-coal plants—making them burn more efficiently and creating higher steam temperatures and water pressures, for instance. These are called "supercritical" and "ultra-supercritical" steam boiler technologies. At best, such plants, intended to operate for at least 50 years, would make coal combustion at least 46 percent efficient, compared with an average 30 percent efficiency rate of conventional pulverized plants. But critics view these plants as a mere band-aid approach to carbon emission reduction and, more important, a diversion from investing in much cleaner coal technology and renewable energy sources. These plants would also require as much water for cooling as conventional generators.

IGCC, mentioned above, is the least polluting coal burning technology being developed today. The advantages of IGCC over pulverized coal technology are that it is more efficient (using less coal to produce each kilowatt-hour of electricity), uses much less water, and emits much less SO_2, particulates, and mercury. Most important, carbon capture can be integrated into an IGCC plant much more easily and cheaply than into a pulverized coal plant. Jana Milford, a senior scientist with the U.S. NGO Environmental Defense, says that IGCC combined with sequestration can cut CO_2 emissions by 90 percent compared with conventional plants.

Sequestration will not be cheap, however. In fact, Environmental Defense, along with Western Resource Advocates (WRA), another environmental research and policy organization, took the unusual step last summer of publicly praising Xcel Energy for proposing the nation's first IGCC plant designed from the beginning to capture and sequester carbon. That is critical, because without capturing and sequestering carbon, IGCC plants do little to reduce carbon emissions beyond their higher efficiency (i.e., slightly lower kilograms of CO_2 per kilowatt-hour produced). (Xcel isn't off the hook, however. Some environmentalists continue opposing the company's ongoing construction of a 750 megawatt supercritical pulverized coal plant in Pueblo, Colorado, next to two existing facilities it operates there.) Currently two other commercial-scale IGCC plants are operating in the United States, one in Indiana and one in Florida, but neither entails capture and sequestration. Several other companies, including Duke Energy, American Electric Power, and TXU, have proposed IGCC plants, but they won't necessarily incorporate capture and sequestration of carbon emissions either.

Utilities often cite the high capital costs of IGCC plants. According to WRA, however, electricity produced with current IGCC technology is estimated to be only 5 to 10 percent more costly than that from a new pulverized coal plant if no carbon capture is included. If carbon capture is considered across all plant types, electricity generated with IGCC will cost 18–32 percent less than pulverized coal technologies, WRA asserts. Retrofitting old-technology power plants to capture carbon is extremely difficult because the entire fluegas stream must be processed to remove the CO_2. With IGCC, the carbon removal occurs early in the generation cycle and involves processing much lower gas volumes.

Many environmentalists are skeptical of IGCC and all other "clean" coal technologies. Perhaps the most vocal and influential of them is former U.S. vice president Al Gore. "It is time to recognize that the phrase 'clean coal technology' is devoid of meaning unless it means 'zero carbon emissions' technology," he said in an address to students at New York University in September. Others, like WRA energy project director John Nielsen, argue that cleaner coal technologies will help some but are not enough by themselves. "We're looking at emissions trajectories over time still going up. They need to go down," says Nielsen. "But eventually we'll need some federal limits on overall CO_2 emissions to bring them down."

True Prices

Whether they dig in their heels or willingly leap forward, utilities will probably face some kind of carbon constraints in the future. The sooner they clean up their act, the lower their future costs will likely be. For many, even politicians, the question is not should we put a price on carbon, but what kind of price, and who pays.

In September the Congressional Budget Office (CBO) published a report on the role of CO_2 pricing. The report notes that human activities are increasing the concentrations of CO_2 and other greenhouse gases in the atmosphere and acknowledges that energy markets fail to capture the "external effects" of emissions from fossil fuel combustion, that is, the costs that are imposed on society by the use of fossil fuels but are not reflected in the prices paid for them. Setting a price (i.e., tax) on carbon emissions would help change industry and consumer behavior, the report says, suggesting that carbon emissions could be assigned prices by taxing fossil fuels in proportion to their carbon content or by establishing a cap-and-trade program. Under such a program, policymakers would set an overall cap on emissions but leave it to carbon emitters to trade rights (called allowances) to those capped emissions.

So far the carbon taxing approach has gained little traction among politicians. Some call it political suicide to even mention the word "tax." But cap-and-trade bills proposed by various politicians are gaining momentum. The one that has enjoyed the most bipartisan support in the US. Senate was introduced in mid-2005 by Jeff Bingaman, a Democrat from New Mexico and a member of the Energy and Natural Resources Committee. Bingaman's proposal called for a cap-and-trade system that would require emitters to gradually stop the growth of their emissions by 2020 but not actually reduce emissions from today's levels. Further, the legislation would place an upper limit for pollution credits of $7 per metric ton of emissions—a modest ceiling aimed at keeping polluters' costs relatively low and predictable. By contrast, carbon credits in the European market, called the European Trading System, have to date traded at substantially higher prices. Some environmentalists have called Bingaman's proposal too soft on businesses.

Although utility executives have been reluctant to publicly endorse any one piece of legislation, as noted above some are increasingly showing support for some form of federally mandated cap on carbon emissions. Executives from Exelon Corporation of Chicago, American Electric Power, Duke Energy, General Electric, and PNM Resources all testified last spring before the Energy and Natural Resources Committee that indeed they want and are ready for carbon regulations. More recently, Xcel Energy chief Richard Kelly voiced his desire for mandatory limits on carbon emissions.

Most utilities that do support federal mandates want any caps to apply to all sectors across the economy rather than just their own, and they prefer waiting for nationwide limits rather than supporting or joining a patchwork of state or regional initiatives. "We believe the science of climate change is real, and we recognize that our industry, the electric sector, contributes roughly one-third of CO_2 emissions in the U.S.," says Helen Howes, vice president of environmental health and safety at Exelon. "But the transportation sector also is a big producer."

As the power companies wait for federal legislation (which few expect to pass before the 2008 elections), they are keeping a close watch on the various state and regional carbon-trading initiatives as well as the privately run and voluntary Chicago Climate Exchange, which now has more than 100 members spanning many industries. Although members' commitments to reduce their greenhouse gas emissions are not enforced, members and industry analysts say the exchange will likely be a model for any federal legislation. The biggest and most-watched regional carbon trading system yet to be crafted is the Regional Greenhouse Gas Initiative. It began in 2005 when seven northeastern states reached an agreement to cap carbon dioxide emissions from their power plants at current levels and to cut those emissions by 10 percent by 2019. The system is slated to go into effect in 2009. (Maryland has since joined the group as well.)

Meanwhile, California has taken the lead among individual states on the climate change front. The state's Democrat-controlled legislature and the Republican governor, Arnold Schwarzenegger, agreed last August on sweeping legislation aimed at reducing CO_2 emissions by 25 percent by 2020. Although this pledge doesn't meet the Kyoto Protocol's mandate to cut carbon emissions to 7 percent below 1990 levels, it is the strongest pledge yet by any state. Further, California has ruled that its electricity distributors will not purchase power unless it meets the lowest possible greenhouse gas emission standards. That ruling will force producers outside the state to accommodate, or lose a huge chunk of their revenues. Pacific Gas and Electric, one of California's regulated electric utilities, already has phased out much of its fossil fuel sources. Forty-six percent of the electricity it delivers to customers comes from fossil fuels, primarily natural gas. Only 1 percent comes from coal, according to utility spokeswoman Darlene Chiu.

Clearly, state and region-wide efforts to cut carbon emissions are grabbing the attention of utilities and the public alike. "What this means is that states are not willing to wait for the Bush administration to catch on," says Dale Bryk, an attorney with the Natural Resources Defense Council, which helped design the northeastern states' program.

Ultimately, many scientists, economists, environmentalists, legislators, and a growing number of utility executives agree that a multi-pronged approach—technology, carbon markets, and public policy—will be necessary to attack global warming. "A responsible approach to solutions would avoid the mistake of trying to find a single magic 'silver bullet'," Gore said in his NYU speech in September. The answer, he added, quoting environmental writer Bill McKibben, lies in a "silver buckshot" approach—an effective blend of several solutions.

EXPLORING THE ISSUE

Should Utilities Burn More Coal?

Critical Thinking and Reflection

1. In what sense does expansion of coal use depend on making carbon capture and storage (CCS) effective and affordable?
2. What are the environmental impacts of using coal for energy?
3. Why is the use of coal for energy likely to expand even if we cannot make carbon capture and storage (CCS) effective and affordable?
4. Why, despite its obvious drawbacks, does coal remain an attractive energy source?

Is There Common Ground?

Civilization requires an ample supply of energy. Both sides agree that coal will continue to be used; the question is whether CCS technology will prove effective and affordable, regulators will require coal-burning power plants to use the technology, and utilities will accept the regulations. Susan Moran, perhaps cynically, says that utilities are rushing to get new coal-burning power plants up and running before such regulations can appear.

If you think new regulations should apply to existing power plants, think again. Owners of existing plants will lobby intensively to have those plants "grandfathered"—meaning only the regulation in force at the time of their construction apply. As an exercise, look up "grandfathering environmental" (even Google will give you plenty of useful material) and answer the following questions.

1. What is grandfathering?
2. Is it likely to be applied to new coal-burning power plants if they can be built before regulators require the use of CCS technology?
3. How might grandfathering be prevented?

Internet References . . .

Intergovernmental Panel on Climate Change (IPCC)

The IPCC was formed by the World Meteorological Organization (WMO) and the United Nations Environment Programme (UNEP) to assess the scientific, technical, and socio-economic information relevant for understanding whether the climate is changing, humans have a role in causing the change, and what the consequences are likely to be.

http://www.ipcc.ch/

Climate Change

The United Nations Environmental Program maintains this site as a central source for substantive work and information resources with regard to climate change.

http://climatechange.unep.net/

International Emissions Trading Association (IETA)

The IETA works to develop an active, flexible, global greenhouse gas market, consistent across national boundaries.

http://www.ieta.org

The U.S. Department of Energy—Carbon Capture

The DOE's Carbon Capture page provides information on current research in this area.

http://www.fossil.energy.gov/programs/sequestration/capture/

Weather Modification Association

The Weather Modification Association promotes research and development of technologies such as cloud seeding for increasing rainfall.

http://www.weathermodification.org/

University Corporation for Atmospheric Research

The University Corporation for Atmospheric Research and the National Center for Atmospheric Research are part of a collaborative community dedicated to understanding the atmosphere—the air around us—and the interconnected processes that make up the Earth system, from the ocean floor to the Sun's core.

http://www.ucar.edu/

UNIT 2

Global Warming

Whatever you call it—the greenhouse effect, global warming, or climate change—it is a major topic of concern and debate around the world today. It is a consequence of the way human activities have added large amounts of carbon dioxide and other "greenhouse gases" to the atmosphere, and among its likely future impacts are rising average temperatures, rising sea level, and spreading disease. Understanding the issues in this unit of the book is essential to understanding what is happening and what society can do about it.

- Is Human Activity Responsible for Global Warming?
- Is Global Warming a Catastrophe That Warrants Immediate Action?
- Will Restricting Carbon Emissions Damage the Economy?
- Is Carbon Capture Technology Ready to Limit Carbon Emissions?
- Is It Time to Think Seriously About "Climate Engineering"?

ISSUE 7

Is Human Activity Responsible for Global Warming?

YES: Mary-Elena Carr, Kate Brash, and Robert F. Anderson, from "Climate Change: Addressing the Major Skeptic Arguments," Deutsche Bank Climate Change Advisors (September 2010)

NO: Alex Newman, from "Global-Warming Alarmism Dying a Slow Death," *New American* (April 12, 2010)

Learning Outcomes

After studying this issue, students will be able to:

- Explain how human activities contribute to future climate change.
- Explain what can be done to cope with future climate change.
- Explain why the nature of science helps critics say the science is not certain.
- Discuss how funding sources and other sources of bias can affect the conclusions a scientist reaches.
- Explain how action to prevent global warming may affect some industries.

ISSUE SUMMARY

YES: Mary-Elena Carr, Kate Brash, and Robert F. Anderson argue that although scientists continue to work on improving our understanding of how carbon emissions affect climate, it is clear that human activities affect climate and that preventive efforts are justified. So-called skeptics misrepresent the science, the adequacy of computer models of climate, the motives of researchers, and the need for action.

NO: Alex Newman argues that critics have revealed so many defects in the science and scientists who support global warming that the climate-crisis crusade is clearly failing, although it is not likely to vanish until after a prolonged battle between the skeptics and alarmists.

Scientists have known for more than a century that carbon dioxide and other "greenhouse gases" (including water vapor, methane, and chlorofluorocarbons) help prevent heat from escaping the Earth's atmosphere. In fact, it is this "greenhouse effect" that keeps the Earth warm enough to support life. Yet, there can be too much of a good thing. Ever since the dawn of the industrial age, humans have been burning vast quantities of fossil fuels, releasing the carbon they contain as carbon dioxide. Because of this, scientists estimate that by the year 2050, the amount of carbon dioxide in the air will be double what it was in 1850. By 1982, an increase was apparent. Less than a decade later, many researchers were saying that the climate had already begun to warm. Today, there is a solid consensus that human-caused climate change is a genuine problem; see Joseph Romm, "The Cold Truth about Climate Change," *Salon* (February 27, 2008) (http://www.salon.com/news/feature/2008/02/27/global_warming_deniers/print.html).

Debate over the reality of the warming trend and its significance for humanity and the environment has been vigorous. Much of this debate has been over the quality of the data and the conclusions drawn from the data, which are crucial in any discussion of global warming (and other issues). If the data and conclusions are not solid, they cannot be used to go further, as in forming public policy designed to ward off disaster that people think the data allow us to predict. If the data *are* solid, however, then moving from the data to public policy may be urgent.

In June 2006, the National Academy of Sciences reported that the Earth is now warmer than it has been in the last 400 years, and perhaps in the last 1000 years (*Surface Temperature Reconstructions for the Last 2,000 Years* (National Academies Press, 2006)). Concerns have been raised about the risks to coastal populations from rising seas and changes in storm patterns; see John Young, "Black Water Rising," *World Watch* (September/October 2006). Nicholas Stern, "Stern Review on the Economics of Climate Change," (October 30, 2006) (www.hm-treasury.gov.uk/independent_reviews/stern_review_economics_climate_change/sternreview_index.cfm), reports that although taking steps now to limit future impacts of global warming would be very expensive, the economic and social impacts of not doing so will be much more expensive. According to Richard A. Kerr, "Pushing the Scary Side of Global Warming," *Science* (June 8, 2007), some climate researchers are concerned that we are seriously underestimating how disastrous global warming will be. Indeed, recent projections suggest sea level may rise as much as 1.5 meters (5 feet) by 2100 (see "Global Sea Levels Set to Rise Above IPCC Forecast," *Geographical* (June 2008)).

On May 29, 2008, the Bush Administration released "Scientific Assessment of the Effects of Global Change on the United States: A Report of the Committee on Environment and Natural Resources, National Science and Technology Council" (available at http://www.ostp.gov/cs/nstc), describing the current and potential impacts of climate change. In sum, it says that the evidence is clear and getting clearer that global warming is "very likely" due to greenhouse gases largely released by human activity and there will be consequent changes in precipitation, storms, droughts, sea level, food production, fisheries, and more. Globally, "poor communities can be especially vulnerable, particularly those concentrated in high-risk areas." Dealing with these effects may require changes in many areas, particularly relating to energy use, but "significant uncertainty exists about the potential impacts of climate change on energy production and distribution, in part because the timing and magnitude of climate impacts are uncertain." See Susan Milius, "Already Feeling the Heat," *Science News* (Web edition) (May 29, 2008).

Consonant with the Bush Administration's previous position that imposing restrictions on fossil fuel use or carbon emissions is a bad idea both because of uncertainty about global warming and because it would harm the economy, mitigation of the risks—which the report says are likely or very likely—is barely mentioned. U.S. goals should be limited to reducing uncertainty in projections of how the Earth's climate and related systems may change in the future. "Reducing uncertainty is crucial to providing decision makers with tools for assessing strategies for adaptation, mitigation, and other forms of risk reduction." Richard Moss of the World Wildlife Foundation, a past director of the CCSP, says that because of its shortcomings, the report fails to meet the needs of the public.

On the other hand, the public may be losing interest. Richard A. Kerr, "Amid Worrisome Signs of Warming, 'Climate Fatigue' Sets In," *Science* (November 13, 2009), notes that many people do not believe the problem is real, perhaps because climate change is too long term and big picture for people to see on a local scale. Long-term, big-picture consequences were highlighted in 2010, when the National Academies of Science, Engineering, and Medicine's Committee on Stabilization Targets for Atmospheric Greenhouse Gas Concentrations released *Climate Stabilization Targets: Emissions, Concentrations, and Impacts over Decades to Millennia (*National Academies Press, 2010), stressing the seriousness of the problem and the need for mitigation efforts.

The price tag for immediate action to prevent or—with luck—reduce the negative impacts of global warming will not be small, but it should be worth it in many ways. Gregg Easterbrook, "Global Warming: Who Loses—and Who Wins?" *The Atlantic* (April 2007), concludes that "Keeping the world economic system and the global balance of power the way they are seems very strongly in the U.S. national interest—and keeping things the way they are requires prevention of significant climate change. That, in the end, is what's in it for us."

Skeptics such as Patrick Frank, "A Climate of Belief," *Skeptic* (vol. 14, no. 1, 2008), argue that the claim that human-caused carbon dioxide emissions are changing climate is insupportable because computer models of climate are imperfect. Roy W. Spencer, "How Serious Is the Global Warming Threat?" *Society* (July 2007), argues that the science of global warming is not as certain as the public is told, but even if predictions of strong global warming are correct, it is not at all clear what the best policy reaction to that threat should be. Massive reductions in greenhouse gas emissions will require new energy technologies, which are most likely to be developed in the countries that can afford massive energy R&D efforts. Therefore, draconian, government-mandated controls on emissions could very well hurt, rather than help, efforts to develop those new technologies. Mike Carey, president of the Ohio Coal

Association, testified before the U.S. House of Representatives' Select Committee on Energy Independence and Global Warming on April 14, 2010, that such problems indicate no need for legislation to combat global warming, and certainly no need for restrictions on the use of fossil fuels such as coal. Unfortunately, many if not most climate-change skeptics have reportedly been funded by the fossil fuel industry, which renders their skepticism highly suspect; see Kate Sheppard, "Inside Koch's Climate Denial Machine," *Mother Jones* (blog; April 1, 2010) (http://motherjones.com/blue-marble/2010/04/inside-kochs-climate-denial-machine), and "Koch Industries: Secretly Funding the Climate Denial Machine," *Greenpeace* (March 2010) (http://www.greenpeace.org/usa/en/media-center/reports/koch-industries-secretly-fund/). See also Union of Concerned Scientists, "Smoke, Mirrors & Hot Air: How ExxonMobil Uses Big Tobacco's Tactics to Manufacture Uncertainty on Climate Science" (January 2007) (http://www.ucsusa.org/global_warming/science/exxonmobil-smoke-mirrors-hot.html), and David Michaels, *Doubt Is Their Product: How Industry's Assault on Science Threatens Your Health* (Oxford University Press, 2008).

In the YES selection, Columbia University's Mary-Elena Carr, Kate Brash, and Robert F. Anderson, with an introductory editorial by Mark Fulton, Global Head of Climate Change Investment Research for Deutsche Bank Group, argue that although scientists continue to work on improving our understanding of how carbon emissions affect climate, it is clear that human activities do affect climate and that preventive efforts are justified. So-called skeptics misrepresent the science, the adequacy of computer models of climate, the motives of researchers, and the need for action. In the NO selection, Alex Newman argues that critics have revealed so many defects in the science and in the character of scientists who support global warming that the climate-crisis crusade is clearly failing, though it is not likely to vanish until after a prolonged battle between the skeptics and alarmists.

Mary-Elena Carr, Kate Brash, and Robert F. Anderson

Climate Change: Addressing the Major Skeptic Arguments

Addressing the Climate Change Skeptics

The purpose of this paper is to examine the many claims and counter-claims being made in the public debate about climate change science.

For most of this year, the volume of this debate has turned way up as the 'skeptics' launched a determined assault on the climate findings accepted by the overwhelming majority of the scientific community. Unfortunately, the increased noise has only made it harder for people to untangle the arguments and form their own opinions. This is problematic because the way the public's views are shaped is critical to future political action on climate change.

For investors in particular, the implications are huge. While there are many arguments in favor of clean energy, water and sustainable agriculture—for instance, energy security, economic growth, and job opportunities—we at DB Climate Change Advisors (DBCCA) have always said that the science is one essential foundation of the whole climate change investment thesis. Navigating the scientific debate is therefore vitally important for investors in this space.

For these reasons, we asked our advisors at the Columbia Climate Center at the Earth Institute, Columbia University, to examine as many as possible of the major skeptic claims in the light of the latest peer reviewed scientific literature and to weigh the arguments of each side in the balance. Although the scientific community has already addressed the skeptic arguments in some detail, there is still a public perception that scientists have been dismissive of the skeptic viewpoint, so the intention in this report is to correct the balance. The result is, we believe, a balanced, expert, and detailed assessment of the scientific case for climate change that will help investors navigate these extremely complex issues.

The paper's clear conclusion is that the primary claims of the skeptics do not undermine the assertion that humanmade climate change is already happening and is a serious long term threat. Indeed, the recent publication on the State of the Climate by the US National Oceanic and Atmospheric Administration (NOAA), analyzing over thirty indicators, or climate variables, concludes that the Earth is warming and that the past decade was the warmest on record. Quantifying cause and effect or projecting future conditions is always incomplete in a system as complex as Earth's climate, where multiple factors impact the observations. Conclusions are thus presented in terms of probabilities rather than dead certainties. This uncertainty is not always adequately explained in the public debate and, when discussed, can appear to be a challenge to the credibility of the field. However, uncertainty is an inevitable component in our understanding of any system for which perfect knowledge is unattainable, be it markets or climate.

To us, the most persuasive argument in support of climate change is that the basic laws of physics dictate that increasing carbon dioxide levels in the earth's atmosphere produce warming. (This will be the case irrespective of other climate events.) The only way that warming can be mitigated by natural processes is if there are countervailing 'feedback mechanisms', such as cooling from increased cloud cover caused by the changing climate. A key finding of the current research is that there has so far been no evidence of such countervailing factors. In fact, most observed and anticipated feedback mechanisms are actually working to amplify the warming process, not reduce it.

Simply put, the science shows us that climate change due to emissions of greenhouse gases is a serious problem. Furthermore, due to the persistence of carbon dioxide in the atmosphere and the lag in response of the climate system, there is a very high probability that we are already heading towards a future where warming will persist for thousands of years. Failing to insure against that high probability does not seem a gamble worth taking.

Introduction

In response to a growing body of research pointing to human-induced warming of Earth's climate, and in recognition of the global nature and potentially sweeping implications of a changing climate, the world's governments have launched national and international efforts to periodically assess the state of knowledge in the many areas of research that bear on climate change. The Intergovernmental Panel on Climate Change (IPCC), a consultative body of scientists from around the world, was established in 1988 under the auspices of the United Nations Environment Program and the World Meteorological Organization. Every six years the IPCC publishes a summary and review of hundreds of peer-reviewed studies relating to the state of knowledge about climate change. The IPCC reports serve as a common, authoritative source and they are a critical tool for enabling an effective international response.

The IPCC's Fourth Assessment Report (hereafter IPCC AR4), released in 2007, states that "[W]arming of the climate system is unequivocal, as is now evident from observations of increases in global average air and ocean temperatures, widespread melting of snow and ice and rising global average sea level." Furthermore, "[M]ost of the observed increase in global average temperatures since the mid-20th century is *very likely* due to the observed increase in anthropogenic GHG [*greenhouse gas*] concentrations." This conclusion is similar to that of the US Global Change Research Program in their latest assessment: "global warming is unequivocal and primarily human-induced." Most recently, the National Academy of Science summarized that "Climate change is occurring, is caused largely by human activities, and poses significant risks for—and in many cases is already affecting—a broad range of human and natural systems."

However, a small but vocal group of individuals and organizations, including some scientists, argue that the current scientific evidence is not sufficient to conclude that human-induced climate change is underway or that it poses a clear and present danger to society. Emails stolen from the University of East Anglia's Climatic Research Unit or CRU and mistakes recently identified in the IPCC AR4 have exacerbated these criticisms.

These critics contend that "the science isn't settled." This is entirely correct in the sense that the scientific process of discovery, testing, peer review and re-assessment carries on. Significant uncertainties about specific dynamics within the climate system persist and scientists continue to pursue research aimed at filling gaps in our knowledge. For example, climate models make poor predictions at the regional scale, and our understanding of cloud dynamics, the distributions and characteristics of aerosols, the rates of glacial melt, and the magnitude of many feedback processes (such as vegetation responses) is still incomplete.

Despite these unanswered questions, the role of carbon dioxide as a driver of Earth's present and future climate is borne out by the increasing body of observations. This is seen most clearly in the annual overviews of climate and weather observations compiled by the National Oceanic and Atmospheric Administration. Over thirty indicators, or climate variables, have been identified to characterize the earth system, including atmospheric levels of greenhouse gases, air temperature, sea level, and extreme weather events. These indicators, which are measured using a wide array of observational techniques, are interpreted in a historical and global context. Some of the variables, like air temperature, are directly related to whether the climate is warming. Other indicators, such as ocean currents or precipitation patterns, need to be monitored over time to understand how the climate system evolves in response to natural and anthropogenic drivers. The latest report, *State of the Climate in 2009*, concludes that Earth is warming and that the past decade was the warmest on record. Even if the thousands of weather stations that report air temperature were suspect, other indicators are consistent with this warming: glaciers continue to lose mass, northern hemisphere snow cover is falling as spring melt occurs earlier, sea levels are rising, and summer Arctic sea ice cover has been on a steadily decreasing trend. Progress in our understanding and more conclusive observational evidence are also manifest in the increased certainty of expressions in successive IPCC reports and of the evolving research. For example, Stott et al. concluded that human influence in regional climate impacts is now discernible throughout the globe based on a review of studies published since the last IPCC report.

Science continues to progress as a discussion among peers and specialists within and across disciplines. Creative and careful research aiming for improved understanding is vital, preferably verified and evaluated via the peer-review process. Although much insight can be gained by scrutiny from people in other fields, the complexity and breadth of climate science do not make it readily or easily accessible in its entirety. Expertise in quantum mechanics does not qualify a person to perform a heart transplant, nor would a surgeon be likely to identify flaws in a finding from particle physics.

Critiques by people from outside the scientific community who aim to "audit" the research process can help identify errors or lead to improved understanding. These criticisms sometimes highlight insufficient transparency within the scientific process. Both the hearing of the UK House of Commons Science and Technology Committee and the Independent Climate Change Email Review into whether the stolen CRU emails indicated misconduct recommended more open sharing of both data and methods. More transparency is always desirable, but making data sets and methods accessible to an untrained public would require significant additional resources for the research endeavor.

Many criticisms, however, center on matters that have been resolved scientifically or on selective use of observations that may be misinterpreted. Cook outlined five characteristics of attacks on science (originally from a study that focused on public health . . . that apply to many claims from those skeptical of climate science: **(1) conspiracy theories**, by which the existence of a large body of accepted evidence is itself purported to be proof of a conspiracy, as has been expressed about the IPCC report; **(2) fake experts**, the presentation as experts of people with scant training in the field; **(3) selectivity**, by which isolated studies or graphs are presented out of context; **(4) impossible expectations**, the practice of demanding research to provide greater certainty than the study system permits, such as complete weather predictability; and **(5) use of misrepresentations and logical fallacies**, including straw man arguments, such as "CO_2 isn't the only driver of climate;" this true statement is integral to our understanding, but is largely irrelevant for the case about anthropogenic change.

This study aims to respond to the most common misconceptions that are presented to challenge the position that GHG emissions are adversely impacting Earth's climate and will continue to do so. . . .

Executive Summary

Periodic summaries and reviews of the state of knowledge about Earth's climate come from several institutions, including the Intergovernmental Panel on Climate Change (IPCC) and national science academies of countries worldwide. These entities have concluded that the increasing body of observations is consistent with the physical principles by which greenhouse gases (GHGs) affect climate: the planet is warming and it will likely continue to warm due to GHG emissions. Although continued research is needed to quantify the timing, location, and extent of climate impacts, many experts are confident that the precautionary principle justifies action to reduce emissions. However, some individuals and organizations dispute this conclusion, asserting that the science isn't settled. These arguments fall into three general categories: Earth is not warming; Earth may be warming but human activity is not responsible; and Earth may be warming, and humans may be responsible, but we don't need to act to stop it. Here we briefly respond to the primary claims made under each of these categories with the current state of scientific knowledge. Each claim is addressed in detail in the body of the report.

Earth Is Not Warming

Claim: Global average temperatures have not risen since 1998. Multiple factors affect global average temperatures, including the long-term warming trend from GHGs. This time-varying interaction of climate drivers can lead to periods of relatively stable temperatures interspersed with periods of warming. The anomalously high global average temperatures in 1998 associated with the El Niño have been followed by comparably high values that reflect a combination of long-term warming and shorter-term natural variability. Periods of relatively constant temperature are not evidence against global warming; in fact, the decade of 2000 to 2009 is the warmest in the instrumental record.

Claim: Climate researchers are engaged in a conspiracy: global warming is a hoax. There is no evidence that scientists have engaged in alleged conspiracies. Four investigations discerned no scientific misconduct in emails stolen from University of East Anglia's Climatic Research Unit. Weather stations have not been deleted purposefully from the global network since the 1990s, as has been claimed; furthermore, the reduction in number of stations reporting data has introduced no detectable bias in the trend of the global average temperature anomaly. The IPCC reports undergo significant scrutiny, but as is inevitable in a 3000-page document, that scrutiny sometimes fails to detect errors. The few errors identified in the latest IPCC report were primarily in referencing and not in content. Their existence does not support a conspiracy to misrepresent climate research.

Claim: Climate models are defective and therefore cannot provide reliable projections of future climate trends. Despite many weaknesses, climate models are increasingly able to represent a range of physical processes and feedbacks and thereby reproduce past and present observations. Consistency between models and observations lends confidence to model projections of future climate change. Models are but one tool, together with theory and observations, to assess and quantify climate processes.

Earth May Be Warming But Human Activity Is Not Responsible

Claim: The greenhouse gas signature is missing. Global observations are consistent with the model-based prediction of GHG-induced cooling in the stratosphere and warming at the surface and throughout the troposphere. Furthermore, new measurements in the tropics suggest greater warming in the upper troposphere than at the surface, as predicted by the models.

Claim: The Medieval Warm Period was just as warm as, or warmer than, today. Scarce records and spotty spatial coverage make estimates of medieval temperatures uncertain. Northern hemisphere average temperatures during the medieval period do not appear to have been higher than those of the late 20th century. Furthermore, a warmer medieval period has no bearing on the conclusion that temperatures have increased in the past half-century, and that temperatures will continue to rise due to GHG emissions.

Claim: Atmospheric CO_2 levels rise hundreds of years after temperature in ice cores. The correlation of records of atmospheric CO_2 and Antarctic temperature over the past 800,000 years indicates that CO_2 amplified the warming attributed to variability in Earth's orbit in the transition out of the ice ages. Different processes can and do affect climate concurrently.

Claim: Earth's climate is driven only by the sun. While the importance of the sun as a driver of Earth's climate is undeniable, the measured changes in solar activity over the last fifty years cannot explain the observed rise in temperature; solar activity has in fact decreased since the 1970s.

Claim: Water vapor is the most prevalent greenhouse gas. Although water vapor plays an important role in the natural greenhouse effect and as a positive feedback, CO_2 and other anthropogenic GHGs are perturbing the natural system.

Claim: CO_2 in the atmosphere is already absorbing all of the infrared radiation that it can. The absorption of infrared radiation by carbon dioxide is an integral part of our understanding of the greenhouse effect and of current climate models which take into account the details of the logarithmic absorption of infrared radiation by CO_2. Adding more CO_2 to the atmosphere will continue to perturb the climate system and warm the planet.

Claim: Climate sensitivity is overestimated in current climate models. Quantifying climate sensitivity, or the change in global mean temperature in response to doubling CO_2, is extremely complex because of the unknown rate and magnitude of feedbacks, such as changes in vegetation or ice cover. Attempts to identify negative feedback processes, which would counter the warming due to GHGs, have not been borne out by observations. Sensitivity values below 2.5°C cannot explain the observed climate changes of the past.

Earth May Be Warming, and Humans May Be Responsible, But We Don't Need to Act to Stop It

Claim: Increasing carbon dioxide will stimulate plant growth and improve agricultural yield. Despite the fertilization effect due to increased CO_2, it is likely that crop yields will be reduced in many regions by rising temperature and shifts in precipitation. Unfortunately, regions that are already food-insecure are expected to suffer the greatest negative impacts. While some locations are expected to benefit from the combination of shifting climate and CO_2 fertilization on the short-term, these yields are unlikely to continue indefinitely,

Claim: Human society and natural systems have adapted to past climate change. Past climate changes have often been accompanied by migration, war, and disease. The growing human population will inevitably make environmental change more disruptive in the future, even in the face of increased technological prowess.

Scientific debate is best carried out within the peer-review literature. However, translation of the scientific literature is a necessary step for the non-expert, as language, results, and implications are often narrowly focused and can be obscure even to those in closely related disciplines. The climate science community must work in a concerted fashion to provide regular state-of-the-art assessments and to answer questions about the current understanding. The present document aims to contribute to the effort by presenting scientific arguments in response to the major claims. . . .

Conclusion

The foundation of the science of human-induced climate change is the understanding of the physical effect of increasing concentrations of greenhouse gases on the planet's heat budget. Observations made throughout the instrumental record are consistent with this understanding: warming, melting land-based glaciers, reduced snowpack and Arctic sea ice, shifts in rainfall, and responses in ecosystems worldwide. The study of past climate through proxy records provides additional insight into the response of the climate system to various forcing mechanisms. Model simulations of present and past changes assist in interpreting these observations while also providinq a means to quantify the climate response to each forcing factor and to project these dynamics into the future. These three avenues of investigation—theory, observations, and modeling—are crucial to build our understanding of Earth's climate and how it changes.

Climate research straddles multiple disciplines, from the physics of light interacting with aerosol particles to the relationship between ecosystems and human societies. It involves processes that occur at a broad range of temporal and spatial scales (e.g., from molecular to astronomical). The complexity and diversity of climate science requires independent bodies that can summarize, assess and integrate results, such as the national academies of countries worldwide (e.g. National Academy of Sciences 2008) or the IPCC. In response to a request from US Congress, the National Academies has released a series of reports on **America's Climate Choices** which address the strategies to reduce human influence on climate, actions to reduce vulnerability and increase adaptive capacity to climate change, and steps to advance scientific understanding of natural and human-induced climate change.

The scientific process relies on skepticism. Results and conclusions are routinely questioned. Existing data and methods are re-evaluated while new data are constantly being acquired. Theory, data, principles and methods are discussed openly. The scientific endeavor with regards to climate change exemplifies this process, as is clearly seen in the successive assessment reports of the IPCC, where uncertainties are reduced, new processes identified, and approaches refined. Key climate indicators that characterize the climate system are consistent with warming. Although continued research is needed to quantify the timing, location, and extent of climate impacts, many experts are confident that "global warming is unequivocal and primarily human-induced."

However, some individuals and organizations dispute this conclusion. These challenges sometimes allege the existence of conspiracies, such as an effort to purposefully remove weather stations from the GHCN or to locate them in close proximity to buildings or airports with the aim of modifying the long-term temperature trend. The claims of conspiracy are not borne out by the facts. Similarly, the recent theft of 15 years of emails from the Climatic Research Unit of University of East Anglia failed to provide any evidence of scientific misconduct or conspiracy. A few errors identified in the IPCC AR4 have added to an atmosphere of criticism. While some assert that these discussions may have had a negative impact on public perception of climate science, the identified errors provide no evidence of conspiracy and have no impact on the scientific conclusions concerning climate change expressed in the IPCC AR4.

In an information-rich world, where sophisticated lay-people question the scientific tenets that inform societal decisions, open access to information and transparency in methods are key to the increased democratization of ideas. The climate community is working to meet this need by facilitating access to observational datasets, method descriptions, and to model code and output. Some examples include the data portal of the National Climate Data Center, the surface temperature datasets maintained by the UK Meteorological Office, the model code of the NASA Goddard Institute of Space Studies, or the output of the models from the IPCC AR4, maintained by the Program for Climate Model Diagnosis and Intercomparison at Lawrence Livermore National Laboratory in the US.

Challenges to scientific understanding are rarely resolved in the mainstream media, as journalists seldom have the training to interpret and present the complex concepts involved. Furthermore, the reader can be easily overwhelmed by journalistic whiplash when competing views are presented. The scientific debate is best addressed within the scientific literature, but translation is required for the non-expert, as language, results, and implications are often narrowly focused and can be obscure even to those in closely related disciplines. The climate science community has the responsibility to provide regular state-of-the-art assessments and to answer questions about the current understanding. Several books, web pages, and blogs work to this end and the present document aims to contribute to the effort by presenting scientific arguments in response to the major claims.

The conclusion that climate is changing in response to human emissions of greenhouse gases does not preclude the existence of other drivers of climate variability, such as solar activity or orbital forcing. Multiple drivers of climate act concurrently today as they have throughout Earth history. The fact that climate has changed in the past without human influence, including a Medieval Warm Period, does not reduce the likelihood of human-induced changes today. Rather, it demonstrates the sensitivity of Earth's climate to perturbation, thereby heightening concern as humans perturb the system by increasing atmospheric carbon dioxide to levels not observed in 800,000 years. Natural variability in climate will continue to occur and the conditions that we experience will result from the interaction of all climate drivers. This is likely to produce periods of warming interspersed with periods of more stable temperatures, as well as trends that vary from one region to another. Regional conditions and short-term variability also can depart from longer-term global trends. Cold conditions observed in parts of the United States and Europe in winter of 2009-2010 were not representative of global temperature anomalies, which were among the highest in the instrumental record.

Science cannot "prove" that a specific feature observed in historical climate records was caused by greenhouse gases, but the scientific method allows general inferences about trends and patterns by rejecting competing hypotheses. For example, the observed long-term warming since the 1950s cannot be reproduced in models that do not include anthropogenic emissions of GHGs. There is no other forcing that could account for the observations. Solar activity declined while Earth warmed and global brightening due to increased regulation of air pollutants is inadequate to explain the warming.

The planet is warming and it is likely to continue to warm as a consequence of increased greenhouse gas emissions. Ultimately, questions revolve around sensitivity: how much will Earth warm, what will be the impacts, and when will they occur. As discussed above, sensitivity estimates below 2.5°C cannot account for past climate changes. Atmospheric carbon dioxide levels are expected to have doubled by the end of the century if we continue with business as usual. Failing to reduce emissions will thus lead the planet beyond the guardrail of 2°C average global warming.

Lower values of sensitivity require the existence of negative feedbacks, processes that would act to counteract warming. The strongest hypotheses of negative feedbacks put forth to date regard shifts in the dynamics of cloud formation; however, they have not been borne out by further research. Instead, as discussed above, warming is accompanied predominantly by positive feedbacks that further heat the planet: a warmer atmosphere holds more water vapor and forests stressed by water and temperature take up less carbon dioxide.

In addition, CO_2 has a direct effect on plant growth, with complex consequences for ecosystems. Crop yields may see a fertilization effect due to increased CO_2 in some regions, but they are likely to be negatively impacted by rising temperature and shifts in water availability elsewhere, especially in regions that are already food-insecure. Our best projections indicate that the most negative impacts of climate change will occur in nations that are already vulnerable to other stressors such as rapid population growth and extreme poverty. Humans have survived climate changes of the past, though never with global populations of the current magnitude. One might ask whether survival of the human species is an adequate standard of success.

Alex Newman

NO

Global-Warming Alarmism Dying a Slow Death

Last December in Copenhagen at the United Nations climate summit, officials and global-warming alarmists seemed confident of their imminent triumph. "There is no doubt in my mind whatsoever that it will yield a success," proclaimed UN global-warming chief Yvo de Boer just weeks before the conference.

But Copenhagen was not the victory de Boer had been anticipating. In fact, most analysts labeled it a significant setback for the alarmist agenda. And since then, problems for the human-caused warming campaign have only grown. After a series of scandals exposed extreme misconduct (if not criminality) by leading climate scientists and errors surrounding the movement's theories, pundits began announcing the inevitable collapse of climate hysteria. But the vested interests will not go down without a long, hard fight.

Scandals

The climate alarmists were already doing poorly in the United States before the Copenhagen failure. An October 2009 Pew poll showed that only 36 percent of Americans even believed in man-made global warming. The issue consistently ranked last among public priorities. Commentators referred to the movement as a "cult," and critics ridiculed the theories and dangerous "solutions" all over the Internet. And that was before the proverbial hitting of the fan late last year.

In November 2009, a scandal now known as Climategate changed everything. Just before the much-touted global-warming conference, incriminating e-mails and data from the University of East Anglia's Climatic Research Unit were revealed to the world. And the picture was not pretty. Prominent climate scientists, including many who were deeply involved with the Intergovernmental Panel on Climate Change (IPCC) report, were exposed plotting to "hide the decline" in global temperatures, conspiring to violate Freedom of Information laws, and scheming to keep contradictory viewpoints excluded. The scandal led to even more distrust of the alarmist narrative.

After Climategate made headlines around the world, obvious factual errors started turning up in the UN's IPCC report as researchers began scrutinizing it more closely. Widely considered the "gospel" of the anthropogenic-warming campaign, the report was rapidly losing credibility.

First came "Glaciergate." In its final report, the IPCC suggested that Himalayan glaciers could melt by 2035 or sooner. It turns out the wild assertion (along with several others in the same paragraph) was lifted from an advocacy group's propaganda literature, which took it from an Indian magazine article that has since been discredited. The claim was totally incorrect. The IPCC has been forced to recant it.

More errors were soon exposed in a flurry of bad press for the alarmists. Amazongate, as it has become known, involved fantastical predictions about global warming's effect on the Amazon rain forest. Up to 40 percent of it could be in danger, according to the report. The IPCC also took this claim from advocacy group literature. But on top of that, it incorrectly attributed it to a report that did not even hint at such a prediction. Critics have correctly labeled the assertion a "fabrication."

There was also Africagate. The IPCC erroneously claimed that rain-dependent agriculture in some African countries could be cut in half by 2020, with the wildly inaccurate claims also taken from an advocacy group report (written by an academic who works with "carbon credits"). Of course, this was also wrong. But this time, IPCC Chair Rajendra Pachauri was the one responsible for allowing the error to be repeated in the condensed "Synthesis Report."

Questions and criticism about the use of temperature data have also been ongoing and continue to plague the UN panel's credibility. "Chinagate," where scientists misused Chinese temperature records, is just one example.

The claims of increased hurricane frequency were also proven fraudulent. And as if those blows were not enough, the Dutch government recently forced the IPCC to retract its claim that 55 percent of the Netherlands was below sea level. It's actually 26 percent.

The rapid loss of public credibility over all the errors has also accompanied numerous calls by prominent voices for official inquiries and even criminal investigations. Several universities involved have already launched reviews. And even previous IPCC chief Professor Robert Watson, for example, is calling for a probe to investigate "warming bias" by the UN panel. "The mistakes all appear to have gone in the direction of making it seem like climate change is more serious by overstating the impact. That is worrying," he told the U.K. Times Online. "The IPCC needs to look at this trend in the errors and ask why it happened." The UN will indeed launch an "independent" inquiry, but critics generally expect a coverup.

U.S. Senator James Inhofe (R-Okla.) went further, proposing criminal investigations to determine if alarmists violated any laws. A Senate report produced for Inhofe's Environment and Public Works Committee concluded that, among other problems, "scientists involved in the CRU controversy violated fundamental ethical principles governing taxpayer-funded research and, in some cases, may have violated federal laws."

Similarly, British authorities were investigating possible criminal activity by Climategate scientists who refused to provide documents and data under lawful Freedom of Information requests. The scientists may reportedly escape prosecution under the FOI law because of a six-month statute of limitations.

Rats and Ships

Even the politicians and officials still pushing the alarmist agenda have distanced themselves from the IPCC report. The analogy of "rats" frantically ditching a "sinking ship" has been used by numerous critics to describe the situation.

Climate chief Yvo de Boer, the executive secretary of the UN Framework Convention on Climate Change who predicted "success" in Copenhagen, announced his resignation in mid-February. He said it was time to pursue "new challenges." But he still advises companies about global warming.

Even a prominent member of the pro-United Nations, internationalist Council on Foreign Relations has thrown in the towel, possibly trying to salvage some credibility by denouncing the scandals. "The global warming movement as we have known it is dead" because of "bad science and bad politics," wrote CFR senior foreign policy fellow Walter Mead in a piece for *The American Interest.* He still believes in human-caused warming, but harshly criticized the movement for its lawbreaking and phony claims. "The global warming meltdown confirms all the populist suspicions out there about an arrogantly clueless establishment invoking faked 'science' to impose cockamamie social mandates on the long-suffering American people, backed by a mainstream media that is totally in the tank," he rightly concluded.

Prominent companies that were once leading the push for "action" on climate change have also been retreating to the shadows. Around the time of de Boer's announcement, three large American firms (including two oil companies) bailed on the U.S. Climate Action Partnership, a powerful lobby pushing for "cap and trade" legislation.

Some of the media have also finally started to report the apparent demise of climate alarmism. "The strategy pursued by activists (including scientists who have crossed the line into advocacy) has turned out to be fatally flawed," declared the *Canadian Globe and Mail* in a recent article entitled "The great global warming collapse: As the science scandals keep coming, the air has gone out of the climate-change movement."

Even a writer for the BBC admitted the campaign was falling apart in a recent piece entitled "The dam is cracking." This same media organization has in recent years issued dire predictions of global warming almost daily and last year sat on the Climategate e-mails for over a month. (The BBC claims it wasn't aware of the significance of the information it was given.)

The "neoconservative" *Weekly Standard*—normally a promoter of the globalist establishment's agenda—actually ran a cover story recently with a cartoon depicting polar bears laughing at a naked and freezing Al Gore. The article, entitled "In Denial—The meltdown of the climate campaign," was written by fellow Steven Hayward with the American Enterprise Institute, an organization that has repeatedly peddled climate propaganda and the desirability of "emission reductions" and a "carbon tax." More rats jumping ship?

Even alarmism ringleader Al Gore seemingly conceded defeat on the impact of his efforts to "educate" the public on human-caused climate change.

"I have thus far failed," he told a Norwegian talk show in early March while promoting his new climate book. But, his fight is far from over.

Alarmists Fight Back

Ironically, some of the news articles announcing the demise of climate hysteria were adorned with government-funded Google ads from the State Department reading "Adapting to a Changing Climate." The link takes readers to America.gov, where articles like "The Need for Action on Climate Change Is Urgent" share the page with a picture of a lonely polar bear and propaganda videos citing the IPCC. (What part of the Constitution authorizes government propaganda ads?) Government has obviously not given up the fight.

Indeed—like an animal backed into a corner—committed alarmists are putting up an increasingly hysterical battle as their movement begins to unravel. The U.K. *Telegraph* ran a piece entitled "Warmists overwhelmed by fear, panic and deranged hatred as their 'science' collapses."

There are many players with a significant stake in making sure the public believes "climate change" is caused by man and carbon emissions. Governments have invested hundreds of billions in it—likely because greenhouse-gas legislation will allow government to monitor and control almost everything every citizen does. Banks, too, have vast sums tied up in the scheme—including Goldman Sachs, one of the most powerful firms on Earth—because they stand to make piles of money trading carbon permits. Gore and Pachauri have their fortunes and their reputations at stake. Many climatologists have their careers to lose, too.

"I, for one, genuinely wish that the climate crisis were an illusion," wrote Gore in a wildly inaccurate editorial temper tantrum printed by the *New York Times* on February 27. Then he proceeded to the usual scaremongering, warning about "the displacement of hundreds of millions of climate refugees" and other calamities. Former Vice President Gore intensified his lobbying efforts, consolidating two of his organizations "to create in one [sic] of the largest non-profit climate change education and advocacy organizations in the world," according to a March 5 press release posted on its website. Will Gore go down with the alarmism ship? As its captain, he probably has no choice at this point—though unlike the captain of the *Titanic,* his action will likely be viewed as less than noble.

Scientists under fire have also attempted to deflect criticism by demonizing skeptics. Stanford climatologist Stephen Schneider, for example, complained of receiving "threatening" e-mails. "They shoot abortion doctors here," he told Tierramérica, a UN-affiliated propaganda organ in Latin America.

And just as global warming was rebranded "climate change" to be more all-encompassing, some Senators are now trying to deviously re-sell the "cap and trade" scheme with linguistic gimmicks. "We will have pollution reduction targets," explained Senator Joseph Lieberman, ignoring the fact that CO_2 is not a pollutant but an essential component of life without which green plants would cease to exist.

And despite the apparent implosion of the alarmist movement, governments are steamrolling ahead with the agenda. The taxation commissioner at the European Commission, for example, recently announced a push for EU-wide carbon taxes.

A recently leaked UN Environment Program document from December 2009 revealed plans to create a "green world order" by 2012, while the head of the International Monetary Fund recently called for the creation of a giant climate-change slush fund. And in the United States, despite bipartisan opposition and the lack of constitutional authority, the Environmental Protection Agency is still moving forward with its anti-CO_2 regulation regime. Other national governments around the world are also marching forward with various climate schemes.

More to Come

The fight is not over yet. But no matter what happens in the coming months and years—trials and convictions for climate swindlers, or taxes on breathing for everyone on Earth—the alarmist campaign will eventually fall.

"There's a lot more to come out yet about the [Climategate] e-mails, and how they cooked the computer models," said Canadian climatologist Dr. Tim Ball, who emphasized that he never received money from oil companies and that he was also against the "global cooling" alarmism of a few decades ago.

He told The New American that the Internet played a pivotal role in exposing the scandals, and that this phenomenon will continue. "It's no coincidence that so much of what was exposed came through the blogs. . . . The mainstream media is ignoring the issues almost completely. And it's because most of them were complicit and bought into the argument." As a consequence, the complicit media is becoming increasingly irrelevant.

But the battle against the alarmist agenda will likely be protracted and difficult. "It won't die—it simply won't die—until the economies start to suffer," Dr. Ball said about the carbon tax and cap-and-trade schemes, citing Spain's experience with new "green jobs" causing an overall loss of jobs as an example of the economic price of alarmist policies. But, he added, the truth will inevitably triumph eventually.

"Reality always comes through, sooner or later, it's just that sometimes it takes a long time," agreed Professor Nils-Axel Mörner, one of the world's foremost experts on sea levels and the head of the Paleogeophysics and Geodynamics Department at Stockholm University until he retired in 2005. He told The New American that as an expert reviewer for the sea-level section of the IPCC report, he had the opportunity to understand the inner workings of the IPCC. And it is doomed to fail eventually.

The sea-level chapter he was supposed to review was "of very poor quality," Mörner said. And the hysteria surrounding sea-level rises, like most of the IPCC scaremongering, "is not grounded in reality." The panel chose authors based on loyalty, not credentials, Mörner explained. And though he warned the IPCC of errors, they mostly ignored the advice. But the anti-science attitude came back to haunt them. Climategate was "wonderful," Mörner exclaimed. He called the scandal an "iceberg of shame," noting that there was still much to be discovered.

"The first thing which has to come now is the restoration of scientific values," he said, explaining that the climate campaign had "autocratically" tried to impose beliefs on the public that were not based on science. "Al Gore is a salesman, not a scientist, and we don't need salesmen." Mörner is optimistic.

Despite all the trouble and wasted resources expended on the movement, there is certainly a bright side emerging as the climate-crisis crusade self-destructs. For one, more people may begin to think twice before blindly trusting governments and the media. Additionally, the whole episode illustrates the crumbling gate-keeping ability of "Big Media" in the age of the Internet. This is an encouraging sign for the future of freedom.

Scaremongering to swindle the public out of money and freedom is an old trick. But hopefully, people will know better than to fall for it again next time.

EXPLORING THE ISSUE

Is Human Activity Responsible for Global Warming?

Critical Thinking and Reflection

1. Why does the fact that it is difficult to predict society's future actions, particularly in the areas of population growth, energy consumption, and energy technologies, make it difficult to pinpoint the magnitude of future climate changes?
2. What measures should be taken in the near future to best prepare for the long-term impacts of global warming?
3. Discuss how the nature of science permits critics to say that the evidence of global warming is not incontrovertible.
4. Discuss the degree to which a scientific report's funding sources affect the report's credibility.
5. What special interest groups can you expect to be harmed by global warming, or by doing something to stop global warming?

Is There Common Ground?

The positions taken in the two essays for this issue leave little room for agreement. But previous issues in this book may help you answer the following questions.

1. Are there actions society can take that make sense whether or not global warming is a real problem?
2. Do fossil fuels come with enough environmental drawbacks to warrant finding replacements, whether or not global warming is a real problem?
3. What technologies seem most helpful for meeting society's future needs, whether or not global warming is a real problem?

ISSUE 8

Is Global Warming a Catastrophe That Warrants Immediate Action?

YES: Global Humanitarian Forum, from *Climate Change—The Anatomy of a Silent Crisis* (May 2009)

NO: Bjorn Lomborg, from "Let's Keep Our Cool About Global Warming," *Skeptical Inquirer* (March/April 2008)

Learning Outcomes

After studying this issue, students will be able to:

- Describe the effects of climate change to date on people.
- Describe how climate change affects different socioeconomic classes of people in different ways.
- Describe what can be done in the near future to prepare for future effects of global warming.
- Explain how population size and growth interact with carbon emissions and global warming.

ISSUE SUMMARY

YES: The Global Humanitarian Forum argues that global warming due to human activities, chiefly the emission of greenhouse gases such as carbon dioxide, is now beyond doubt. Impacts on the world's poorest people are already severe and will become much worse. Immediate action is essential to tackle climate change, increase funding for adaptation to its effects, and end the suffering it causes.

NO: Bjorn Lomborg argues that although global warming has genuine impacts on people, the benefits of continuing to use fossil fuels are so much greater than the costs that the best approach to a solution is not to demand draconian cuts in carbon emissions, but to invest globally in research and development of non-carbon-emitting energy technologies and thereby "recapture the vision of delivering both a low-carbon and a high-income world."

Exactly what will global warming do to the world and its people? Projections for the future have grown steadily worse; see Eli Kintisch, "Projections of Climate Change Go from Bad to Worse, Scientists Report," *Science* (March 20, 2009). Effects include rising sea level, more extreme weather events, reduced global harvests (Constance Holden, "Higher Temperatures Seen Reducing Global Harvests," *Science* (January 9, 2009)), and threats to the economies and security of nations (Michael T. Klare, "Global Warming Battlefields: How Climate Change Threatens Security," *Current History* (November 2007) and Scott G. Bergerson, "Arctic Meltdown: The Economic and Security Implications of Global Warming," *Foreign Affairs* (March/April 2008)). As rainfall patterns change and the seas rise, millions of people will flee their homelands; see Alex de Sherbinin, Koko Warner, and Charles Erhart, "Casualties of Climate Change," *Scientific American* (January 2011). Perhaps worst of all, even if we somehow stopped emitting greenhouse gases today, the effects would continue for 1000 years or more, during which sea-level rise may exceed "several meters." See Susan Solomon et al., "Irreversible Climate Change Due to Carbon Dioxide Emissions," *Proceedings of the National Academy of Sciences* (February 10, 2009).

It seems clear that something must be done, but what? How urgently? And with what aim? Should we be trying to reduce or prevent human suffering? Or to avoid political conflicts? Or to protect the global economy—meaning standards of living, jobs, and businesses? The humanitarian and economic approaches are obviously connected, for protecting jobs certainly has much to do with easing or preventing suffering. However, these approaches can also conflict. In October 2009, the Government Accountability Office (GAO) released "Climate Change Adaptation: Strategic Federal Planning Could Help Government Officials Make More Informed Decisions" (GAO-10-113; www.gao.gov/products/GAO-10-113), which noted the need for multiagency coordination and strategic (long-term) planning, both of which are often resisted by bureaucrats and politicians. Robert Engelman, *Population, Climate Change, and Women's Lives* (Worldwatch Institute, 2010), notes that addressing population size and growth would help but "Despite its key contribution to climate change, population plays little role in current discussions on how to address this serious challenge."

The U.S. Climate Change Science Program's "Scientific Assessment of the Effects of Global Change on the United States. A Report of the Committee on Environment and Natural Resources, National Science and Technology Council" (May 29, 2008; available at www.ostp.gov/galleries/NSTC%20Reports/Scientific%20Assessment%20FULL%20Report.pdf) describes the current and potential impacts of climate change. In sum, it says that the evidence is clear and getting clearer that global warming is "very likely" due to greenhouse gases largely released by human activity and there will be consequent changes in precipitation, storms, droughts, sea level, food production, fisheries, and more. Dealing with these effects may require changes in many areas, particularly relating to energy use, but "significant uncertainty exists about the potential impacts of climate change on energy production and distribution, in part because the timing and magnitude of climate impacts are uncertain."

See Susan Milius, "Already Feeling the Heat," *Science News* (Web edition) (May 29, 2008).

Consonant with the Bush Administration's previous position that imposing restrictions on fossil fuel use or carbon emissions is a bad idea both because of uncertainty about global warming and because it would harm the economy, mitigation of the risks—which the report says are likely or very likely—is barely mentioned. The Bush Administration's position was that there are wrong ways and right ways to go about solving the global warming problem. The wrong way is raising energy taxes and prices (which hurts consumers and business), imposing restrictions, and abandoning the use of nuclear power and coal. The right way is setting realistic goals, adopting policies that spur investment in new technologies, encouraging the use of nuclear power and "clean" coal, and enhancing international cooperation through free trade in clean energy technologies. Such measures rely on the market to stimulate voluntary reductions in emissions. It is possible that increasing funding for alternative energy technologies would lead to solutions that greatly ease the problem, but many observers think it folly to rely on future breakthroughs.

President Barack Obama has indicated that his administration will take global warming more seriously. In June 2009, the U.S. House of Representatives passed an Energy and Climate bill that promised to cap carbon emissions and stimulate use of renewable energy. The Senate version of the bill failed to pass; see Daniel Stone, "Who Killed the Climate and Energy Bill?" *Newsweek* (September 15, 2010). The Obama Administration also said it was committed to negotiating seriously at the Copenhagen Climate Change Conference in December 2009. Unfortunately, the Copenhagen meeting ended with little accomplished except agreements to limit global temperature increases to two degrees Celsius by 2100, but only through voluntary cuts in carbon emissions, to have developed nations report their cuts, to have developed nations fund mitigation and adaptation in developing nations, and to continue talking about the problem (see Elizabeth Finkel, "Senate Looms as Bigger Hurdle after Copenhagen," *Science* (January 1, 2010)). There were few signs that the world is ready to take the extensive actions deemed necessary by many; see, for example, Janet L. Sawin and William R. Moomaw, "Renewing the Future and Protecting the Climate," *World Watch* (July/August 2010).

In May 2010, the National Research Council released three books, *Advancing the Science of Climate Change* (www.nap.edu/catalog.php?record_id=12782), *Limiting the Magnitude of Future Climate Change* (www.nap.edu/catalog.php?record_id=12785), and *Adapting to the Impacts of Climate Change* (www.nap.edu/catalog.php?record_id=12783). Together, they stress the reality of the problem, the need for immediate action to keep the problem from getting worse, and the need for advance planning and preparation to deal with the impacts. However, their message may not be enough to convince Congress to pass necessary legislation; see Richard A. Kerr and Eli Kintisch, "NRC Reports Strongly Advocate Action on Global Warming," *Science* (May 28, 2010).

In the YES selection, the Global Humanitarian Forum argues that global warming due to human activities, chiefly the emission of greenhouse gases such as carbon dioxide, is now beyond doubt. Impacts on the world's poorest people are already severe and will become much worse. Immediate action is essential to tackle climate change, increase funding for adaptation to its effects, and end the suffering caused by it. In the NO selection, economist Bjorn Lomborg argues that although global warming has genuine impacts on people, the benefits of continuing to use fossil fuels are so much greater than the costs that the best approach to a solution is not to demand draconian cuts in carbon emissions, but to invest globally in research and development of non-carbon-emitting energy technologies and thereby "recapture the vision of delivering both a low-carbon and a high-income world."

YES  Global Humanitarian Forum

Climate Change—The Anatomy of a Silent Crisis

Science is now unequivocal as to the reality of climate change. Human activities, in particular emissions of greenhouse gases like carbon dioxide are recognized as its principle cause. This report clearly shows that climate change is already causing widespread devastation and suffering around the planet today. Furthermore, even if the international community is able to contain climate change, over the next decades human society must prepare for more severe climate change and more dangerous human impacts.

This report documents the full impact of climate change on human society worldwide today. It covers in specific detail the most critical areas of the global impact of climate change, namely on food, health, poverty, water, human displacement, and security. The third section of this report highlights the massive socio-economic implications of those impacts, in particular, that the worst affected are the world's poorest groups, who cannot be held responsible for the problem. The final section examines how sustainable development and the millennium development goals are in serious danger, the pressures this will exert on humanitarian assistance, and the great need to integrate efforts in adapting to climate change.

Based on verified scientific information, established models, and, where needed, the best available estimates, this report represents the most plausible narrative of the human impact of climate change. It reports in a comprehensive manner the adverse effects people already suffer today due to climate change within a single volume, encompassing the full spectrum of the most important impacts evidenced to date.

The findings of the report indicate that every year climate change leaves over 300,000 people dead, 325 million people seriously affected, and economic losses of US$125 billion. Four billion people are vulnerable, and 500 million people are at extreme risk. These figures represent averages based on projected trends over many years and carry a significant margin of error. The real numbers could be lower or higher. The different figures are each explained in more detail and in context in the relevant sections of the report. Detailed information describing how these figures have been calculated is also included in the respective sections and in the end matter of the report.

These already alarming figures may prove too conservative. Weather-related disasters alone cause significant economic losses. Over the past five

From *Climate Change—The Anatomy of a Silent Crisis,* May 2009. Published by Global Humanitarian Forum.

years this toll has gone as high as $230 billion, with several years around $100 billion and a single year around $50 billion. Such disasters have increased in frequency and severity over the past 30 years in part due to climate change. Over and above these costs are impacts on health, water supply, and other shocks not taken into account. Some would say that the worst years are not representative and they may not be. But scientists expect that years like these will be repeated more often in the near future.

Climate Change Through the Human Lens

Climate change already has a severe human impact today, but it is a silent crisis—it is a neglected area of research as the climate change debate has been heavily focused on physical effects in the long-term. This human impact report on climate change, therefore, breaks new ground. It focuses on human impact rather than physical consequences. It looks at the increasingly negative consequences that people around the world face as a result of a changing climate. Rather than focusing on environmental events in 50–100 years, the report takes a unique social angle. It seeks to highlight the magnitude of the crisis at hand in the hope to steer the debate towards urgent action to overcome this challenge and reduce the suffering it causes.

The human impact of climate change is happening right now—it requires urgent attention. Events like weather-related disasters, desertification and rising sea levels, exacerbated by climate change, affect individuals and communities around the world. They bring hunger, disease, poverty, and lost livelihoods—reducing economic growth and posing a threat to social and, even, political stability. Many people are not resilient to extreme weather patterns and climate variability. They are unable to protect their families, livelihoods and food supply from negative impacts of seasonal rainfall leading to floods or water scarcity during extended droughts. Climate change is multiplying these risks.

Today, we are at a critical juncture—just months prior to the Copenhagen summit where negotiations for a post-2012 climate agreement must be finalized. Negotiators cannot afford to ignore the current impact of climate change on human society. The responsibility of nations in Copenhagen is not only to contain a serious future threat, but also to address a major contemporary crisis. The urgency is all the more apparent since experts are constantly correcting their own predictions about climate change, with the result that climate change is now considered to be occurring more rapidly than even the most aggressive models recently suggested. The unsettling anatomy of the human impact of climate change cannot be ignored at the negotiating tables.

Climate Change Is a Multiplier of Human Impacts and Risks

Climate change is already seriously affecting hundreds of millions of people today and in the next twenty years those affected will likely more than double—making it the greatest emerging humanitarian challenge of our time. Those seriously affected are in need of immediate assistance either following a weather-related disaster, or because livelihoods have been severely compromised by climate change. The number of those severely affected by climate change is more than ten times greater than for instance those injured in traffic accidents each year, and more than the global annual number of new malaria cases. Within the next 20 years, one in ten of the world's present population could be directly and seriously affected.

Already today, hundreds of thousands of lives are lost every year due to climate change. This will rise to roughly half a million in 20 years. Over nine in ten deaths are related to gradual environmental degradation due to climate change—principally malnutrition, diarrhoea, [and] malaria, with the remaining deaths being linked to weather-related disasters brought about by climate change.

Economic losses due to climate change currently amount to more than one hundred billion US dollars per year, which is more than the individual national GDPs of three quarters of the world's countries.

This figure constitutes more than the total of all Official Development Assistance in a given year. Already today, over half a billion people are at extreme risk to the impacts of climate change, and six in ten people are vulnerable to climate change in a physical and socio-economic sense. The majority of the world's population does not have the capacity to cope with the impact of climate change without suffering a potentially irreversible loss of wellbeing or risk of loss of life. The populations most gravely and immediately at risk live in some of the poorest areas that are also highly prone to climate change—in particular, the semi-arid dry land belt countries from the Sahara to the Middle East and Central Asia, as well as sub-Saharan Africa, South Asian waterways and Small Island Developing States.

A Question of Justice

It is a grave global justice concern that those who suffer most from climate change have done the least to cause it. Developing countries bear over nine-tenths of the climate change burden: 98 percent of the seriously affected and 99 percent of all deaths from weather-related disasters, along with over 90 percent of the total economic losses. The 50 Least Developed Countries contribute less than 1 percent of global carbon emissions.

Climate change exacerbates existing inequalities faced by vulnerable groups, particularly women, children, and the elderly. The consequences of climate change and poverty are not distributed uniformly within communities. Individual and social factors determine vulnerability and capacity to adapt to the effects of climate change. Women account for two-thirds of the world's poor and comprise about seven in ten agricultural workers. Women and children are disproportionately represented among people displaced by extreme weather events and other climate shocks.

The poorest are hardest hit, but the human impact of climate change is a global issue. Developed nations are also seriously affected, and increasingly so. The human impact of recent heat waves, floods, storms and forest fires in rich countries has been alarming. Australia is perhaps the developed nation most vulnerable to the direct impacts of climate change and also to the indirect impact from neighbouring countries that are stressed by climate change.

The Time to Act Is Now

Climate change threatens sustainable development and all eight Millennium Development Goals. The international community agreed at the beginning of the new millennium to eradicate extreme hunger and poverty by 2015. Yet, today, climate change is already responsible for forcing some fifty million additional people to go hungry and driving over ten million additional people into extreme poverty. Between one-fifth and one-third of Official Development Assistance is in climate-sensitive sectors and thereby highly exposed to climate risks.

To avert the worst outcomes of climate change, adaptation efforts need to be scaled up by a factor of more than 100 in developing countries. The only way to reduce the present human impact is through adaptation. But funding for adaptation in developing countries is not even one percent of what is needed. The multilateral funds that have been pledged for climate change adaptation funding currently amount to under half a billion US dollars.

Despite the lack of funding, some cases of successful adaptation do provide a glimmer of hope. Bangladesh is one such example. Cyclone Sidr, which struck Bangladesh in 2007, demonstrates how well adaptation and prevention efforts can pay off. Disaster preparation measures, such as early warning systems and storm-proof houses, minimized damage and destruction. Cyclone Sidr's still considerable death toll of 3,400, and economic damages of $1.6 billion, nevertheless compare favourably to the similar scale cyclone Nargis, which hit Myanmar in 2008, resulting in close to 150,000 deaths and economic losses of around $4 billion.

Solutions do also exist for reducing greenhouse gas emissions, some even with multiple benefits. For instance, black carbon from soot, released by staple energy sources in poor communities, is likely causing as much as 18 percent of warming. The provision of affordable alternative cooking stoves to the poor can, therefore, have both positive health results, since smoke is eliminated, and an immediate impact on reducing emissions, since soot only remains in the atmosphere for a few weeks. Integrating strategies between adaptation, mitigation, development, and disaster risk reduction can and must be mutually reinforcing. Climate change adaptation, mitigation, humanitarian assistance and development aid underpin each other, but are supported by different sets of institutions, knowledge centres, policy frameworks, and funding mechanisms. These policies are essential to combat the human impact of climate change, but their links to one another have received inadequate attention.

A key conclusion of this report is that the global society must work together if humanity is to overcome this shared challenge: nations have to realize their common interest at Copenhagen, acting decisively with one voice; humanitarian and development actors of all kinds have to pool resources, expertise, and efforts in order to deal with the rapidly expanding challenges brought about by climate change; and in general, people, businesses, and communities everywhere should become engaged and promote steps to tackle climate change and end the suffering it causes.

Bjorn Lomborg

Let's Keep Our Cool About Global Warming

There is a kind of choreographed screaming about climate change from both sides of the debate. Discussion would be on much firmer ground if we could actually hear the arguments and the facts and then sensibly debate long-term solutions.

Man-made climate change is certainly a problem, but it is categorically not the end of the world. Take the rise in sea levels as one example of how the volume of the screaming is unmatched by the facts. In its 2007 report, the United Nations estimates that sea levels will rise about a foot over the remainder of the century. While this is not a trivial amount, it is also important to realize that it is not unknown to mankind: since 1860, we have experienced a sea level rise of about a foot without major disruptions. It is also important to realize that the new prediction is lower than previous Intergovernmental Panel on Climate Change (IPCC) estimates and much lower than the expectations from the 1990s of more than two feet and the 1980s, when the Environmental Protection Agency projected more than six feet.

We dealt with rising sea levels in the past century, and we will continue to do so in this century. It will be problematic, but it is incorrect to posit the rise as the end of civilization.

We will actually lose very little dry land to the rise in sea levels. It is estimated that almost all nations in the world will establish maximal coastal protection almost everywhere, simply because doing so is fairly cheap. For more than 180 of the world's 192 nations, coastal protection will cost less than 0.1 percent GDP and approach 100 percent protection.

The rise in sea level will be a much bigger problem for poor countries. The most affected nation will be Micronesia, a federation of 607 small islands in the West Pacific with a total land area only four times larger than Washington, D.C. If nothing were done, Micronesia would lose some 21 percent of its area by the end of the century. With coastal protection, it will lose just 0.18 percent of its land area. However, if we instead opt for cuts in carbon emissions and thus reduce both the sea level rise and economic growth, Micronesia will end up losing a *larger* land area. The increase in wealth for poor nations is more important than sea levels: poorer nations will be less able to defend themselves against rising waters, even if they rise more slowly. This is the same for other vulnerable nations: Tuvalu, the Maldives, Vietnam, and Bangladesh.

From *Skeptical Inquirer,* March/April 2008, pp. 42–45.

The point is that we cannot just talk about CO_2 when we talk about climate change. The dialogue needs to include *both* considerations about carbon emissions and economics for the benefit of humans and the environment. Presumably, our goal is not just to cut carbon emissions, but to do the best we can for people and the environment.

We should take action on climate change, but we need to be realistic. The U.K has arguably engaged in the most aggressive rhetoric about climate change. Since the Labour government promised in 1997 to cut emissions by a further 15 percent by 2010, emissions have *increased* 3 percent. American emissions during the Clinton/Gore administration increased 28 percent.

Look at our past behavior: at the Earth Summit in Rio in 1992, nations promised to cut emissions back to 1990 levels by 2000. The member countries of the Organisation for Economic Cooperation and Development (OECD) overshot their target in 2000 by more than 12 percent.

Many believe that dramatic political action will follow if people only knew better and elected better politicians. Despite the European Union's enthusiasm for the Kyoto Protocol on Climate Change—and a greater awareness and concern over global warming in Europe than in the United States—emissions per person since 1990 have remained stable in the U.S. while E.U. emissions have *increased* 4 percent.

Even if the wealthy nations managed to reign in their emissions, the majority of this century's emissions will come from developing countries—which are responsible for about 40 percent of annual carbon emissions; this is likely to increase to 75 percent by the end of the century.

In a surprisingly candid statement from Tony Blair at the Clinton Global Initiative, he pointed out:

> I think if we are going to get action on this, we have got to start from the brutal honesty about the politics of how we deal with it. The truth is no country is going to cut its growth or consumption substantially in the light of a long-term environmental problem. What countries are prepared to do is to try to work together cooperatively to deal with this problem in a way that allows us to develop the science and technology in a beneficial way.

Similarly, one of the top economic researchers tells us: "Deep cuts in emissions will only be achieved if alternative energy technologies become available at reasonable prices." We need to engage in a sensible debate about how to tax CO_2. If we set the tax too low, we emit too much. If we set it too high, we end up much poorer without doing enough to reduce warming.

In the largest review of all of the literature's 103 estimates, climate economist Richard Tol makes two important points. First, the really scary, high estimates typically have not been subjected to peer review and published. In his words: "studies with better methods yield lower estimates with smaller uncertainties." Second, with reasonable assumptions, the cost is very unlikely to be higher than \$14 per ton of CO_2 and likely to be much smaller. When I specifically asked him for his best guess, he wasn't too enthusiastic about shedding his cautiousness—as few true researchers invariably are—but gave the estimate of $2 per ton of CO_2.

Therefore, I believe that we should tax CO_2 at the economically feasible level of about $2/ton, or maximally $14/ton. Yet, let us not expect this will make any major difference. Such a tax would cut emissions by 5 percent and reduce temperatures by 0.16°F. And before we scoff at 5 percent, let us remember that the Kyoto protocol, at the cost of 10 years of political and economic toil, will reduce emissions by just 0.4 percent by 2010.

Neither a tax nor Kyoto nor draconian proposals for future cuts move us closer toward finding better options for the future. Research and development in renewable energy and energy efficiency is at its lowest for twenty-five years. Instead, we need to find a way that allows us to "develop the science and technology in a beneficial way," a way that enables us to provide alternative energy technologies at reasonable prices. It will take the better part of a century and will need a political will spanning parties, continents, and generations. We need to be in for the long haul and develop cost-effective strategies that won't splinter regardless of overarching ambitions or false directions.

This is why one of our generational challenges should be for *all nations to commit themselves to spending 0.05 percent of GDP in research and development of noncarbon emitting energy technologies*. This is a tenfold increase on current expenditures, yet would cost a relatively minor $25 billion per year (seven times cheaper than Kyoto and many more times cheaper than Kyoto II). Such a commitment could include all nations, with wealthier nations paying the larger share, and would let each country focus on its own future vision of energy needs, whether that means concentrating on renewable sources, nuclear energy, fusion, carbon storage, conservation, or searching for new and more exotic opportunities.

Funding research and development globally would create a momentum that could recapture the vision of delivering both a low-carbon and high-income world. Lower energy costs and high spin-off innovation are potential benefits that possibly avoid ever stronger temptations to free-riding and the ever tougher negotiations over increasingly restrictive Kyoto Protocol-style treaties. A global financial commitment makes it plausible to envision stabilizing climate changes at reasonable levels.

I believe it would be the way to bridge a century of parties, continents, and generations, creating a sustainable, low-cost opportunity to create the alternative energy technologies that will power the future.

To move toward this goal we need to create sensible policy dialogue. This requires us to talk openly about priorities. Often there is strong sentiment in any public discussion that we should do *anything* required to make a situation better. But clearly we don't actually do that. When we talk about schools, we know that more teachers would likely provide our children with a better education. Yet we do not hire more teachers simply because we also have to spend money in other areas. When we talk about hospitals, we know that access to better equipment is likely to provide better treatment, yet we don't supply an infinite amount of resources. When we talk about the environment, we know tougher restrictions will mean better protection, but this also comes with higher costs.

Consider traffic fatalities, which are one of the ten leading causes of deaths in the world. In the U.S., 42,600 people die in traffic accidents and 2.8 million people are injured each year. Globally, it is estimated that 1.2 million people die from traffic accidents and 50 million are injured every year.

About 2 percent of all deaths in the world are traffic-related and about 90 percent of the traffic deaths occur in third world countries. The total cost is a phenomenal $512 billion a year. Due to increasing traffic (especially in the third world) and due to ever better health conditions, the World Heath Organization estimates that by 2020, traffic fatalities will be the second leading cause of death in the world, after heart disease.

Amazingly, we have the technology to make all of this go away. We could instantly save 1.2 million humans and eliminate $500 billion worth of damage. We would particularly help the third world. The answer is simply lowering speed limits to 5 mph. We could avoid almost all of the 50 million injuries each year. But of course we will not do this. Why? The simple answer that almost all of us would offer is that the benefits from driving moderately fast far outweigh the costs. While the cost is obvious in terms of those killed and maimed, the benefits are much more prosaic and dispersed but nonetheless important—traffic interconnects our society by bringing goods at competitive prices to where we live and bringing people together to where we work, and lets us live where we like while allowing us to visit and meet with many others. A world moving only at 5 mph is a world gone medieval.

This is not meant to be flippant. We really could solve one of the world's top problems if we wanted. We know traffic deaths are almost entirely caused by man. We have the technology to reduce deaths to zero. Yet, we persist in exacerbating the problem each year, making traffic an ever-bigger killer.

I suggest that the comparison with global warming is insightful; we have the technology to reduce it to zero, yet we seem to persist in going ahead and exacerbating the problem each year, causing temperatures to continue to increase to new heights by 2020. Why? Because the benefits from moderately using fossil fuels far outweigh the costs. Yes, the costs are obvious in the "fear, terror, and disaster" we read about in the papers every day.

But the benefits of fossil fuels, though much more prosaic, are nonetheless important. Fossil fuels provide us with low-cost electricity, heat, food, communication, and travel. Electrical air-conditioning means that people in the U.S. no longer die in droves during heat waves. Cheaper fuels would have avoided a significant number of the 150,000 people that have died in the UK since 2000 due to cold winters.

Because of fossil fuels, food can be grown cheaply, giving us access to fruits and vegetables year round, which has probably reduced cancer rates by at least 25 percent. Cars allow us to commute to city centers for work while living in areas that provide us with space and nature around our homes, whereas communication and cheap flights have given ever more people the opportunity to experience other cultures and forge friendships globally.

In the third world, access to fossil fuels is crucial. About 1.6 billion people don't have access to electricity, which seriously impedes human development. Worldwide, about 2.5 billion people rely on biomass such as wood and waste (including dung) to cook and keep warm. For many Indian women, searching for wood takes about three hours each day, and sometimes they walk more than 10 kilometers a day. All of this causes excessive deforestation. About 1.3 million people—mostly women and children—die each year due to heavy indoor air pollution. A switch from biomass to fossil fuels would dramatically improve 2.5 billion lives; the cost of $1.5 billion annually would be greatly superseded by benefits of about $90 billion. Both for the developed and the developing world, a world without fossil fuels—in the short or medium term—is, again, a lot like reverting back to the middle ages.

This does not mean that we should not talk about how to reduce the impact of traffic and global warming. Most countries have strict regulation on speed limits—if they didn't, fatalities would be much higher. Yet, studies also show that lowering the average speed in Western Europe by just 5 kilometers per hour could reduce fatalities by 25 percent—with about 10,000 fewer people killed each year. Apparently, democracies in Europe are not willing to give up on the extra benefits from faster driving to save 10,000 people.

This is parallel to the debate we are having about global warming. We can realistically talk about $2 or even a $14 CO_2 tax. But suggesting a $140 tax, as Al Gore does, seems to be far outside the envelope. Suggesting a 96 percent carbon reduction for the OECD by 2030 seems a bit like suggesting a 5 mph speed limit in the traffic debate. It is technically doable, but it is very unlikely to happen.

One of the most important issues when it comes to climate change is that we cool our dialogue and consider the arguments for and against different policies. In the heat of a loud and obnoxious debate, facts and reason lose out.

EXPLORING THE ISSUE

Is Global Warming a Catastrophe That Warrants Immediate Action?

Critical Thinking and Reflection

1. How is climate change killing people today?
2. In what sense is dealing with climate change a matter of social justice?
3. What measures should be taken in the near future to best prepare for the long-term impacts of global warming?
4. How do population size and growth affect carbon emissions and climate change?

Is There Common Ground?

Both sides in this issue agree that global warming is already affecting people. They differ in how concerned we should be, or in how much we should be worrying about "social justice." Visit the Social Justice Training Institute site http://www.sjti.org/ to find out more.

1. What is social justice?
2. How does the concept apply to environmental problems such as global warming?
3. What other aspects of life can you apply the concept to?

ISSUE 9

Will Restricting Carbon Emissions Damage the Economy?

YES: Paul Cicio, from "Competitiveness and Climate Policy: Avoiding Leakage of Jobs and Emissions," testimony before the House Committee on Energy and Commerce, Subcommittee on Energy and Environment (March 18, 2009)

NO: Aaron Ezroj, from "How Cap and Trade Will Fuel the Global Economy," *Environmental Law Reporter* (July 2010)

Learning Outcomes

After studying this issue, students will be able to:

- Explain how a cap-and-trade system can reduce pollution levels over time.
- Explain how cap-and-trade systems motivate corporate polluters to pollute less.
- Explain why restricting carbon emissions would stimulate the development of new energy-related technology.
- Discuss what factors are most important in making policy decisions such as whether to restrict emissions of greenhouse gases.

ISSUE SUMMARY

YES: Paul Cicio argues that lacking global agreements, capping greenhouse gas emissions of the industrial sector will make domestic production less competitive in the global market, drive investment and jobs offshore, increase exports, and damage the economy. The real greenhouse gas problem lies with other sectors of the economy, and that is where attention should be focused.

NO: Aaron Ezroj argues that although restricting emissions (as in a cap-and-trade program) may increase costs for some businesses, it will create many business opportunities in the financial sector, low-carbon technologies, carbon capture-and-storage projects, advanced-technology vehicles, and legal and nonlegal consulting. The overall effect will be to fuel the global economy.

Following World War II, the United States and other developed nations experienced an explosive period of industrialization accompanied by an enormous increase in the use of fossil fuel energy sources and a rapid growth in the manufacture and use of new synthetic chemicals. In response to growing public concern about pollution and other forms of environmental deterioration resulting from this largely unregulated activity, the U.S. Congress passed the National Environmental Policy Act of 1969. This legislation included a commitment on the part of the government to take an active and aggressive role in protecting the environment. The next year the Environmental Protection Agency (EPA) was established to coordinate and oversee this effort. During the next two decades, an unprecedented series of legislative acts and administrative rules were promulgated, placing numerous restrictions on industrial and commercial activities that might result in the pollution, degradation, or contamination of land, air, water, food, and the workplace.

Such forms of regulatory control have always been opposed by the affected industrial corporations and developers as well as by advocates of a free-market policy, who prefer reliance on voluntary measures. More moderate critics of the government's regulatory program recognize that adequate environmental protection will not result from completely voluntary policies. They suggest that a new set of strategies is needed. Arguing that "top-down, federal, command and control legislation" is not an appropriate or effective means of preventing ecological degradation, they propose a wide range of alternative tactics, many of which are designed to operate through the economic marketplace. The first significant congressional response to these proposals was the incorporation of tradable pollution emission rights into the 1990 Clean Air Act amendments as a means for achieving the set goals for reducing acid-rain-causing sulfur dioxide emissions. More recently, the 1997 international negotiations on controlling global warming in Kyoto, Japan, resulted in a protocol that includes emissions trading as one of the key elements in the plan to limit the atmospheric buildup of greenhouse gases.

Charles W. Schmidt, "The Market for Pollution," *Environmental Health Perspectives* (August 2001), argues that emissions trading schemes represent "the most significant developments" in the use of economic incentives to motivate corporations to reduce pollution. In "A Low-Cost Way to Control Climate Change," *Issues in Science and Technology* (Spring 1998), Byron Swift argues that the "cap-and-trade" feature of the U.S. Acid Rain Program has been so successful that a similar system for implementing the Kyoto Protocol's emissions trading mandate as a cost-effective means of controlling greenhouse gases should work. In March 2001, the U.S. Senate Committee on Agriculture, Nutrition, and Forestry held a "Hearing on Biomass and Environmental Trading Opportunities for Agriculture and Forestry," in which witnesses urged Congress to encourage trading for both its economic and its environmental benefits. Richard L. Sandor, chairman and chief executive officer of Environmental Financial Products LLC, said that "200 million tons of CO_2 could be sequestered through soils and forestry in the United States per year. At the most conservative prices of $20–$30 per ton, this could potentially generate $4–$6 billion in additional agricultural income."

Cap-and-trade systems work by setting a limit (the cap) on how much of a pollutant can be emitted per year. Permits to emit a portion of the pollutant are then made available—either free or as the result of an auction process—to businesses that emit that pollutant. Businesses that do not emit as much as their permits allow can then sell their unused permits. This provides an incentive to reduce emissions by spending money to improve efficiency. But a cap-and-trade system does require spending money for initial permits, for improved technology, and/or for additional permits when a company cannot or does not control emissions. This means added expenses, with an impact on profitability. Many businesses are concerned that this will affect their competitive position and even their ability to stay in business. A crucial question thus becomes whether environmental or economic protection should come first.

Europe is already implementing its own Greenhouse Gas Emissions Trading Scheme, although Marianne Lavelle, "The Carbon Market Has a Dirty Little Secret," *U.S. News and World Report* (May 14, 2007), reports that in Europe the value of tradable emissions allowances fell so low at one point, partly because too many allowances were issued, that it was cheaper to burn more fossil fuel and emit more carbon than to burn and emit less. Future trading schemes will need to be designed to avoid the problem, and the U.S. Congress is actively considering ways to address the issue (see "Support Grows for Capping and Trading Carbon Emissions," *Issues in Science and Technology* (Summer 2007)). David G. Victor and Danny Cullenward, "Making Carbon Markets Work," *Scientific American* (December 2007), argue that what is needed is to combine a trading program with limits on emissions and careful management. Bill McKibben, "The Greenback Effect," *Mother Jones* (May/June 2008), agrees, noting that although markets may not be perfect, when they work, they work fast.

Meanwhile, there is also interest in what is known as "carbon offsets," by which corporations, governments, and even individuals compensate for carbon dioxide emissions by investing in activities that remove carbon dioxide from the air or reduce emissions from a different source. See Anja Kollmuss, "Carbon Offsets 101," *World Watch* (July/August 2007). Unfortunately, present carbon-offset schemes contain loopholes that mean they may do little to reduce overall emissions; see Madhusree Mukerjee, "A Mechanism of Hot Air," *Scientific American* (June 2009).

On May 21, 2009, the House Energy and Commerce Committee approved H.R. 2454, "The American Clean Energy and Security Act." The goal of the Act, said Committee Chair Henry A. Waxman (D-CA), was to "break our dependence on foreign oil, make our nation the world leader in clean energy jobs and technology, and cut global warming pollution. I am grateful to my colleagues who supported this legislation and to President Obama for his outstanding leadership on these critical issues." Among other things, the Act called for Title VII of the Clean Air Act to provide a declining limit on global warming pollution (a "cap" as in "cap-and-trade") and to hold industries accountable for pollution reduction under the limit. The aim was to cut global warming pollution by 17 percent compared with 2005 levels in 2020, by 42 percent in 2030, and by

83 percent in 2050. (See the summary of the Act at http://energycommerce.house.gov/Press_111/20090515/hr2454_summary.pdf). In June 2009, the House of Representatives passed the bill, which also called for utilities to use more renewable energy sources. Unfortunately, the Senate refused to pass a corresponding bill, and "cap and trade" is dead, at least for now; see Jeff Johnson, "Cap and Trade Dies in Senate," *Chemical & Engineering News* (August 2, 2010). Part of the reason was the perception of the bill as imposing additional taxes in a time of economic difficulty; see C. Boyden Gray, "The Problem with Cap and Trade," *American Spectator* (June 2010).

Many people feel that it is about time that the United States took such action. According to the Global Humanitarian Forum's "Human Impact Report: Climate Change—The Anatomy of a Silent Crisis" (May 29, 2009) global warming is already affecting more than 300 million people and is responsible for 300,000 deaths per year. Action now is clearly appropriate, even though it does seem inevitable that this deadly impact of global warming must grow worse for many years before it can be stopped. However, the debate over the proper actions to take is by no means over. Some analysts argue that a carbon tax would be more effective; see Bettina B. F. Wittneben, "Exxon Is Right: Let Us Re-Examine Our Choice for a Cap-and-Trade System over a Carbon Tax," *Energy Policy* (June 2009). William B. Bonvillian, "Time for Climate Plan B," *Issues in Science and Technology* (Winter 2011), calls for a major effort to stimulate the development of new technologies.

In the YES selection, Paul Cicio, president of the Industrial Energy Consumers of America, argues that lacking global agreements, capping greenhouse gas emissions of the industrial sector will make domestic production less competitive in the global market, drive investment and jobs offshore, increase exports, and damage the economy. The real greenhouse gas problem lies with other sectors of the economy, and that is where attention should be focused. In the NO selection, Aaron Ezroj, Law Fellow at Adams Broadwell Joseph & Cardozzo, argues that although restricting emissions (as in a cap-and-trade program) may increase costs for some businesses, it will create many business opportunities in the financial sector, low-carbon technologies, carbon capture-and-storage projects, advanced-technology vehicles, and legal and nonlegal consulting. The overall effect will be to fuel the global economy.

YES 

Paul Cicio

Competitiveness and Climate Policy: Avoiding Leakage of Jobs and Emissions

Key Points

Capping the greenhouse gas (GHG) emissions of the industrial sector will drive investment and jobs offshore and increase imports. It will not bring major developing countries to the table but they will benefit through increased exports to the US. Even the third phase of the EU Emissions Trading Scheme (ETS) contains a provision to ensure their trade exposed industries receive compensation in order to prevent job loss and emissions leakage. Regulating the US industrial sector "before" negotiating an international agreement undermines our ability to achieve a fair and effective GHG reduction agreement for US industry.

For the industrial sector, climate policy is also trade, energy, [and] economic and employment policy. They are all intrinsically linked and inseparable. It is for this reason that regulating GHG emissions for the industrial sector be negotiated with both developed and developing countries in the context of a fair trade and productivity.

The US industrial sector is not the problem. In the US, the industrial sector's GHG emissions have risen only 2.6% above 1990 levels while emissions from the residential sector are up 29%, commercial up 39%, transportation up 27% and electricity generation up 29%.

The industrial sector competes globally and requires a global GHG policy solution that is based on productivity, something that the developing countries industrial sector can potentially agree to. A GHG cap is an unacceptable policy alternative for them and for us.

The US cannot grow the economy without using more volume of our products. The only question is whether the product will be supplied from domestic sources or imports. In fact a cap limits economic efficiency because it even limits the ability to maximize production from existing facilities that are not running at installed capacity. Since 2000, US manufacturing has been losing ground. From 2000 to 2008, imports are up 29% and manufacturing unemployment fell 22%, losing 3.8 million jobs, a direct statistical correlation.

The use of energy by the industrial sector is value-added. Our products enable GHG emission reductions. Lifecycle studies show that they save much

U.S. Senate, March 18, 2009.

more energy and GHG emissions than what is used/emitted in their production. Raising energy costs raises the cost of these valuable products.

The industrial sector already has a price signal for GHG emissions, it is called global competition and because we are energy intensive, we either drive down our energy costs or go out of business.

Under cap and trade, the industrial sector pays twice: [t]hrough the additional cost of carbon embedded in energy purchases and through the higher cost of natural gas and electricity. Higher demand for natural gas will result in higher prices for all consumers. Since natural gas power generation sets the marginal price of electricity, higher natural gas prices will mean higher electricity prices for all consumers.

A cap will damage the ability of the US industrial sector to take back market share from imports and increase exports.

Cap and trade does not address our country's fundamental need to significantly increase the availability, affordability and reliability of low carbon sources of supply.

Carbon trading and market manipulation is of great concern. The US government has proven unable to prevent market manipulation for mature energy and food commodities and credit default swaps—carbon markets will be much harder to regulate.

If the US proceeds to cap GHGs, it must provide to industry free allowances equal to the resulting increased direct and indirect costs due to GHG regulation until major competing developing countries have similar cost increases.

Congressional Justification for Not Capping GHG Emissions of the Industrial Sector

Congress has a choice to make and it is a decision it cannot afford to make incorrectly. It must decide whether to maintain and possibly increase US manufacturing jobs by not capping GHG emissions on the industrial sector—or create jobs in foreign countries by importing manufacturing products to supply the needs of our economy.

The Industrial Energy Consumers of America is an association of leading manufacturing companies with $510 billion in annual sales and with more than 850,000 employees nationwide. It is an organization created to promote the interests of manufacturing companies IECA membership represents a diverse set of industries including plastics, cement, paper, food processing, brick, chemicals, fertilizer, insulation, steel, glass, industrial gases, pharmaceuticals, aluminum and brewing.

The decision should not be hard because there is very sound economic and environmental justification for Congress to not act in the short term to cap GHG reductions on the industrial sector but to forge a different policy path that will provide sustained GHG reductions globally by harnessing real market forces called competition.

The industrial sector needs a globally level playing field that lets the best companies win. Adding costs by unilateral action helps "all" of our competitors in other countries take our business and our jobs. **We need US leadership to forge a global effort to address industrial sector GHG emission reductions that is focused on "fair trade" and "productivity." This is the only way to potentially bring developing nations to the table.**

Productivity is a language that all manufacturers understand and fundamental to competition. We believe that all governments want increased productivity by their industrial sector. We urge you to take action in this more realistic direction.

The world in which the industrial manufacturing company operates is diverse and business is often won or lost on the difference between pennies per unit of product. Competitiveness is everything. Some segments of industry, such as the power producers, may support cap and trade, but that's because they don't compete globally and they simply pass through their increased costs, we don't have that luxury.

Unlike that vision that many Americans have of China building coal-fired power plants using antiquated technology, it is vitally important that the Congress understand that a great number of companies that we compete with from developing countries are top-in-class competitors. They are utilizing the latest, world class technology. Some of these facilities are state owned or supported. Many also have subsidized energy costs. Energy costs most often determine our competitiveness and it can be our largest non-controllable cost.

The Congress can act in the public interest to consider both the cost and benefits of not imposing the cap on the industrial sector. The benefits of not imposing a GHG cap include good paying jobs, exports that reduce our balance of payments and the domestic production of products that are solutions to our climate challenges.

So far, only the environmental costs have been debated. We caution you to consider that your policy decision can lead to a further acceleration of the loss of the industrial sector. Just look at the facts. Due to the loss of competitiveness since 2000, the manufacturing sector has lost 3.8 million jobs thru 2008. During this same time period, imports rose 29%, a direct statistical correlation.

President Obama rightfully points to the disappearing middle class as troubling. We agree. The US began to lose the middle class when the industrial sector began to lose competitiveness along with our high paying jobs that most often pay benefits. The timing is consistent. We encourage the president and Congress to work with us to put new industrial policies in place that will increase competitiveness and grow the industrial sector and greatly restore the middle class.

To their credit, Representative Inslee and Doyle have rightfully recognized the need to protect manufacturing competitiveness. They are well intentioned but their solution is not really a solution for an industry that competes globally. We will still be burdened with costs and uncertainty. Most importantly, it does not do anything to bring the industrial sectors of developing countries into a climate agreement. Instead, a global solution is warranted that puts us on equal footing with our competitors. The international agreement should be negotiated first, not second. Regulating the US industrial sector in advance of negotiations completely removes our negotiation leverage.

The global reality is that developing nations place a significant priority on their manufacturing sector for both domestic economic growth and exports. They have a long history of providing all types of subsidies that include energy and trade credits. If they subsidize energy costs for their manufacturers, why wouldn't they also subsidize the cost of GHG reductions to enable exports to the US? US industry needs a level playing field—and then let us compete.

The justification is obvious and in the best interests of the country. The industrial sector's absolute GHG emissions are only 2.6% above 1990 levels and the rate of change has been flat due to energy efficiency improvements and a declining manufacturing presence. In contrast, according to the EPA, the transportation sector emissions are up 27%, residential up 29%, commercial up 39% and power generation up 29%. The point is that the industrial sector is *not* a contributor to growing GHG emissions and should not be a high priority for GHG reduction mandates.

Secondly, the products we produce are essential for economic growth of the country and a vibrant opportunity to create new high paying jobs. As the economy rebounds, our country will require significant volumes of the products that we produce such as cement, steel, aluminum, chemicals, plastics, paper, glass, and fertilizer which are all energy intensive. You can't produce renewable energy without our products. The question Congress must answer is whether it wants these products to be supplied by production facilities in the US or imported from foreign countries.

If Congress places a declining GHG cap on the industrial sector, you can be pretty confident that US companies will "not" invest their capital nor create jobs in the US. The reason is obvious. There is a lack of confidence that other countries will place a GHG cap on their manufacturers any time soon which would place US industry at a significant competitive disadvantage. Setting a starting date of 2012 for a GHG cap will result in industrial companies making pre-emptive capital decisions on where to locate and increase the production of their products that anticipates these assumptions.

Third, products from the manufacturing sector provide the "enabling solutions" to the challenges of climate change and it is important that GHG regulation does not increase the cost of these products to deter consumer purchases.

It takes energy to save energy. Insulation can be made from glass, plastic or paper, all of which are energy intensive. Double pane windows use twice the amount of glass but save an enormous amount of energy over the life of a building. Reducing the weight of autos, trucks and aircraft is an essential solution but requires greater use of aluminum, composite plastics and different grades of steel. More steel and plastics are needed for wind turbines. The production of solar silicon used to make solar panels is energy intensive. There are literally a thousand examples of how manufacturing products contribute to the climate solution and it is important to keep the cost of these products low.

The industrial sector is the "green sector." Manufacturing has a remarkable track record of reducing energy while continuing to increase the output of product. They predominantly use natural gas as a fuel versus coal. They are the largest consumer of biomass that is used for making paper and as a fuel for producing energy efficient steam and power. They utilize combined heat and power extensively and substantial quantities of recycled steel, aluminum, glass and paper which is extraordinarily energy efficient.

Fourth, placing a GHG cap on manufacturing makes it much more difficult for our sector to reclaim domestic market share and increase exports. The US has a significant trade deficit in part due to declining manufacturing product exports that accelerated in 2000 as US natural gas prices rose and imports increased.

A lot of these imports are from China, a country that values its manufacturing sector. And now, the US is dependent upon China to finance its burgeoning debt. Improving the competitive health of our manufacturing sector can help reduce this dependency. Increasing competitiveness of the industrial sector and increasing exports is an important matter of public policy that needs [to be] addressed.

The decision is yours to make. Company CEOs have a responsibility to their shareholders to protect the company's interest and they will. The manufacturing sector is agile and mobile to survive and thrive—it is just a question of where.

Climate Policy and Manufacturing Competitiveness

IECA has not attempted to gain consensus by the industrial sector on what is the best way to regulate GHG emissions for the US economy or for the manufacturing sector. However, there is little question how the majority view policy options.

Every discussion begins and ends with "competitiveness." Manufacturers compete globally and for many, the cost of energy and carbon will determine whether they will successfully compete in domestic and global markets.

The "absolute" cost of energy and carbon does not matter so long as all of our competitors around the world have the same increased costs. What matters to manufacturers is the "relative" cost of energy and carbon compared to our major global competitors regardless of whether they are in Europe or a developing country.

For that reason, US climate policy must not increase our relative costs. This means that manufacturing competitiveness must also be dealt with at the international level. While this presents a challenge for policy makers, it also provides a wonderful opportunity.

Those of us from the industry believe that more GHG emission reductions can be achieved globally when industrial climate policy instruments are focused on productivity that is, increasing production while reducing energy consumption. It's a win-win and recognizes that all players can only manage the energy use inside their plant and often have little control on the type of energy available.

There is general agreement by US manufacturers that other countries will not knowingly sacrifice their manufacturing jobs in response to climate policy. Since China tends to be a policy lynchpin, it is important to note that they especially will not sacrifice their manufacturing competitiveness to address climate change.

It is China's manufacturing sector that has raised its status to a world power by creating jobs and exports that have provided a significant and unequaled trade surplus. Now, the US is dependent upon them to buy our treasury bills and finance our debt. This is not an enviable position for the US nor is it necessary.

To its credit, the Chinese government has a history of emphasizing the importance of the manufacturing sector which is in contrast to the US government. China has also provided export tax credits, subsidies for energy costs and manages its currency. Some US government officials claim that currency control gives China a 40% competitive advantage over US manufacturers. Whether it's the currency or not, China's manufacturing sector is winning and US manufacturing is losing.

Any US climate policy option must hold manufacturing harmless until major competitors in both developed and countries in transition have comparable energy and carbon cost increases. Comparable reduction requirements do not meet the test. Without this protection, US manufacturers will protect their shareholders and move production facilities to countries that offer a competitive environment.

Well intentioned members of Congress have proposed a cap and trade system that would provide manufacturers with "some" free allowances that would decline over time and would cover "some" of the resulting higher energy costs. While appreciated, these provisions are not adequate to allow the industrial sector to compete, grow domestic production and exports. Many US industries have been working on energy efficiency for decades and simply don't have technology available to make step changes needed to meet these ratcheting targets.

Under these provisions we will still have a declining GHG cap that reduces our production; unpredictable costs for energy, carbon and transaction costs; and un-necessary cost increases. It also does not do anything to help our domestic customers who will be asked to absorb higher costs for our products.

Economy-wide cap and trade is simply the wrong policy platform for the manufacturing sector. IECA wants a climate policy that will allow US manufacturing to: invest in the US; does not create winners and losers; does not penalize those who have already invested in energy efficiency; and transparency so that the system cannot be manipulated or gamed.

Relatively few manufacturers in the industry support cap and trade. The ones that do have either inherent special circumstances that allow them to gain a relative competitive advantage; have already moved their energy intensive manufacturing offshore; will significantly benefit from increased product sales or are simply not energy intensive and are not measurably impacted.

We do not know any manufacturing companies who support carbon cap and trade with auction. This is completely understandable because the manufacturing sector needs predictability over long time horizons for capital investment. The auction of carbon allowances does not give price certainty plus manufacturers are disadvantaged in competing for the auctioned carbon with regulated utilities who can afford to pay any price and then pass the cost on to consumers to pay.

If the government lets Wall Street participate, the auction option gets even worse. In general, manufacturers believe that only companies who are required to reduce GHG emissions should be allowed to purchase carbon allowances or offsets. This leaves Wall Street out.

Auctioning is the quickest way to lose manufacturing jobs and they will go silently, one at a time and without an announcement. Each manufacturing production unit has a cost break—even that varies significantly from plant to plant and from company to company. As the cost of carbon rises, the manufacturer will not have any choice but to shut it down.

Very few companies support cap and trade even if allowances are initially provided free of charge because they recognize that these temporary allowances are not a safety net and their economic viability is in jeopardy long term. The engineering limitations of their manufacturing facilities leave little room for imagination—just realism.

A carbon tax is better than a cap and trade program because it does not constrict our ability to increase the volume of product produced, it is superior in transparency, and more easily adjusted at the border. Nonetheless, it is a cost that is not welcomed and un-necessary for the industrial sector to reduce carbon intensity. Clearly, a high carbon tax will be just as effective of putting us out of business.

There are about 350,000 manufacturing facilities in the US. It is estimated that about 7,800 facilities would emit 10,000 tons of CO_2 per year. By itself, regulating the industrial sector presents a significant regulatory challenge for the federal government. While only 7,800 would be regulated, the other 342,200 facilities and the American consuming public would be asked to absorb higher resulting product costs. . . .

Aaron Ezroj

How Cap and Trade Will Fuel the Global Economy

Cap-and-trade programs and related measures will spark a wide range of business opportunities. The financial sector will grow to facilitate hundreds of billions of dollars worth of climate change-related exchanges. By 2050, markets for low-carbon technologies are likely to be worth at least $500 billion annually, and possibly much more. Numerous carbon capture-and-storage projects will emerge. Plug-in and other advanced technology vehicles will become the norm. Moreover, a plethora of legal and nonlegal consulting agencies will be advising government agencies and companies on climate change. Countries and companies should position themselves now in order to take full advantage of these opportunities.

The European Union (EU) has done the most to position itself and companies within it. Although it initially disfavored cap and trade, the EU has implemented the world's most expansive cap-and-trade program: the EU Emission Trading System (EU ETS). Following the implementation of the EU ETS, financial markets in London are overseeing the trading of billions of dollars in carbon allowances. Diplomats in Brussels are negotiating guidelines for offset projects in China. Moreover, Copenhagen is becoming the global center for the development of wind turbine technology. These efforts have provided the EU and companies within it with a significant head start in positioning themselves for the transition to a low-emissions global economy.

Eying the success of cap and trade in Europe, New Zealand is moving forward with the New Zealand Emissions Trading Scheme (NZ ETS), Australia is moving forward with the Australian Carbon Pollution Reduction Scheme (CPRS), and individual states and provinces within the United States and Canada are moving forward with their programs and joining larger regional collectives, such as the Regional Greenhouse Gas Initiative (RGGI) in the northeastern United States and the Western Climate Initiative (WCI) in the western United States and Canadian Provinces. The U.S. federal government is also contemplating putting together a program of enormous proportions, involving agencies such as the U.S. Environmental Protection Agency (EPA), the U.S. Department of Agriculture, the Federal Energy Regulatory Commission, and the Commodities Futures Trading Commission, among others. As more countries and regions enact cap-and-trade programs, these programs will fuel the global economy.

Financial Markets

In 2008, transactions on the global carbon market amounted to $92 billion. There were over three billion spot, future, and option contracts traded for a variety of reasons, including compliance, risk management, and arbitrage. Between 2007 and 2008, the value of transactions nearly doubled for the EU ETS. As current cap-and-trade programs expand to cover more sectors of the economy and other countries and regions develop cap-and-trade programs, the global carbon market will continue to grow rapidly. While it is impossible to forecast the growth of this emerging sector with exact accuracy, it has been projected that if all developed countries had carbon markets covering all fossil fuels, the global carbon market would grow by 200%. Moreover, if markets were established in all the top 20 emitting countries, the global carbon market would grow by 400%. According to Louis Redshaw, the head of environmental markets at Barclays Capital: "Carbon will be the world's biggest commodity market, and it could become the world's biggest market overall."

Allowances are the basic unit traded within the global carbon market. A single allowance provides a compliance entity with the right to emit one ton of carbon dioxide (CO_2) or CO_2 equivalent. Allowances are introduced into the market through a distribution made by a government agency or an auction, and resulting revenue funds policy mandates. Allowances are then traded between compliance entities through market exchanges or over-the-counter transactions. At the end of each compliance period, compliance entities surrender allowances to a designated regulator for each ton of CO_2 or CO_2 equivalent that they emitted during the period.

Emissions trading enables a compliance entity to emit more than permitted by its current holding of allowances if it can obtain spare allowances from another compliance entity. The overall environmental outcome is the same as if both compliance entities used their allowances exactly, but with the important difference that both buying and selling companies benefited from the flexibility allowed by trading. Moreover, emissions trading encourages compliance entities to find cost-effective ways to reduce their emissions, which allows compliance entities to purchase fewer allowances.

However, the price for allowances is not static. The price of an allowance can spike upward as a result of weather fluctuations. For instance, a low water year affects the generation of hydroelectricity. The price of an allowance can spike downward if there is a recession, there is less demand for energy, and in turn compliance entities require less allowances to emit greenhouse gases (GHGs). Moreover, the price of an allowance can change as a result of market speculation from investment banks who themselves have no compliance obligations but are still active in carbon markets. Or, the price of allowances can change as a result of environmental nongovernmental organizations (NGOs) purchasing allowances to retire, and thus restricting the ability of compliance entities to emit GHGs.

Furthermore, if compliance entities do not have enough allowances to surrender at the end of a compliance period, they will be forced to pay heavy penalties. The EU ETS fines compliance entities 100 euros for each ton of

CO_2 or CO_2 equivalent emitted for which the operator has not surrendered an allowance. RGGI has a 3x allowance penalty for each ton of CO_2 or CO_2 equivalent emitted for which the operator has not surrendered an allowance. Additionally, each day and each excess ton of emissions is considered separate violations of state law for which the source can be subject to administrative or civil fines and proceedings.

Because the price for allowances fluctuates and compliance entities need to make sure that they will have enough allowances or else they will face heavy penalties, carbon markets currently include or are likely to include a number of financial instruments to manage risk. These instruments, referred to as derivatives, include forward contracts, futures contracts, option contracts, and swaps. Forward contracts allow buyers and sellers to agree upon the delivery of allowances at a specified date. Futures contracts give the holder the right to sell a specified quantity of allowances at a specific price within a specified time, regardless of the market price for allowances. Option contracts give the holder the right to buy a specified quantity of allowances at a specific price within a specified period of time, regardless of the market price for allowances. Lastly, swaps allow for the exchange of one asset or liability for another asset or liability.

In the EU ETS, the majority of allowance-based instruments are traded as derivatives, rather than allowances. Concerns over allowance price volatility or a low volume of allowances may also make this the case in a U.S. carbon market.

Robust carbon markets, involving allowances and derivatives, will lead to numerous employment and investment opportunities. Large investment firms are well aware of the enormous potential that cap and trade presents for the financial industry and are already active in markets for carbon emissions and other climate-related commodities. In 2006, Goldman Sachs made a minority equity investment in Climate Exchange PLC, which owns the European Climate Exchange, the Chicago Climate Exchange, and the newly created California Climate Exchange.

Alternative Energy

In 2007, the size of the market for renewable energy products was approximately $38 billion and employed approximately 1.7 million people. Overall, the market grew by 25% in 2005. Within the overall total, some renewable energy technologies grew at an even faster rate. The global install-capacity of solar photovoltaic rose by 55% in 2005. The market for wind power grew by nearly 50%. In the year prior to August 2006, the market capitalization of solar companies grew 38-fold to $27 billion. Growth in the biofuel sector only rose by 15% in 2005, but still the total market for the sector is worth over $15 billion. It has been predicted that by the year 2050, the annual market for low-carbon technologies could be worth hundreds of billions of dollars and employ over 25 million people.

A study conducted by the United Kingdom's (U.K.'s) Secretary of State for Energy and Climate Change, explained: "Climate change is not only one of the most significant challenges of our generation; it also presents a huge opportunity. Supplying the demands of a low-carbon economy offers a significant potential contribution to the economic growth and job creation in the U.K." Further, the study projected that whole new industrial sectors may emerge and will provide around 100 billion pound sterling worth of investment opportunities and up to 500,000 U.K. renewable energy jobs.

Moreover, the study emphasized:

> The current economic difficulties make this even more important: now is the not the time to scale back our ambitions on tackling climate change and securing our energy supplies. The increased levels of investment in renewable energy in the U.K. and across Europe over the next decade and beyond will involve significant adjustment costs, but the high investment in renewable energy has the potential to boost our economy in the short term and will help kick-start our long-term transition to a low-carbon economy.

The best way for a country and companies within it to capitalize on opportunities in the renewable energy market is to adopt a cap-and-trade program and related measures. Doing so allows for: (A) auction revenue to be allocated to alternative energy projects; (B) offsets enabling projects that would otherwise not be feasible; and (C) renewable energy certificates that subsidize the development of alternative energy projects.

Auction Revenue Will Be Allocated to Alternative Energy Projects

When allowances are auctioned, the revenue from the auction is used to fund specific policy mandates. In H.R. 2454, which details the U.S. House of Representative's proposal for a cap-and-trade program, allowance revenue is used to provide rebates for low- and moderate-income families; to offset increased costs faced by consumers of electricity, natural gas, and heating oil; to subsidize GHG capture and storage; to support other domestic and international technology programs; to safeguard the competitiveness of energy-intensive, trade-exposed industries; and to support domestic and international adaptation programs. From 2012 to 2050, 15% of allowance revenue will go to renewables, efficiency, GHG capture and storage, autos, and other green technologies.

California's cap-and-trade program may also devote auction revenue to carbon reduction technologies, such as alternative energy projects. The Economic and Technology Advancement Advisory Committee, which has been advising the California Air Resources Board, recommended using allowance revenue to fund research and development and to support a green technology workforce training program.

Offset Credits Enable Projects That Would Otherwise Not Be Feasible

Carbon offset credits are awarded for GHG reductions from renewable energy projects that are "additional," meaning that they would not have been financially viable without the prospect of revenue from the sale of offsets.

Renewable energy projects are unlikely to qualify for offset credits within a U.S. cap-and-trade program because project developers would have difficulty demonstrating that a renewable energy project would not have been financially viable without the prospect of offset revenues. In a U.S. cap-and-trade program, energy sector emissions will likely be capped. This will make fossil fuels more expensive and thus make renewable energy sources more attractive.

However, while domestic renewable energy projects are unlikely to qualify for offset credits in a national emissions reduction program, renewable energy projects will still qualify for offset credits in countries without GHG emissions controls on their energy sectors. The EU encourages companies within it to support capacity-building activities in developing countries to help them take advantage of the Clean Development Mechanism (CDM) in a manner that supports sustainable development in the host's country. Indeed, the Kyoto Protocol reads:

> The purpose of the CDM *shall be to assist Parties not included in Annex I in achieving sustainable development* and in contributing to the ultimate objective of the Convention, and to assist Parties included in Annex I in achieving compliance with their quantified emission limitation and reduction commitments under Article 3.

Currently, a large percentage of offset credits are generated from renewable energy projects in countries without GHG emissions controls. Thirty-five percent of the projects in the CDM pipeline are renewable energy projects. There are 399 wind CDM projects in China, producing 22,209 megawatts (MW) of electricity, and 320 wind CDM projects in India, producing 5,915 MW of electricity. There are 819 hydro CDM projects in China, producing 25,896 MW of electricity, and 130 hydro CDM projects in India, producing 5,737 MW of electricity. Moreover, 30% of the projects in the Joint Implementation (JI) pipeline are renewable energy projects.

H.R. 2454 lays out a framework for international offset-crediting and emphasizes that credits can only be issued for projects in developing countries. Thus, assuming that a large percentage of offset credits are generated from renewable energy projects in countries without GHG emissions controls, offset credits will continue to enable projects that would otherwise not be feasible.

Renewable Energy Certificates Will Subsidize the Development of Alternative Energy Projects

Renewable Energy Certificates (RECs) are awarded for each MW-hour of renewable energy generated from qualifying renewable energy projects, such as wind, solar, geothermal, and certain hydropower projects. Typically, RECs are unbundled and sold separately from the underlying electricity generated by renewable energy projects, allowing renewable energy generators to sell both RECs and the wholesale electricity they produce. Overall, RECs act as a subsidy for the development of alternative energy projects.

Currently, REC registries are being set up in many states, regions, and countries. California requires utilities to meet part of their electricity demand through renewable energy sources. The California Energy Commission estimates that renewable energy sources generate 12% of California's retail electricity load. California S.B. 107 requires investor-owned utilities to increase the share of renewables in their electricity portfolios to 20% by 2010. At the same time, public-owned utilities are encouraged to meet the same target. Recently, Gov. Arnold Schwarzenegger called for renewables to make up 33% of electricity portfolios and accordingly, it is anticipated that renewable energy sources will generate 33% of California's electricity by 2020.

H.R. 2454 amends the Public Utility Regulatory Policies Act of 1978 to require retail electricity suppliers to meet 20% of their electricity demand through renewable energy sources and energy efficiency by 2020. Each retail electricity supplier that annually sells four million MW-hours of electricity or more would need to submit RECs equal to at least three-quarters of their allotted requirement.

Other countries are also enacting systems involving mechanisms similar to RECs. For example, a Renewables Obligation (RO) was introduced in the U.K. in 2002. Under the RO, generators receive Renewable Obligation Certificates (ROCs) for renewable electricity. Electricity suppliers are incentivized to buy ROCs from generators, and ROCs provide renewable generators with financial support in addition to what they receive from selling their electricity. The RO has so far increased RO-eligible renewable electricity generation in the U.K. from 1.8% of the country's electricity load in 2002 to 5.3% of the country's electricity load in 2008.

It is unclear whether RECs will be part of a cap-and-trade program or be a related measure in a larger climate change legislative package. Principally, RECs serve as proof that one MW-hour of electricity was generated and delivered to the grid from a qualifying renewable energy source, but the definition of RECs has been extended to imply or explicitly claim that RECs also offset GHG emissions and should be treated as offset credits. There are, however, serious problems with treating RECs as offset credits. First, it is difficult to prove ownership with RECs. Operation of the electric power grid is complex, and it is difficult to establish linkage between renewable energy generation and changes in generation at other power plants on the grid. Second, REC programs have eligibility requirements that do not necessarily consider additionality. While some renewable energy projects may not have been implemented without RECs, other projects may have been implemented without them.

However, whether RECs are directly incorporated into a cap-and-trade program or whether they are supplemental as part of a larger climate change legislative package, they will certainly encourage the development of renewable energy projects.

GHG Capture and Storage

Cap and trade will also lead to the development of numerous GHG capture-and-storage projects. Domestic and international offset projects are rapidly increasing in number. In 2007, the value of transactions in the primary market for offset projects grew 34% to $8.2 billion. Currently, the market is dominated by the main offset mechanisms of the Kyoto Protocol: the CDM and the JI. In 2007, CDM transactions accounted for 87% of project-based transaction volumes and JI transactions doubled in volume and tripled in value. The remaining market activity was split among other compliance mechanisms and voluntary purchases.

The offset market is highly sophisticated, involving a number of government, quasi-government, and private-sector participants. At its 21st meeting, the CDM Executive Board discussed work on the registration of CDM project activities as part of the CDM Management Plan. The CDM Executive Board decided to: "Make publicly available relevant information, submitted to it for this purpose, on proposed CDM project activities in need of funding and on investors seeking opportunities, in order to assist in arranging funding of CDM project activities, as necessary." Subsequently, the United Nations Framework Convention on Climate Change (UNFCCC) established the UNFCCC CDM Bazaar, which is a web-based facility serving as a platform for the exchange of information on CDM project opportunities.

There are offset retailers, such as Climate Trust, TerraPass, NativeEnergy, and Myclimate. Offset prices vary by factors, such as project type, location, and stringency of offset program requirements. Certified emission reductions (CERs), awarded for CDM projects, and emission reduction units (ERUs), awarded for JI projects, can be valued at upward of 80% of the trading price of EU allowances. Prices for voluntary offset credits vary significantly based on project types, project locations, standards used, offset quality, delivery guarantees, and contract terms.

Recently, JP Morgan agreed to purchase EcoSecurities, an offset aggregator, for $204 million. EcoSecurities has been involved in the carbon market for over 10 years and has offices and representatives in more than 20 countries and five continents. It sources, develops, and trades emissions reductions credits and manages a diverse portfolio of credits, including different project types, project locations, volumes, technologies, methodologies, risk profiles, contract terms, volumes, and sustainability co-benefits.

As more countries and regions enact cap-and-trade programs and the demand for offsets increases, two project types are likely to expand rapidly: (1) methane capture and destruction projects; and (2) biological sequestration projects.

Methane Capture and Destruction Projects Will Become Extremely Popular

Methane capture and destruction projects are likely to become extremely popular because methane has 25 times the heat-trapping ability of CO_2 and the global warming potential of GHGs influences the volume of offsets generated by a project. A developer who reduces one ton of methane gas receives 25 times the credits that they would receive for reducing one ton of CO_2.

Current practices could be changed to curb emissions. For instance, in the United States, coal mines account for about 10% of all man-made methane emissions. Because the gas can present a safety risk for miners, methane released during the extraction of coals is removed through ventilation fans and vented into the atmosphere. Through an offset project, the methane could instead be recovered and burned to produce energy or flared to reduce its heat-trapping ability when it is released into the atmosphere.

In 2007, about 49% of the U.S. offset supply was produced from projects that capture and destroy methane from coal mines, agricultural operations, or landfills. Ninety-three of the 211 projects that produced U.S. offsets were methane projects. The Clean Energy Jobs and American Power Act considers: "methane collection and combustion projects at active underground coal mines"; "methane collection and combustion projects at landfills"; and, "nonlandfill methane collection, combustion and avoidance projects involving organic waste streams that would have otherwise emitted methane in the atmosphere, including manure management and biogas capture and combustion."

Biological Sequestration Will Result in Numerous Forestry Projects

Forestry and other land use projects aimed at sequestering GHGs will also increase in popularity. These projects can reduce and avoid the atmospheric buildup of GHGs in a number of ways. First, tree biomass and soils can act as carbon sinks, removing and storing CO_2. Statistics released in an EPA study explained that afforestation can sequester 2.2-9.5 tons of CO_2 per acre per yearj and reforestation can sequester 1.1-7.7 tons of CO_2 per acre per year for 90 to more than 120 years before saturation occurs. Second, CO_2 emissions can be avoided by using biofuels rather than fossil fuels. Third, agricultural emissions from fertilizers can be reduced by changing livestock management and fertilizer applications.

The Kyoto Protocol vaguely promised to include emissions reductions for forestry and other land use projects because these projects have the potential to sequester CO_2. However, there are concerns about how much CO_2 these projects actually remove from the atmosphere, how to measure the CO_2 that they remove, and whether the removal from the atmosphere is permanent. For these reasons, the use of forestry and other land use projects in meeting emissions targets has been controversial. However, the United States and other economies with high energy-intensity and population growth, Australia and Canada, have pushed for a maximum of flexibility in achieving emissions reductions. The United States, specifically, has insisted on the inclusion of sinks from forestry and other land use projects.

In 2007, 17% of the U.S. offset supply was generated from forestry and other land use projects. This includes 52 forestry projects that produced about 7% of the total U.S. supply. The Clean Energy Jobs and American Power Act considers awarding offset credits for: "projects involving afforestation or reforestation of acreage not forested as of January 1, 2009"; "forest management resulting in an increase in forest carbon stores, including harvested wood products"; "agricultural, grassland, and rangeland sequestration and management" practices"; and, "changes in carbon stocks attributed to land use change and forestry activities." Moreover, H.R. 2454 gives financial incentives to farmers and ranchers to plant trees. According to an EPA analysis of H.R. 2454, about 18 million acres of new trees would be planted by 2020. With the implementation of a U.S. cap-and-trade program, afforestation efforts would be even greater than those carried out by the Civilian Conservation Corps between 1933 and 1942, which planted 3 billion trees.

Plug-in and Other Advanced Technology Vehicles

Transportation is one of the largest sources of GHG emissions. In California, it is the largest source. Cap and trade and related measures are taking steps to decrease emissions in this sector by encouraging the production and purchasing of plug-in and other advanced technology vehicles.

Auction Revenue Will Be Allocated to Plug-in and Other Advanced Technology Vehicles

Fuel providers are compliance entities in the proposed U.S. cap-and-trade program, California's cap-and-trade program, and New Zealand's cap-and-trade program. Within these programs, auction revenue may be allocated to plug-in and other advanced technology vehicles. For example, under H.R. 2454, the U.S. cap-and-trade program contains significant incentives for automakers to produce plug-in and other advanced technology vehicles. In the beginning of the program, 3% of allowances would be allocated to the automotive sector to provide grants to refit or establish plants to build plug-ins and other advanced vehicles. Depending on the price of allowances, this allocation could end up being worth billions of dollars each year.

Consumers Will Switch to Plug-in and Other Advanced Technology Vehicles Because of Increased Fuel Costs

Because fuel providers will have to purchase allowances in certain cap-and-trade programs, in these programs, their costs will increase and these costs will be passed on to consumers who will have to pay more for gasoline at the pump. Facing increased costs in gasoline, consumers will be incentivized to purchase plug-in and other advance technology vehicles. This trend was demonstrated in the 1980s, for instance, when consumers responded to high gasoline prices by driving smaller, more fuel-efficient cars.

A study conducted by the Center for the Study of Energy Markets postulates that this trend may not be incredibly strong, especially with today's consumers, who may be less likely to curb their gas consumption with increased fuel prices than consumers in earlier decades. This could be because incomes have grown and consumers are now less sensitive to price increases because gasoline consumption is a smaller share of their budget. The study, however, looks at short-run rather than long-run gasoline price increases, and acknowledges that consumers may respond to higher gasoline prices in the long run by purchasing more fuel-efficient vehicles.

Low-Carbon Fuel Standards Will Encourage the Production of Alternative Energy Vehicles

Additionally, programs aimed at reducing the carbon intensity of transportation fuels, such as a low-carbon fuel standard (LCFS), will encourage the production of alternative energy vehicles. The California Air Resources Board developed an LCFS that requires fuel providers to track the average life-cycle GHG intensity of their products, including production, transportation, storage, and fuel use, and reduce the average life-cycle GHG intensity of transportation fuels they sell in California by at least 10% by 2020.

Following California's lead, the EU, several other U.S. states, and some Canadian provinces are developing LCFS proposals. The U.K. has also introduced the U.K. Renewable Transport Fuel Obligation Programme, which includes reporting requirements and methodologies for calculating life-cycle GHG emissions and requires fuel providers to ensure biofuels constitute 2.5% of total road transport fuels in 2008-2009, 3.75% in 2009-2010, and 5% after 2009-2010.

It is unclear how LCFS and other programs aimed at reducing the carbon intensity of transportation fuels will fit into a cap-and-trade program. A study prepared by the Center for the Study of Energy Markets recommended that California's LCFS should be kept separate from California's cap-and-trade program for at least the first 10 years to ensure innovation and investment in low global warming-intensive fuel technologies. However, whether these standards are directly incorporated into a cap-and-trade program or whether they are supplemental as part of a larger climate change legislative package, they will certainly encourage the production of alternative energy vehicles.

Legal and Nonlegal Consulting

Because of the introduction of cap-and-trade programs and related measures, a plethora of legal and nonlegal consulting agencies will be advising companies and government agencies on climate change [and related issues]. . . .

Conclusion

Cap and trade and related measures are not just environmental efforts that will curb the effects of global warming. They also present a wide range of business opportunities that will fuel the global economy. This includes growth in the financial sector, the development of low-carbon technologies, numerous carbon capture-and-storage projects, increased production of plug-in and other advanced technology vehicles, and a plethora of legal and nonlegal consulting opportunities. Countries and companies should position themselves now in order to take full advantage of these opportunities.

EXPLORING THE ISSUE

Will Restricting Carbon Emissions Damage the Economy?

Critical Thinking and Reflection

1. If the regulatory bodies in charge of assigning permitted pollution limits or caps reduce those caps over time, what will happen to overall pollution levels? What will happen if the caps stay the same over time?
2. How do cap-and-trade systems motivate companies to reduce carbon emissions?
3. How would controlling carbon emissions stimulate the development of advanced technology vehicles?
4. Should important decisions—such as whether to restrict emissions of greenhouse gases—be made primarily in terms of near-term effects on the economy, or in terms of long-term effects?

Is There Common Ground?

Both sides in this issue agree that restricting carbon emissions will increase costs for some businesses. One important question is whether the benefits to other businesses will compensate. Another is whether any business has the right to expect the conditions under which it operates not to change.

1. What businesses or industries seem most likely to experience increased costs if carbon emissions are restricted?
2. What businesses or industries seem most likely to benefit economically if carbon emissions are restricted?
3. When the conditions under which a business operates change, what should the business do?

ISSUE 10

Is Carbon Capture Technology Ready to Limit Carbon Emissions?

YES: David G. Hawkins, from "Carbon Capture and Sequestration," testimony before the Committee on House Energy and Commerce, Subcommittee on Energy and Air Quality (March 6, 2007)

NO: Charles W. Schmidt, from "Carbon Capture & Storage: Blue-Sky Technology or Just Blowing Smoke?" *Environmental Health Perspectives* (November 2007)

Learning Outcomes

After studying this issue, students will be able to:

- Explain what carbon capture and sequestration or storage (CCS) is.
- Describe several methods of CCS.
- Explain why long-term stability of carbon storage is of concern.
- Explain why areas subject to earthquakes and volcanic eruptions may be poor choices for carbon storage locations.

ISSUE SUMMARY

YES: David G. Hawkins, director of the Climate Center of the Natural Resources Defense Council, argues that we know enough to implement large-scale carbon capture and sequestration for new coal plants.

NO: Charles W. Schmidt argues that the technology is not yet technically and financially feasible, research is stuck in low gear, and the political commitment to reducing carbon emissions is lacking.

It is now well established that burning fossil fuels is a major contributor to global warming, and thus a hazard to the future well-being of human beings and ecosystems around the world. The reason lies in the release of carbon dioxide, a major "greenhouse gas." It follows logically that if we reduce the amount of carbon dioxide we release to the atmosphere, we must prevent or ease global warming. Such a reduction would of course follow if we shift away from fossil fuels as an energy source. Another option is to capture the carbon dioxide before it reaches the atmosphere—perhaps directly from power plant exhaust streams—and put it somewhere else. Still another option is to remove carbon dioxide from the atmosphere. One method is to plant trees, or even just to refrain from cutting forests; at the December 2010 United Nations Framework Convention on Climate Change, held in Cancun, Mexico, forest preservation became a dominant theme; see Eli Kintisch and Antonio Regalado, "Cancun Delegates See the Trees Through a Forest of Hot Air," *Science* (December 17, 2010). There is also a technological approach, according to Klaus S. Lackner, "Washing Carbon Out of the Air," *Scientific American* (June 2010).

All of these possibilities require that millions of tons of captured carbon dioxide be stored someplace safe. Lal Rattan, "Carbon Sequestration," *Philosophical Transactions: Biological Sciences* (February 2008), describes a number of techniques and notes that all are expensive and have leakage risks. Forests, for instance, can catch fire. One proposal is that supplying nutrients to ocean waters could stimulate the growth of algae, which remove carbon dioxide from the air; when the algae die, the carbon should settle to the ocean floor. The few experiments that have been done so far indicate that although fertilization does in fact stimulate algae growth, carbon does not always settle deep enough to keep it out of the air for long. The experiments also fail to say whether the procedure would damage marine ecosystems. See Eli Kintisch, "Should Oceanographers Pump Iron?" *Science* (November 30, 2007), and Sandra Upson, "Algae Bloom Climate-Change Scheme Doomed," *IEEE Spectrum* (January 2008). Another proposal is that carbon dioxide captured from power plant exhaust or the atmosphere be concentrated, liquefied, and pumped to the deep ocean, where it would remain for thousands of years. Concern that that storage time is not long enough helped shift most attention to underground storage. Carbon dioxide, in either gas or liquid form, can be pumped into porous rock layers deep beneath the surface. Such layers are accessible in the form of depleted oil deposits, and in fact carbon dioxide injection can be used to force residual oil out of the deposits. See Robert H. Sokolow, "Can We Bury Global Warming?" *Scientific American* (July 2005), and Valerie Brown, "A Climate Change Solution?" *High Country News* (September 3, 2007). Jennie C. Stephens and Bob Van Der Zwann, "The Case for Carbon Capture and Storage," *Issues in Science and Technology* (Fall 2005), note that the technology exists but that industry lacks incentives to implement it; such incentives could be supplied if the federal government established limits for carbon emissions.

One of the great concerns about carbon capture and sequestration (CCS) is that once immense amounts of carbon dioxide have been stored underground, it will leak out again, either slowly or—perhaps after an earthquake—suddenly. Study of past eras has suggested that sudden releases of carbon dioxide from volcanoes have led to rapid greenhouse warming, which reduced oxygen levels in the ocean and caused the buildup of toxic hydrogen sulfide, which in turn reached the air and killed plants and animals on land, resulting in mass extinctions such as the one 250 million years ago. See Peter D. Ward, "Impact from the Deep," *Scientific American* (October 2006). A smaller scale threat, exemplified by Cameroon's Lake Nyos, which released so much dissolved carbon dioxide in 1986 that it flowed downhill and suffocated almost 2000 people, along with their domestic animals, has been cited by environmental justice groups protesting CCS legislation in California. See Valerie J. Brown, "Of Two Minds: Groups Square Off on Carbon Mitigation," *Environmental Health Perspectives* (November 2007).

Such threats should concern us, but so should the threat of global warming itself. In the long run, we must move to non–fossil fuel sources of energy, for even coal, as plentiful as it is, will not last forever. In the short run, we have coal-burning power plants that continue to emit carbon dioxide, and we are planning to build more. As the Royal Society of Chemistry (RSC) notes in "Can We Bury Our Carbon Dioxide Problem?" *Bulletin 3* (Spring 2006), CCS will require the use of energy and will therefore increase the burning of fossil fuels and the price of energy to the consumer. The RSC also notes that researchers are not sure that there is enough underground capacity for all the carbon dioxide that CCS would endeavor to keep out of the atmosphere. More research is needed in this area, as well as in finding better, more efficient methods of capturing carbon dioxide, which can account for three quarters of the cost of CCS. Fortunately, researchers are developing new materials that may lower that cost significantly. See Kevin Bullis, "A Better Way to Capture Carbon," *Technology Review* (online) (February 15, 2008) (www.technologyreview.com/Energy/20295/?a=f), and Sid Perkins, "Down with Carbon," *Science News* (May 10, 2008). H. Jesse Smith, Julia Fahrenkamp-Uppenbrink, and Robert Coontz note in "Clearing the Air," their introduction to a special section on carbon capture and sequestration in *Science* (September 25, 2009), that "The prospects of CCS are uncertain, but its promise is great. . . . there are abundant reasons to hope that CCS can be implemented effectively."

It is worth stressing that those who favor CCS also believe, as David G. Hawkins says in his last paragraph below, that although CCS "is an important strategy to reduce CO_2 emissions from fossil fuel . . . it is not the basis for a climate protection program by itself. Increased reliance on low-carbon energy resources is the key to protecting the climate." Recognizing the value of CCS, U.S. Senator Jeff Bingaman (D-NM) proposed legislation to stimulate development and implementation of the technology; a hearing was held by the Senate Energy and Natural Resources Committee on April 20, 2010 (frwebgate.access.gpo.gov/cgi-bin/getdoc.cgi?dbname=111_senate_hearings&docid=f:61699.pdf). Unfortunately, the bill died when the Senate failed to pass energy policy legislation at the end of the year.

In the YES selection, David G. Hawkins, director of the Climate Center of the Natural Resources Defense Council, argues that we already know enough to implement large scale carbon capture and sequestration for new coal plants. The technology is ready to do so safely and effectively. In the NO selection, Charles W. Schmidt argues that the technology is not yet technically and financially feasible, research is stuck in low gear, and the political commitment to reducing carbon emissions is lacking. In addition, it has not been shown that carbon dioxide stored in underground reservoirs will stay in place indefinitely, and it has not been decided who will monitor such storage and take responsibility if it fails.

YES 

David G. Hawkins

Carbon Capture and Sequestration

Today, the U.S. and other developed nations around the world run their economies largely with industrial sources powered by fossil fuel and those sources release billions of tons of carbon dioxide (CO_2) into the atmosphere every year. There is national and global interest today in capturing that CO_2 for disposal or sequestration to prevent its release to the atmosphere. To distinguish this industrial capture system from removal of atmospheric CO_2 by soils and vegetation, I will refer to the industrial system as carbon capture and disposal or CCD.

The interest in CCD stems from a few basic facts. We now recognize that CO_2 emissions from use of fossil fuel result in increased atmospheric concentrations of CO_2, which along with other so-called greenhouse gases, trap heat, leading to an increase in temperatures, regionally and globally. These increased temperatures alter the energy balance of the planet and thus our climate, which is simply nature's way of managing energy flows. Documented changes in climate today along with those forecasted for the next decades, are predicted to inflict large and growing damage to human health, economic well-being, and natural ecosystems.

Coal is the most abundant fossil fuel and is distributed broadly across the world. It has fueled the rise of industrial economies in Europe and the U.S. in the past two centuries and is fueling the rise of Asian economies today. Because of its abundance, coal is cheap and that makes it attractive to use in large quantities if we ignore the harm it causes. However, per unit of energy delivered, coal today is a bigger global warming polluter than any other fuel: double that of natural gas; 50 per cent more than oil; and, of course, enormously more polluting than renewable energy, energy efficiency, and, more controversially, nuclear power. To reduce coal's contribution to global warming, we must deploy and improve systems that will keep the carbon in coal out of the atmosphere, specifically systems that capture carbon dioxide (CO_2) from coal-fired power plants and other industrial sources for safe and effective disposal in geologic formations. . . .

The Need for CCD

Turning to CCD, my organization supports rapid deployment of such capture and disposal systems for sources using coal. Such support is not a statement about how dependent the U.S. or the world should be on coal and for how

From U.S. House of Representatives Committee on House Energy and Commerce by David G. Hawkins, (March 6, 2007).

long. Any significant additional use of coal that vents its CO_2 to the air is fundamentally in conflict with the need to keep atmospheric concentrations of CO_2 from rising to levels that will produce dangerous disruption of the climate system. Given that an immediate world-wide halt to coal use is not plausible, analysts and advocates with a broad range of views on coal's role should be able to agree that, if it is safe and effective, CCD should be rapidly deployed to minimize CO_2 emissions from the coal that we do use.

Today coal use and climate protection are on a collision course. Without rapid deployment of CCD systems, that collision will occur quickly and with spectacularly bad results. The very attribute of coal that has made it so attractive—its abundance—magnifies the problem we face and requires us to act now, not a decade from now. Until now, coal's abundance has been an economic boon. But today, coal's abundance, absent corrective action, is more bane than boon.

Since the dawn of the industrial age, human use of coal has released about 150 billion metric tons of carbon into the atmosphere—about half the total carbon emissions due to fossil fuel use in human history. But that contribution is the tip of the carbon iceberg. Another 4 *trillion* metric tons of carbon are contained in the remaining global coal resources. That is a carbon pool nearly seven times greater than the amount in our pre-industrial atmosphere. Using that coal without capturing and disposing of its carbon means a climate catastrophe. And the die is being cast for that catastrophe today, not decades from now. Decisions being made today in corporate board rooms, government ministries, and congressional hearing rooms are determining how the next coal-fired power plants will be designed and operated. Power plant investments are enormous in scale, more than $1 billion per plant, and plants built today will operate for 60 years or more. The International Energy Agency (IEA) forecasts that more than $5 trillion will be spent globally on new power plants in the next 25 years. Under IEA's forecasts, over 1800 gigawatts (GW) of new coal plants will be built between now and 2030—capacity equivalent to 3000 large coal plants, or an average of ten new coal plants every month for the next quarter century. This new capacity amounts to 1.5 times the total of all the coal plants operating in the world today.

The astounding fact is that under IEA's forecast, 7 out of every 10 coal plants that will be operating in 2030 don't exist today. That fact presents a huge opportunity—many of these coal plants will not need to be built if we invest more in efficiency; additional numbers of these coal plants can be replaced with clean, renewable alternative power sources; and for the remainder, we can build them to capture their CO_2, instead of building them the way our grandfathers built them.

If we decide to do it, the world could build and operate new coal plants so that their CO_2 is returned to the ground rather than polluting the atmosphere. But we are losing that opportunity with every month of delay—10 coal plants were built the old-fashioned way last month somewhere in the world and 10 more old-style plants will be built this month, and the next and the next. Worse still, with current policies in place, none of the 3000 new plants projected by IEA are likely to capture their CO_2.

Each new coal plant that is built carries with it a huge stream of CO_2 emissions that will likely flow for the life of the plant—60 years or more. Suggestions that such plants might be equipped with CO_2 capture devices later in life might come true but there is little reason to count on it. As I will discuss further in a moment, while commercial technologies exist for pre-combustion capture from gasification-based power plants, most new plants are not using gasification designs and the few that are, are not incorporating capture systems. Installing capture equipment at these new plants after the fact is implausible for traditional coal plant designs and expensive for gasification processes.

If all 3000 of the next wave of coal plants are built with no CO_2 controls, their lifetime emissions will impose an enormous pollution lien on our children and grandchildren. Over a projected 60-year life these plants would likely emit 750 billion tons of CO_2, a total, from just 25 years of investment decisions, that is 30% greater than the total CO_2 emissions from all previous human use of coal. Once emitted, this CO_2 pollution load remains in the atmosphere for centuries. Half of the CO_2 emitted during World War I remains in the atmosphere today. In short, we face an onrushing train of new coal plants with impacts that must be diverted without delay. What can the U.S. do to help? The U.S. is forecasted to build nearly 300 of these coal plants, according to reports and forecasts published by the U.S. EIA. By taking action ourselves, we can speed the deployment of CO_2 capture here at home and set an example of leadership. That leadership will bring us economic rewards in the new business opportunities it creates here and abroad and it will speed engagement by critical countries like China and India.

To date our efforts have been limited to funding research, development, and limited demonstrations. Such funding can help in this effort if it is wisely invested. But government subsidies—which are what we are talking about—cannot substitute for the driver that a real market for low-carbon goods and services provides. That market will be created only when requirements to limit CO_2 emissions are adopted. This year in Congress serious attention is finally being directed to enactment of such measures and we welcome your announcement that you intend to play a leadership role in this effort.

Key Questions about CCD

I started studying CCD in detail ten years ago and the questions I had then are those asked today by people new to the subject. Do reliable systems exist to capture CO_2 from power plants and other industrial sources? Where can we put CO_2 after we have captured it? Will the CO_2 stay where we put it or will it leak? How much disposal capacity is there? Are CCD systems "affordable"? To answer these questions, the Intergovernmental Panel on Climate Change (IPCC) decided four years ago to prepare a special report on the subject. That report was issued in September, 2005 as the IPCC Special Report on Carbon Dioxide Capture and Storage. I was privileged to serve as a review editor for the report's chapter on geologic storage of CO_2.

CO_2 Capture

The IPCC special report groups capture or separation of CO_2 from industrial gases into four categories: post-combustion; pre-combustion; oxyfuel combustion; and industrial separation. I will say a few words about the basics and status of each of these approaches. In a conventional pulverized coal power plant, the coal is combusted using normal air at atmospheric pressures. This combustion process produces a large volume of exhaust gas that contains CO_2 in large amounts but in low concentrations and low pressures. Commercial post-combustion systems exist to capture CO_2 from such exhaust gases using chemical "stripping" compounds and they have been applied to very small portions of flue gases (tens of thousands of tons from plants that emit several million tons of CO_2 annually) from a few coal-fired power plants in the U.S. that sell the captured CO_2 to the food and beverage industry. However, industry analysts state that today's systems, based on publicly available information, involve much higher costs and energy penalties than the principal demonstrated alternative, pre-combustion capture.

New and potentially less expensive post-combustion concepts have been evaluated in laboratory tests and some, like ammonia-based capture systems, are scheduled for small pilot-scale tests in the next few years. Under normal industrial development scenarios, if successful such pilot tests would be followed by larger demonstration tests and then by commercial-scale tests. These and other approaches should continue to be explored. However, unless accelerated by a combination of policies, subsidies, and willingness to take increased technical risks, such a development program could take one or two decades before post-combustion systems would be accepted for broad commercial application.

Pre-combustion capture is applied to coal conversion processes that gasify coal rather than combust it in air. In the oxygen-blown gasification process coal is heated under pressure with a mixture of pure oxygen, producing an energy-rich gas stream consisting mostly of hydrogen and carbon monoxide. Coal gasification is widely used in industrial processes, such as ammonia and fertilizer production around the world. Hundreds of such industrial gasifiers are in operation today. In power generation applications as practiced today this "syngas" stream is cleaned of impurities and then burned in a combustion turbine to make electricity in a process known as Integrated Gasification Combined Cycle or IGCC. In the power generation business, IGCC is a relatively recent development—about two decades old and is still not widely deployed. There are two IGCC power-only plants operating in the U.S. today and about 14 commercial IGCC plants are operating, with most of the capacity in Europe. In early years of operation for power applications a number of IGCC projects encountered availability problems but those issues appear to be resolved today, with Tampa Electric Company reporting that its IGCC plant in Florida is the most dispatched and most economic unit in its generating system.

Commercially demonstrated systems for pre-combustion capture from the coal gasification process involve treating the syngas to form a mixture of hydrogen and CO_2 and then separating the CO_2, primarily through the use of solvents. These same techniques are used in industrial plants to separate CO_2 from natural gas and to make chemicals such as ammonia out of gasified coal. However, because CO_2 can be released to the air in unlimited amounts under today's laws, except in niche applications, even plants that separate CO_2 do not capture it; rather they release it to the atmosphere. Notable exceptions include the Dakota Gasification Company plant in Beulah, North Dakota, which captures and pipelines more than one million tons of CO_2 per year from its lignite gasification plant to an oil field in Saskatchewan, and ExxonMobil's Shute Creek natural gas processing plant in Wyoming, which strips CO_2 from sour gas and pipelines several million tons per year to oil fields in Colorado and Wyoming.

Today's pre-combustion capture approach is not applicable to the installed base of conventional pulverized coal in the U.S. and elsewhere. However, it is ready today for use with IGCC power plants. The oil giant BP has announced an IGCC project with pre-combustion CO_2 capture at its refinery in Carson, California. When operational the project will gasify petroleum coke, a solid fuel that resembles coal more than petroleum to make electricity for sale to the grid. The captured CO_2 will be sold to an oil field operator in California to enhance oil recovery. The principal obstacle for broad application of pre-combustion capture to new power plants is not technical, it is economic: under today's laws it is cheaper to release CO_2 to the air rather than capturing it. Enacting laws to limit CO_2 can change this situation, as I discuss later.

While pre-combustion capture from IGCC plants is the approach that is ready today for commercial application, it is not the only method for CO_2 capture that may emerge if laws creating a market for CO_2 capture are adopted. I have previously mentioned post-combustion techniques now being explored. Another approach, known as oxyfuel combustion, is also in the early stages of research and development. In the oxyfuel process, coal is burned in oxygen rather than air and the exhaust gases are recycled to build up CO_2 concentrations to a point where separation at reasonable cost and energy penalties may be feasible. Small scale pilot studies for oxyfuel processes have been announced. As with post-combustion processes, absent an accelerated effort to leapfrog the normal commercialization process, it could be one or two decades before such systems might begin to be deployed broadly in commercial application.

Given, the massive amount of new coal capacity scheduled for construction in the next two decades, we cannot afford to wait until we see if these alternative capture systems prove out, nor do we need to. Coal plants in the design process today can employ proven IGCC and pre-combustion capture systems to reduce their CO_2 emissions by about 90 per cent. Adoption of policies that set a CO_2 performance standard now for such new plants will not anoint IGCC as the technological winner since alternative approaches can be employed when they are ready. If the alternatives prove superior to IGCC and pre-combustion capture, the market will reward them accordingly. As I will discuss later, adoption of CO_2 performance standards is a critical step to improve today's capture methods and to stimulate development of competing systems.

I would like to say a few words about so-called "capture-ready" or "capture-capable" coal plants. I will admit that some years ago I was under the impression that some technologies like IGCC, initially built without capture equipment could be properly called "capture-ready." However, the implications of the rapid build-out of new coal plants for global warming and many conversations with engineers since then have educated me to a different view. An IGCC unit built without capture equipment can be equipped later with such equipment and at much lower cost than attempting to retrofit a conventional pulverized coal plant with today's demonstrated post-combustion systems. However, the costs and engineering reconfigurations of such an approach are substantial. More importantly, we need to begin capturing CO_2 from new coal plants without delay in order to keep global warming from becoming a potentially runaway problem. Given the pace of new coal investments in the U.S. and globally, we simply do not have the time to build a coal plant today and think about capturing its CO_2 down the road.

Implementation of the Energy Policy Act of 2005 approach to this topic needs a review in my opinion. The Act provides significant subsidies for coal plants that do not actually capture their CO_2 but rather merely have carbon "capture capability." While the Act limits this term to plants using gasification processes, it is not being implemented in a manner that provides a meaningful substantive difference between an ordinary IGCC unit and one that genuinely has been designed with early integration of CO_2 capture in mind. Further, in its FY2008 budget request, the administration seeks appropriations allowing it to provide $9 billion in loan guarantees under Title XVII of the Act, including as much as $4 billion in loans for "carbon sequestration optimized coal power plants." The administration request does not define a "carbon sequestration optimized" coal power plant and it could mean almost anything, including, according to some industry representatives, a plant that simply leaves physical space for an unidentified black box. If that makes a power plant "capture-ready" Mr. Chairman, then my driveway is "Ferrari-ready." We should not be investing today in coal plants at more than a billion dollars apiece with nothing more than a hope that some kind of capture system will turn up. We would not get on a plane to a destination if the pilot told us there was no landing site but options were being researched.

Geologic Disposal

We have a significant experience base for injecting large amounts of CO_2 into geologic formations. For several decades oil field operators have received high pressure CO_2 for injection into fields to enhance oil recovery, delivered by pipelines spanning as much as several hundred miles. Today in the U.S. a total of more than 35 million tons of CO_2 are injected annually in more than 70 projects. (Unfortunately, due to the lack of any controls on CO_2 emissions, about 80 per cent of that CO_2 is sources from natural CO_2 formations rather than captured from industrial sources. Historians will marvel that we persisted so long in pulling CO_2 out of holes in the ground in order to move it hundreds of miles and stick it back in holes at the same time we were recognizing the harm being caused by emissions of the same molecule from nearby large industrial sources.) In addition to this enhanced oil recovery experience, there are several other large injection projects in operation or announced. The longest running of these, the Sleipner project, began in 1996.

But the largest of these projects injects on the order of one million tons per year of CO_2, while a single large coal power plant can produce about five million tons per year. And of course, our experience with man-made injection projects does not extend for the thousand year or more period that we would need to keep CO_2 in place underground for it to be effective in helping to avoid dangerous global warming. Accordingly, the public and interested members of the environmental, industry and policy communities rightly ask whether we can carry out a large scale injection program safely and assure that the injected CO_2 will stay where we put it.

. . . In its 2005 report the IPCC concluded the following with respect to the question of whether we can safely carry out carbon injection operations on the required scale:

> "With appropriate site selection based on available subsurface information, a monitoring programme to detect problems, a regulatory system and the appropriate use of remediation methods to stop or control CO_2 releases if they arise, the local health, safety and environment risks of geological storage would be comparable to the risks of current activities such as natural gas storage, EOR and deep underground disposal of acid gas."

The knowledge exists to fulfill all of the conditions the IPCC identifies as needed to assure safety. While EPA has authority regulate large scale CO_2 injection projects its current underground injection control regulations are not designed to require the appropriate showings for permitting a facility intended for long-term retention of large amounts of CO_2. With adequate resources applied, EPA should be able to make the necessary revisions to its rules in two to three years. We urge this Committee to act to require EPA to undertake this effort this year.

Do we have a basis today for concluding that injected CO_2 will stay in place for the long periods required to prevent its contributing to global warming? The IPCC report concluded that we do, stating:

> "Observations from engineered and natural analogues as well as models suggest that the fraction retained in appropriately selected and managed geological reservoirs is very likely to exceed 99% over 100 years and is likely to exceed 99% over 1,000 years."

Despite this conclusion by recognized experts there is still reason to ask what are the implications of imperfect execution of large scale injection projects, especially in the early years before we have amassed more experience? Is this reason enough to delay application of CO_2 capture systems to new power plants until we gain such experience from an initial round of multi-million ton "demonstration" projects? To sketch an answer to this question, my colleague Stefan Bachu, a geologist with the Alberta Energy and Utilities Board, and I wrote a paper for the Eighth International Conference on Greenhouse Gas Control Technologies in June 2006. The obvious and fundamental point we made is that without CO_2 capture, new coal plants built during any "delay and research" period will put 100 per cent of their CO_2 into the air and may do so for their operating life if they were "grandfathered" from retrofit requirements. Those releases need to be compared to hypothetical leaks from early injection sites.

Our conclusions were that even with extreme, unrealistically high hypothetical leakage rates from early injection sites (10% per year), a long period to leak detection (5 years) and a prolonged period to correct the leak (1 year), a policy that delayed installation of CO_2 capture at new coal plants to await further research would result in cumulative CO_2 releases twenty times greater than from the hypothetical faulty injection sites, if power plants built during the research period were "grandfathered" from retrofit requirements. If this wave of new coal plants were all required to retrofit CO_2 capture by no later than 2030, the cumulative emissions would still be four times greater than under the no delay scenario. I believe that any objective assessment will conclude that allowing new coal plants to be built without CO_2 capture equipment on the ground that we need more large scale injection experience will always result in significantly greater CO_2 releases than starting CO_2 capture without delay for new coal plants now being designed.

The IPCC also made estimates about global storage capacity for CO_2 in geologic formations. It concluded as follows:

> "Available evidence suggests that, worldwide, it is likely that there is a technical potential of at least about 2,000 $GtCO_2$ (545 GtC) of storage capacity in geological formations. There could be a much larger potential for geological storage in saline formations, but the upper limit estimates are uncertain due to lack of information and an agreed methodology."

Current CO_2 emissions from the world's power plants are about 10 Gt (billion metric tons) per year, so the IPCC estimate indicates 200 years of capacity if power plant emissions did not increase and 100 years capacity if annual emissions doubled.

Policy Actions to Speed CCD

As I stated earlier, research and development funding is useful but it cannot substitute for the incentive that a genuine commercial market for CO_2 capture and disposal systems will provide to the private sector. The amounts of capital that the private sector can spend to optimize CCD methods will almost certainly always dwarf what Congress will provide with taxpayer dollars. To mobilize those private sector dollars, Congress needs a stimulus more compelling than the offer of modest handouts for research. Congress has a model that works:

intelligently designed policies to limit emissions cause firms to spend money finding better and less expensive ways to prevent or capture emissions.

Where a technology is already competitive with other emission control techniques, for example, sulfur dioxide scrubbers, a cap and trade program like that enacted by Congress in 1990, can result in more rapid deployment, improvements in performance, and reductions in costs. Today's scrubbers are much more effective and much less costly than those built in the 1980s. However, a CO_2 cap and trade program by itself may not result in deployment of CCD systems as rapidly as we need. Many new coal plant design decisions are being made literally today. Depending on the pace of required reductions under a global warming bill, a firm may decide to build a conventional coal plant and purchase credits from the cap and trade market rather than applying CCD systems to the plant. While this may appear to be economically rational in the short term, it is likely to lead to higher costs of CO_2 control in the mid and longer term if substantial amounts of new conventional coal construction leads to ballooning demand for CO_2 credits. Recall that in the late 1990's and the first few years of this century, individual firms thought it made economic sense to build large numbers of new gas-fired power plants. The problem is too many of them had the same idea and the resulting increase in demand for natural gas increased both the price and volatility of natural gas to the point where many of these investments are idle today.

Moreover, delaying the start of CCD until a cap and trade system price is high enough to produce these investments delays the broad demonstration of the technology that the U.S. and other countries will need if we continue substantial use of coal as seem likely. The more affordable CCD becomes, the more widespread its use will be throughout the world, including in rapidly growing economies like China and India. But the learning and cost reductions for CCD that are desirable will come only from the experience gained by building and operating the initial commercial plants. The longer we wait to ramp up this experience, the longer we will wait to see CCD deployed here and in countries like China.

Accordingly, we believe the best policy package is a hybrid program that combines the breadth and flexibility of a cap and trade program with well-designed performance measures focused on key technologies like CCD. One such performance measure is a CO_2 emissions standard that applies to new power investments. California enacted such a measure in SB1368 last year. It requires new investments for sale of power in California to meet a performance standard that is achievable by coal with a moderate amount of CO_2 capture.

Another approach is a low-carbon generation obligation for coal-based power. Similar in concept to a renewable performance standard, the low-carbon generation obligation requires an initially small fraction of sales from coal-based power to meet a CO_2 performance standard that is achievable with CCD. The required fraction of sales would increase gradually over time and the obligation would be tradable. Thus, a coal-based generating firm could meet the requirement by building a plant with CCD, by purchasing power generated by another source that meets the standard, or by purchasing credits from those who build such plants. This approach has the advantage of speeding the deployment of CCD while avoiding the "first mover penalty." Instead of causing the first builder of a commercial coal plant with CCD to bear all of the incremental costs, the tradable low-carbon generation obligation would spread those costs over the entire coal-based generation system. The builder of the first unit would achieve far more hours of low-carbon generation than required and would sell the credits to other firms that needed credits to comply. These credit sales would finance the incremental costs of these early units. This approach provides the coal-based power industry with the experience with a technology that it knows is needed to reconcile coal use and climate protection and does it without sticker shock.

A bill introduced in the other body, S. 309, contains such a provision. It begins with a requirement that one-half of one per cent of coal-based power sales must meet the low-carbon performance standard starting in 2015 and the required percentage increases over time according to a statutory minimum schedule that can be increased in specified amounts by additional regulatory action.

A word about costs is in order. With today's off the shelf systems, estimates are that the production cost of electricity at a coal plant with CCD could be as much as 40% higher than at a conventional plant that emits its CO_2. But the impact on average electricity prices of introducing CCD now will be very much smaller due to several factors. First, power production costs represent about 60% of the price you and I pay for electricity; the rest comes from transmission and distribution costs. Second, coal-based power represents just over half of U.S. power consumption. Third, and most important, even if we start now, CCD would be applied to only a small fraction of U.S. coal capacity for some time. Thus, with the trading approach I have outlined, the incremental costs on the units equipped with CCD would be spread over the entire coal-based power sector or possibly across all fossil capacity depending on the choices made by Congress. Based on CCD costs available in 2005 we estimate that a low-carbon generation obligation large enough to cover all forecasted new U.S. coal capacity through 2020 could be implemented for about a two per cent increase in average U.S. retail electricity rates.

Conclusions

To sum up, since we will almost certainly continue using large amounts of coal in the U.S. and globally in the coming decades, it is imperative that we act now to deploy CCD systems. Commercially demonstrated CO_2 capture systems exist today and competing systems are being researched. Improvements in current systems and emergence of new approaches will be accelerated by requirements to limit CO_2 emissions. Geologic disposal of large amounts of CO_2 is viable and we know enough today to conclude that it can be done safely and effectively. EPA must act without delay to revise its regulations to provide the necessary framework for efficient permitting, monitoring and operational practices for large scale permanent CO_2 repositories.

Finally CCD is an important strategy to reduce CO_2 emissions from fossil fuel use but it is not the basis for a climate protection program by itself.

Increased reliance on low-carbon energy resources is the key to protecting the climate. The lowest carbon resource of all is smarter use of energy; energy efficiency investments will be the backbone of any sensible climate protection strategy. Renewable energy will need to assume a much greater role than it does today. With today's use of solar, wind and biomass energy, we tap only a tiny fraction of the energy the sun provides every day. There is enormous potential to expand our reliance on these resources.

We have no time to lose to begin cutting global warming emissions. Fortunately, we have technologies ready for use today that can get us started.

Charles W. Schmidt

Carbon Capture & Storage: Blue-Sky Technology or Just Blowing Smoke?

Towering 650 feet over the sea surface and spouting an impressive burning flare, it would be easy to mistake the Sleipner West gas platform for an environmental nightmare. Its eight-story upper deck houses 200 workers and supports drilling equipment weighing 40,000 tons. Located off the Norwegian coast, it ranks among Europe's largest natural gas producers, delivering more than 12 billion cubic feet of the fuel annually to onshore terminals by pipeline. Roughly 9% of the natural gas extracted here is carbon dioxide (CO_2), the main culprit behind global warming. But far from a nightmare, Sleipner West is actually a bellwether for environmental innovation. Since 1996, the plant's operators have stripped CO_2 out of the gas on-site and buried it 3,000 feet below the sea floor, where they anticipate it will remain for at least 10,000 years.

Operated by StatoilHydro, Norway's largest company, Sleipner is among the few commercial-scale facilities in the world today that capture and bury CO_2 underground. Many experts believe this practice, dubbed carbon capture and storage (sometimes known as carbon capture and sequestration, but in either case abbreviated CCS), could be crucial for keeping industrial CO_2 emissions out of the atmosphere. Sleipner injects 1 million tons of CO_2 annually into the Utsira Formation, a saline aquifer big enough to store 600 years' worth of emissions from all European power plants, company representatives say.

With mounting evidence of climate change—and predictions that fossil fuels could supply 80% of global energy needs indefinitely—the spotlight on CCS is shining as brightly as the Sleipner flare. A panel of experts from the Massachusetts Institute of Technology (MIT) recently concluded that CCS is "the critical enabling technology to reduce CO_2 emissions significantly while allowing fossil fuels to meet growing energy needs." The panel's views were presented in *The Future of Coal,* a report issued by MIT on 14 March 2007.

Environmental groups are split on the issue. Speaking for the Natural Resources Defense Council (NRDC), David Hawkins, director of the council's Climate Center and a member of the MIT panel's external advisory committee, says, "We believe [CCS] is a viable way to cut global warming pollution. . . . We have the knowledge we need to start moving forward." Other environmental groups, including the World Resources Institute, Environmental Defense, and

From *Environmental Health Perspectives,* 115(11), November 2007, pp. A538–A545.

the Pew Center on Global Climate Change, have also come out in support of CCS. These groups view CCS as one among many alternatives (including renewable energy) for reducing CO_2 emissions.

Greenpeace is perhaps the most vocal critic of CCS. Truls Gulowsen, Greenpeace's Nordic climate campaigner, stresses that CCS deflects attention from renewable energy and efficiency improvements, which, he says, offer the best solutions to the problem of global warming. "Companies are doing a lot of talking about CCS, but they're doing little to actually put it into place," he says. "So, they're talking about a possible solution that they don't really want to implement now, and at the same time, they're trying to push for more coal, oil, and gas development instead of renewables, which we already know can deliver climate benefits."

Coal Use Drives CCS Adoption

The pressure to advance on CCS has been fueled by soaring coal use worldwide. China, which is building coal-fired power plants at the rate of two per week, surpassed the United States as the world's largest producer of greenhouse gases in June 2007, years earlier than predicted. Coal use in India and other developing nations is also on the rise, while the United States sits on the largest coal reserves in the world, enough to supply domestic energy needs for 300 years, states the MIT report. Coal already supplies more than 50% of U.S. electricity demand and could supply 70% by 2025, according to the International Energy Agency. Meanwhile, coal-fired power plants already account for nearly 40% of CO_2 emissions worldwide, a figure that—barring some dramatic advance in renewable energy technology—seems poised to rise dramatically. During a 6 September 2007 hearing of the House Select Committee on Energy Independence and Global Warming, Chairman Edward Markey (D–MA) noted that more than 150 new coal plants are being planned in the United States alone, with another 3,000 likely to be built worldwide by 2030.

A mature CCS system would capture, transport, and inject those emissions underground to depths of at least 1 km, where porous rock formations in geologically favorable locations absorb CO_2 like a sponge. At those depths, high pressures and temperatures compress the gas into a dense, liquid-like "supercritical" state that displaces brine and fills the tiny pores between rock grains. Three types of geological formations appear especially promising for sequestration: saline (and therefore nonpotable) aquifers located beneath freshwater deposits; coal seams that are too deep or thin to be extracted economically; and oil and gas fields, where CO_2 stripped from fuels on-site can be injected back underground to force dwindling reserves to the surface, a process called "enhanced recovery." Using CO_2 for enhanced recovery has a long history, particularly in southwestern Texas, where oil yields have been declining for decades.

Of these three options, saline aquifers—with their large storage capacity and broad global distribution—are considered the most attractive. Thomas Sarkus, director of the Applied Science and Energy Technology Division of the DOE National Energy Technology Laboratory (NETL), suggests saline aquifers in the central United States could conceivably store 2,000 years' worth of domestic CO_2 emissions.

Apart from Sleipner, only two other industrial-scale CCS projects are in operation today. In Algeria, a joint venture involving three energy companies—Statoil-Hydro, BP, and Sonatrach—stores more than 1 million tons of CO_2 annually under a natural gas platform near In Salah, an oasis town in the desert. And in Weyburn, Canada, comparable volumes are being used by EnCana Corporation, a Canadian energy company, for enhanced recovery at an aging oil field. The CO_2 sequestered at Weyburn comes by pipeline from a coal gasification plant in Beulah, North Dakota, 200 miles away. Unlike other enhanced recovery projects—wherein the ultimate fate of CO_2 is not the primary concern—Weyburn combines fossil fuel recovery with research to study sequestration on a large scale.

What's needed now, says Jim Katzer, a visiting scholar at MIT's Laboratory for Energy and the Environment, are more large-scale demonstrations of CCS in multiple geologies, integrated with policies that address site selection, licensing, liability, and other issues. Katzer says there are a number of investigations that are investigating storage in the 5,000- to 20,000-ton-capacity range, and they're generating some useful information. "But," he says, "none of them are getting us to the answer we really need: how are we going to manage storage in the millions of tons over long periods of time?"

Paying for Storage

The task of managing carbon storage is nothing if not daunting: in the United States alone, coal plants produce more than 1.5 billion tons of CO_2 every year. Sequestering that amount of gas will require not only a vast new infrastructure of pipelines and storage sites but also that the country's coal plants adopt costly technologies for carbon capture. Most existing U.S. plants—indeed, most of the world's 5,000 coal-fired power plants, including the ones now being built in China—burn pulverized coal (PC) using technologies essentially unchanged since the Industrial Revolution. CO_2 can be extracted from PC plants only after the fuel has been burned, which is inefficient because the combustion emissions are highly diluted with air.

A more efficient approach is to capture highly concentrated streams of CO_2 from coal before it's burned. Precombustion capture is usually applied at integrated gasification combined cycle (IGCC) coal plants, which are extremely rare, numbering just five worldwide, according to Sarkus. IGCC plants cost roughly 20% more to operate because the gasification process requires additional power, which explains why there are so few of them.

Although they don't rule out the possibility, none of the industry sources interviewed for this article welcome the prospect of retrofitting traditional PC plants for carbon capture. That would require major plant modifications and could potentially double the cost of electricity to consumers, they say. But by ignoring existing facilities, industry will set back CCS expansion by decades—most PC plants in use today have been designed for lifetimes of 30 to 40 years.

Whatever path it takes, the transition to CCS will require enormous sums of money. When used for enhanced recovery, CO_2 is a commodity that pays for its own burial. But only a small fraction of the CO_2 generated by coal plants and other industrial processes is used for that purpose. Creating a broad CCS infrastructure will ultimately require a charge on carbon emissions that, according to calculations described in *The Future of Coal,* should total at least $30 per ton—$25 per ton for CO_2 capture and pressurization and $5 per ton for transportation and storage—with this figure rising annually in accordance with inflation.

Sally Benson, a professor of energy resources engineering at Stanford University, points to different ways to pay that charge. One is a tax on CO_2 emissions, an option she concedes has little political support. Funds could also be raised with a "cap-and-trade" system, which sets area-wide limits on CO_2 emissions that industries can meet by trading carbon credits on the open market. A cap-and-trade system for CO_2 has already been established by the European Union, which regulates the greenhouse gas to meet obligations under the Kyoto Protocol. Jeff Chapman, chief executive officer of the Carbon Capture and Storage Association, a trade group based in London, suggests the European cap-and-trade system could ultimately raise €62 billion.

In the United States, a national cap-and-trade system likely won't appear until the federal government regulates CO_2 as a pollutant, says Luke Popovich, vice president of external communications with the National Mining Association, a coal industry trade group in Washington, DC. In the meantime, individual states—for instance, California, which sets its own air quality standards per a waiver under the Clean Air Act—are planning for their own cap-and-trade systems. California regulates CO_2 under a state law called AB32, which directs industries to reduce all greenhouse gas emissions by 25% over the next 13 years. CCS may ultimately emerge on a state-by-state basis in this country, where charges on carbon emissions allow it, Benson suggests.

Technical Questions Remain

Until the early 1990s, most researchers involved in CCS worked in isolation. But in March 1992, more than 250 gathered for the first International Conference on Carbon Dioxide Removal in Amsterdam. Howard Herzog, a principal research engineer at the MIT Laboratory for Energy and the Environment and a leading expert on CCS, says attendees arrived as individuals but left as a research community that now includes funding agencies, industries, and nongovernmental organizations throughout the world. Unfortunately, that community doesn't have nearly the resources it needs to study CCS on a realistic scale, Katzer says. Indeed, *The Future of Coal* states emphatically that "government and private-sector programs to implement on a timely basis the large-scale integrated demonstrations needed to confirm the suitability of carbon sequestration are completely inadequate."

Absent sufficient evidence, most experts simply assume that vast amounts of sequestered CO_2 will stay in place without leaking to the atmosphere. They base that assumption on available monitoring data from the big three industrial projects—none of which have shown any evidence of CO_2 leakage from their underground storage sites, according to *The Future of Coal*—and also on expectations that buried CO_2 will behave in essentially the same way as underground fossil fuel deposits. "We're optimistic it will work," says Jeffrey Logan, a senior associate in the Climate, Energy, and Transport Program at the World Resources Institute. "The general theory is that if oil and gas resources can remain trapped for millions of years, then why not CO_2?"

Franklin Orr, director of the Global Climate and Energy Project at Stanford University, says monitoring data show that CO_2 injected underground for enhanced oil and gas recovery remains trapped there by the same geological structures that trapped the fuels for millions of years; specifically overlying shale deposits through which neither fossil fuels nor CO_2 can pass. Decades of research by the oil and gas industries, in addition to basic research in geology, have revealed the features needed for CO_2 sequestration, he says: "You're looking for deep zones with highly porous rocks—for instance, sandstone—capped by shale seals with low permeability. Sleipner and Weyburn are both good examples; both have thick shale caps that keep the CO_2 from getting out."

But Orr concedes that questions remain about how large amounts of CO_2 might behave underground. A key risk to avoid, he says, is leakage through underlying faults or abandoned wells that provide conduits to the atmosphere. Yousif Kharaka, a research hydrologist with the USGS in Menlo Park, California, says an unknown but possibly large number of orphaned or abandoned wells in the United States could pose a risk of leakage to the atmosphere. And that, he warns, would negate the climate benefits of sequestration.

The likelihood that CO_2 levels could accumulate and cause health or ecological injuries is minimal, Kharaka says, echoing the conclusions reached in *The Future of Coal*. He says CO_2 in air only becomes harmful to humans at concentrations of 3% or above, which is far higher than might be expected from slow leaks out of the ground. Nonetheless, the possibility that CO_2 leaking from underground storage sites might accumulate to harmful levels in basements or other enclosed spaces can't be discounted entirely, cautions Susan Hovorka, a senior research scientist at the Bureau of Economic Geology, a state-sponsored research unit at the University of Texas at Austin. "It's important that we manage this substance correctly," she says. "If you determine that there's a risk to confined places, then you have to provide adequate ventilation. But we have a high level of confidence that CO_2 will be retained at depth."

The greater concern says Kharaka, is that migrating CO_2 might mix with brine, forming carbonic acid that could leach metals such as iron, zinc, or lead from the underlying rock. In some cases, acidified brine alone could migrate and mix with fresh groundwater, posing health risks through drinking or irrigation water, he says. Results from an investigation conducted near Houston, Texas, led by Hovorka as principal investigator along with Kharaka and other scientists from 21 organizations, indicate that CO_2 injected into saline aquifers produced sharp drops in brine pH, from 6.5 to around 3.5. These results were published in the September 2007 report *Water–Rock Interaction: Proceedings of the 12th International Symposium on Water–Rock Interaction,*

Kunming, China, 31 July–5 August, 2007. Chemical analyses showed the brine contained high concentrations of iron and manganese, which suggests toxic metal contamination can't be ruled out, Hovorka says. "I'd describe this as a nonzero concern," she adds. "It's not something we should write off, but it's not a showstopper."

Experts in this area consistently point to the need for more detailed investigations of CO_2 movements at depth and their geochemical consequences. Hovorka's investigation was among the first of this kind, but its scale—just 1,600 tons—paled in comparison to realistic demands for CO_2 mitigation to combat climate change.

Constrained by inadequate funding, the DOE has put much of its CCS investment into a project dubbed "FutureGen." This initiative seeks to build a prototype coal-fired power plant that will integrate all three features of a CCS system, namely, carbon capture (achieved with IGCC technology), CO_2 transportation, and sequestration. Supported by the DOE and an alliance of industry partners, the four-year, $1.5 billion project was announced formally by President Bush in his 2002 State of the Union Address. Once operational, the plant will supply 275 megawatts of power (compared with the 600–1,300 megawatts supplied by typical U.S. coal plants), enough for 275,000 households. Sarkus, who is also the FutureGen director, says four potential sites for the plant and its CO_2 reservoirs—including two in Illinois and two in Texas—are under consideration. Final site selection, he says, will depend on community support, adequate transportation lines, and proximity to underground storage reservoirs.

The Bush administration's stance is that FutureGen will promote CCS advancements throughout the coal and utility industries. But many stakeholders don't think it goes far enough toward meeting existing needs; the project is "too much 'future' and not enough 'generation,'" quips Hawkins. "What we need is legislation that specifies future power plants must be outfitted with CCS, period." To that, Katzer adds, "FutureGen was announced in 2002, and they still haven't settled on site selection, nor have they resolved key design issues. Operations were set to begin in 2012, and now that's slipping back even further. Assuming you start in 2012 and operate for four years, you're looking at 2016 before you complete a single demonstration project. That stretches things out too far, and speaks to the need for several demonstration projects funded now by the U.S. government so we can deal with CO_2 emissions in a timely fashion."

The Developing Country Factor

With U.S. research efforts stuck in low gear, concerns over a comparable lack of progress in the developing world are growing. China already obtains more than 80% of its domestic electricity from coal. And with a relentless push for economic growth, lowering CO_2 emissions from its coal plants is a low priority. It's likely that none of China's coal-fired plants are outfitted for carbon capture, says Richard Lester, a professor of nuclear science and engineering at MIT. "Given the scale and expansion of China's electric power sector, the eventual introduction of CCS there is going to be absolutely critical to global efforts to abate or reduce the atmospheric carbon burden," he says.

Meanwhile, India lags just a decade or less behind China in terms of its own economic growth, which is increasingly fueled by coal use, Katzer says. The key difference between the two countries, he says, has to do with planning for environmental and energy development. In China, Katzer explains, growth and environment strategies seem to be dictated at regional levels without any central coordination, which is ironic considering the country's socialist political structure. India, on the other hand, seems to have what Katzer calls a "master plan" for growth. "But they have no clue how to move forward in terms of CO_2 reductions," Katzer says. "What officials in India say to me is, 'We'll manage CO_2 if it doesn't cost too much.' That's the downside in all of this."

In the end, CCS seems to be stuck in a catch-22: In the view of the developing world, the United States and other wealthier nations should take the lead with respect to emissions reduction technology. Governments in wealthier nations, meanwhile—particularly the United States—look to industries in the free market for solutions to the problem. But U.S. industries say they can't afford large-scale research; in industry's opinion, the government should pay for additional studies that lay the groundwork for industry research and the technology's future implementation. The government, however, doesn't fund the DOE and other agencies at nearly the amounts required to achieve this. And at the same time, the two mechanisms that could possibly generate sufficient revenues for CCS—carbon taxes and cap-and-trade systems for CO_2 emissions—are trapped by perpetual political gridlock.

Leslie Harroun, a senior program officer at the Oak Foundation, a Geneva-based organization that funds social and environmental research, warns that industry might leverage the promise of CCS as a public relations strategy today while doing little to ensure its broad-based deployment tomorrow. "The coal industry's many proposals to build 'clean' coal plants that are 'capture ready' across the U.S. is a smokescreen," she asserts. "Coal companies are hoping to build new plants before cap-and-trade regulations go into effect—and they will, soon—with the idea that the plants and their greenhouse gas emissions will be grandfathered in until sequestration is technically and financially feasible. This is an enormously risky investment decision on their part, and morally irresponsible, but maybe they think there is power in numbers."

In a sense, the inertia surrounding CCS might reflect a collective wilt in the face of a seemingly overwhelming technical and social challenge. To make a difference for climate change, a CCS infrastructure will have to capture and store many billions of tons of CO_2 throughout the world for hundreds of years. Those buried deposits will have to be monitored by unknown entities far into the future. Many questions remain about who will "own" these deposits and thereby assume responsibility for their long-term storage. Meanwhile, industry and the government are at an impasse, with neither taking a leading role toward making large-scale CCS a reality. How this state of affairs ultimately plays out for health of the planet remains to be seen.

Whatever Happened to Deep Ocean Storage?

One CCS option that appears to have fallen by the wayside is deep ocean storage. Scientists have long speculated that enormous volumes of CO_2 could be stored in the ocean at depths of 3 km or more. High pressure would compress the CO_2, making it denser than seawater and thus enabling it to sink. So-called CO_2 lakes would hover over the sea floor, suggests Ken Caldeira, a Stanford University professor of global ecology.

"A coal-fired power plant produces a little under one kilogram of CO_2 for each kilowatt-hour of electricity produced," says Caldeira. "An individual one-gigawatt coal-fired power plant, . . . if completely captured and the CO_2 stored on the sea floor, would make a lake ten meters deep and nearly one kilometer square—and it [would grow] by that much each year."

But Caldeira and others acknowledge that deep ocean storage doesn't offer a permanent solution. Unless the gas is somehow physically confined, over time—perhaps 500 to 1,000 years—up to half the CO_2 would diffuse through the ocean and be released back into the atmosphere. Moreover, most life within CO_2 lakes would be extinguished. However, Caldeira believes this consequence would be balanced by the benefits of keeping the greenhouse gas out of the atmosphere, where under global warming scenarios it acidifies and endangers sea life at the surface.

No one knows precisely what would happen during deep ocean storage because it's never been tested. A planned experiment off the coast of Hawaii in the late 1990s, with participation of U.S., Norwegian, Canadian, and Australian researchers, was canceled because of opposition of local environmental activists. According to Caldeira, who previously co-directed the DOE's now-defunct Center for Research on Ocean Carbon Sequestration, government program managers who backed the Hawaiian study were laterally transferred, sending a signal that advocating for this type of research was politically dangerous for career bureaucrats. "Today, there's zero money going into it," Caldeira says. "Right now, ocean sequestration is dead in the water."

EXPLORING THE ISSUE

Is Carbon Capture Technology Ready to Limit Carbon Emissions?

Critical Thinking and Reflection

1. What is carbon capture and sequestration or storage (CCS)?
2. Describe three ways in which captured carbon can be stored.
3. What events may lead to the release of captured and stored carbon back to the atmosphere?
4. What factors contribute to the definition of an ideal carbon storage location?

Is There Common Ground?

The contributors to this issue agree that carbon capture and sequestration or storage (CCS) is a potential solution to the problem of what to do with carbon dioxide, if we must generate it by burning fossil fuels or if we remove it from the air. The differences lie in judgments of technical, economic, and political ability to do CCS.

1. As noted in the introduction to the issue, legislation to stimulate CCS failed to pass in December 2010. This relates to the "political will" Charles W. Schmidt says is lacking. Why is that will lacking?
2. What can be done to improve the technical ability to do CCS?
3. How would a carbon tax improve the economic ability to do CCS?

ISSUE 11

Is It Time to Think Seriously About "Climate Engineering"?

YES: Kevin Bullis, from "The Geoengineering Gambit," *Technology Review* (January/February 2010)

NO: James R. Fleming, from "The Climate Engineers," *Wilson Quarterly* (Spring 2007)

Learning Outcomes

After studying this issue, students will be able to:

- Compare the benefits of addressing global warming by reducing greenhouse gas emissions with the benefits of addressing global warming by blocking incoming solar radiation.
- Explain the drawbacks of addressing global warming by blocking incoming solar radiation.
- Explain why even if the better approach to controlling global warming involves reducing emissions of greenhouse gases, studying geoengineering remains a good idea.
- Explain the political hazards of geoengineering.

ISSUE SUMMARY

YES: Kevin Bullis, Energy Editor of *Technology Review,* reviews the latest thinking about "geoengineering" as a solution to the global warming problem, and concludes that despite potential side-effects and the risk of unknown impacts on the environment, it may be time to consider technologies that offset global warming.

NO: James R. Fleming, professor of science, technology, and society, argues that climate engineers fail to consider both the risks of unintended consequences to human life and political relationships and the ethics of the human relationship to nature.

It has been known for a very long time that natural events such as volcanic eruptions can cool the climate, sometimes dramatically, by injecting large quantities of dust and sulfates into the stratosphere, where they serve as a "sunshade" that reflects a portion of solar heat back into space before it can warm the Earth. In 1815, the Tambora volcano on Sumbawa Island, Indonesia, put so much material (especially sulfates) into the atmosphere that 1816 was known in the United States, Canada, and Europe as the "year without a summer." There was crop-killing frost, snow, and ice all summer long, which gave the year its other name of "eighteen-hundred-and-froze-to-death." See Clive Oppenheimer, "Climatic, Environmental and Human Consequences of the Largest Known Historic Eruption: Tambora Volcano (Indonesia) 1815," *Progress in Physical Geography* (June 2003). In 1992, Mount Pinatubo, in the Philippines, had a similar, if smaller, effect and hid for a time the climate warming otherwise produced by increasing amounts of greenhouse gases. See Alan Robock, "The Climatic Aftermath," *Science* (February 15, 2002). Changes in solar activity can also have effects. Periods of climate chilling and climate warming have been linked to decreases and increases in the amount of energy released by the sun and reaching the Earth; see Caspar M. Ammann et al., "Solar Influence on Climate During the Past Millennium: Results from Transient Simulations with the NCAR Climate System Model," *Proceedings of the National Academy of Sciences of the United States of America* (March 6, 2007).

Such effects have prompted many researchers to think that global warming is not just a matter of increased atmospheric content of greenhouse gases such as carbon dioxide (which slow the loss of heat to space and thus warm the planet) but also of the amount of sunlight that reaches Earth from the sun. So far, most attempts to find a solution to global warming have focused on reducing human emissions of greenhouse gases. But it does not seem unreasonable to consider the other side of the problem, the energy that reaches Earth from the sun. After all, if you are too warm in bed at night, you can remove the blanket *or* turn down the furnace. Suggestions that something similar might be done on a global scale go back more than 40 years; see Robert Kunzig, "A Sunshade for Planet Earth," *Scientific American* (November 2008).

Paul Crutzen suggested in "Albedo Enhancement by Stratospheric Sulfur Injections: A Contribution to Resolve a Policy Dilemma?" *Climate Change* (August 2006) that adding sulfur compounds to the stratosphere (as volcanoes have done) could reflect some solar energy and help relieve the problem. According to Bob Henson, "Big Fixes for Climate?" *UCAR Quarterly* (Fall 2006) (www.ucar.edu/communications/quarterly/fall06/bigfix.jsp), the National Center for Atmospheric Research is currently testing the idea with computer simulations; one conclusion is that a single Pinatubo-sized stratospheric injection could buy 20 years of time before we would have to cut back carbon dioxide emissions in a big way. Such measures would not be cheap, and at present there is no way to tell whether they would have undesirable side-effects, although G. Bala, P. B. Duffy, and K. E. Taylor, "Impact of Goengineering Schemes on the Global Hydrological Cycle," *Proceedings of the National Academy of Sciences of the United States of America* (June 3, 2008), suggest it is likely that precipitation would be significantly reduced.

Roger Angel, "Feasibility of Cooling the Earth with a Cloud of Small Spacecraft near the Inner Lagrange Point (L1)," *Proceedings of the National Academy of Sciences of the United States of America* (November 14, 2006), argues that if dangerous changes in global climate become inevitable, despite greenhouse gas controls, it may be possible to solve the problem by reducing the amount of solar energy that hits the Earth by using reflective spacecraft. He does not suggest that climate engineering solutions such as this or injecting sulfur compounds into the stratosphere should be tried *instead of* reducing greenhouse gas emissions. Rather, he suggests that such solutions should be evaluated for use *in extremis*, if greenhouse gas reductions are not sufficient or if global warming runs out of control. This position may make a good deal of sense. In January 2009, a survey of climate scientists found broad support for exploring the geoengineering approach and even developing techniques; see "What Can We Do to Save Our Planet?" *The Independent* (January 2, 2009) (www.independent.co.uk/environment/climate-change/what-can-we-do-to-save-our-planet-1221097.html). However, any such program will require much more international cooperation than is usually available; see Jason J. Blackstock and Jane C. S. Long, "The Politics of Geoengineering," *Science* (January 29, 2010). Jamais Cascio, "The Potential and Risks of Geoengineering," *Futurist* (May/June 2010), argues that there will be a place in international diplomacy for efforts "to control climate engineering technologies and deal with their consequences." Among those consequences may be major changes in the amount and distribution of rain and snow; see Gabriele C. Hegerl and Susan Solomon, "Risks of Climate Engineering," *Science* (August 21, 2009).

Chris Mooney, "Climate Repair Made Simple," *Wired* (July 2008), notes that geoengineering proposals might actually work for climate control, but increasing carbon dioxide levels have other consequences too, such as increasing acidity of the oceans. Alan Robock lists this among "20 Reasons Why Geoengineering May Be a Bad Idea," *Bulletin of the Atomic Scientists* (May/June 2008). Alternative methods such as ocean fertilization to stimulate algae to remove carbon dioxide from the atmosphere do not seem likely to be successful; see Noreen Parks, "Fertilizing the Seas for Climate Mitigation—Promising Strategy or Sheer Folly?" *Bioscience* (February 2008). Among other approaches are spraying seawater into the stratosphere and removing carbon dioxide from the atmosphere; see Erik Sofge, "A Geo-Engineered World," *Popular Mechanics* (February 2010), and Erica Engelhaupt, "Engineering a Cooler Earth," *Science News* (June 5, 2010). Robert B. Jackson and James Salzman, "Pursuing Geoengineering for Atmospheric Restoration," *Issues in Science and Technology* (Summer 2010), favors approaches that remove carbon dioxide from the atmosphere. Awareness of potential problems has already lead to one major meeting that concluded that geoengineering research was essential but should be conducted with "humility"; see Eli Kintisch, "'Asilomar 2' Takes Small Steps Toward Rules for Geoengineering," *Science* (April 2, 2010).

James R. Fleming, "The Pathological History of Weather and Climate Modification: Three Cycles of Promise and Hype," *Historical Studies in the Physical and Biological Sciences* (vol. 37, no. 1, 2006), finds the history of attempts to modify weather and climate so marred by excessive optimism that we should doubt the rationality of present climate engineering proposals. Thomas Sterner et al., "Quick Fixes for the Environment: Part of the Solution or Part of the Problem?" *Environment* (December 2006), say that "Quick fixes are sometimes appropriate because they work sufficiently well and/or buy time to design longer term solutions. . . . [but] When quick fixes are deployed, it is useful to tie them to long-run abatement measures." Fundamental solutions (such as reducing emissions of greenhouse gases to solve the global warming problem) are to be preferred, but they may be opposed because of "lack of understanding of ecological mechanisms, failure to recognize the gravity of the problem, vested interests, and absence of institutions to address public goods and intergenerational choices effectively."

There are two basic problems with "sunshades" solutions such as stratospheric sulfate injections, seawater sprays, or reflective spacecraft. One is that they are difficult to test short of full-scale implementation; see Alan Robock et al., "A Test for Geoengineering?" *Science* (January 29, 2010). The second is that even if they work as intended, with few or no undesirable side effects, they must be maintained indefinitely, while the underlying problem continues. If the maintenance falters, the underlying problem will still be there. In fact, atmospheric levels of greenhouse gases will be higher than before because emissions will have continued to rise.

Such solutions also fail to recognize that we face more than one problem. Fossil fuels are finite in supply. We will eventually run out of them. At that time, if we have failed to develop alternative energy sources, we will face a tremendous crisis. But if we reduce greenhouse gas emissions in part by developing alternative energy sources, that crisis will never arrive or will not be as severe when it does. One question with which human society is presently struggling is whether we can shift away from fossil fuels despite the vested interests mentioned by Sterner et al.

It is worth noting that much of the discussion of climate engineering presumes that we will try as a global society to reduce carbon emissions but that our efforts will be insufficient. However, the United Nations Global Climate Change Conference, held in Copenhagen in December 2009, failed to produce a binding agreement to start reducing emissions. Many people feel that if we do not move faster, geoengineering may be our only option. Robert B. Jackson and James Salzman, "Pursuing Geoengineering for Atmospheric Restoration," *Issues in Science and Technology* (Summer 2010), argue that we need to move toward increased energy efficiency and greater use of renewable energy and explore ways to remove carbon from the atmosphere. If we are successful, we will be less likely to need "sunshades" approaches to geoengineering; however, research into such approaches is essential. Ken Caldeira and David W. Keith, "The Need for Climate Engineering Research," *Issues in Science and Technology* (Fall 2010), say that climate engineering may prove to be "the only affordable and fast-acting option to avoid a global catastrophe."

In the YES selection, Kevin Bullis, reviews the latest thinking about "geoengineering" as a solution to the global warming problem, and concludes that despite potential risks it may be time to consider technologies that can rapidly cool the planet and offset global warming. In the NO selection, Roger Fleming, argues that climate engineers fail to consider both the risks of unintended consequences to human life and political relationships and the ethics of the human relationship to nature.

YES 

Kevin Bullis

The Geoengineering Gambit

Rivers fed by melting snow and glaciers supply water to over one-sixth of the world's population—well over a billion people. But these sources of water are quickly disappearing: the Himalayan glaciers that feed rivers in India, China, and other Asian countries could be gone in 25 years. Such effects of climate change no longer surprise scientists. But the speed at which they're happening does. "The earth appears to be changing faster than the climate models predicted," says Daniel Schrag, a professor of earth and planetary sciences at Harvard University, who advises President Obama on climate issues.

Atmospheric levels of carbon dioxide have already climbed to 385 parts per million, well over the 350 parts per million that many scientists say is the upper limit for a relatively stable climate. And despite government-led efforts to limit carbon emissions in many countries, annual emissions from fossil-fuel combustion are going up, not down: over the last two decades, they have increased 41 percent. In the last 10 years, the concentration of carbon dioxide in the atmosphere has increased by nearly two parts per million every year. At this rate, they'll be twice preindustrial levels by the end of the century. Meanwhile, researchers are growing convinced that the climate might be more sensitive to greenhouse gases at this level than once thought. "The likelihood that we're going to avoid serious damage seems quite low," says Schrag. "The best we're going to do is probably not going to be good enough."

This shocking realization has caused many influential scientists, including Obama advisors like Schrag, to fundamentally change their thinking about how to respond to climate change. They have begun calling for the government to start funding research into geoengineering—large-scale schemes for rapidly cooling the earth.

Strategies for geoengineering vary widely, from launching trillions of sun shields into space to triggering vast algae blooms in oceans. The one that has gained the most attention in recent years involves injecting millions of tons of sulfur dioxide high into the atmosphere to form microscopic particles that would shade the planet. Many geoengineering proposals date back decades, but until just a few years ago, most climate scientists considered them something between high-tech hubris and science fiction. Indeed, the subject was "forbidden territory," says Ronald Prinn, a professor of atmospheric sciences at MIT. Not only is it unclear how such engineering feats would be accomplished and whether they would, in fact, moderate the climate, but most scientists worry that they could have disastrous unintended consequences. What's more, relying on geoengineering to cool the earth, rather than cutting greenhouse-gas emissions, would commit future generations to maintaining these schemes indefinitely. For these reasons, mere discussion of geoengineering was considered a dangerous distraction for policy makers considering how to deal with global warming. Prinn says that until a few years ago, he thought its advocates were "off the deep end."

It's not just a fringe idea anymore. The United Kingdom's Royal Society issued a report on geoengineering in September that outlined the research and policy challenges ahead. The National Academies in the United States are working on a similar study. And John Holdren, the director of the White House Office of Science and Technology Policy, broached the idea soon after he was appointed. "Climate change is happening faster than anyone previously predicted," he said during one talk. "If we get sufficiently desperate, we may try to engage in geoengineering to try to create cooling effects." To prepare ourselves, he said, we need to understand the possibilities and the possible side effects. Even the U.S. Congress has now taken an interest, holding its first hearings on geoengineering in November.

Geoengineering might be "a terrible idea," but it might be better than doing nothing, says Schrag. Unlike many past advocates, he doesn't think it's an alternative to reducing greenhouse-gas emissions. "It's not a techno-fix. It's not a Band-Aid. It's a tourniquet," he says. "There are potential side effects, yes. But it may be better than the alternative, which is bleeding to death."

Sunday Storms

The idea of geoengineering has a long history. In the 1830s, James Espy, the first federally funded meteorologist in the United States, wanted to burn large swaths of Appalachian forest every Sunday afternoon, supposing that heat from the fires would induce regular rainstorms. More than a century later, meteorologists and physicists in the United States and the Soviet Union separately considered a range of schemes for changing the climate, often with the goal of warming up northern latitudes to extend growing seasons and clear shipping lanes through the Arctic.

In 1974 a Soviet scientist, Mikhail Budyko, first suggested what is today probably the leading plan for cooling down the earth: injecting gases into the upper reaches of the atmosphere, where they would form microscopic particles to block sunlight. The idea is based on a natural phenomenon. Every few decades a volcano erupts so violently that it sends several millions of tons of sulfur—in the form of sulfur dioxide—more than 10 kilometers into the upper reaches of the atmosphere, a region called the stratosphere. The resulting sulfate particles spread out quickly and stay suspended for years. They reflect and diffuse sunlight, creating a haze that whitens blue skies and causes dramatic sunsets. By decreasing the amount of sunlight that reaches the surface, the haze also lowers its temperature. This is what happened after the 1991 eruption of Mount Pinatubo in the Philippines, which released about 15 million tons of sulfur dioxide into the stratosphere. Over the next 15 months, average temperatures dropped by half a degree Celsius. (Within a few years, the sulfates settled out of the stratosphere, and the cooling effect was gone.)

Scientists estimate that compensating for the increase in carbon dioxide levels expected over this century would require pumping between one million and five million tons of sulfur into the stratosphere every year. Diverse strategies for getting all that sulfur up there have been proposed. Billionaire investor Nathan Myhrvold, the former chief technology officer at Microsoft and the founder and CEO of Intellectual Ventures, based in Bellevue, WA, has thought of several, one of which takes advantage of the fact that coal-fired power plants already emit vast amounts of sulfur dioxide. These emissions stay close to the ground, and rain washes them out of the atmosphere within a couple of weeks. But if the pollution could reach the stratosphere, it would circulate for years, vastly multiplying its impact in reflecting sunlight. To get the sulfur into the stratosphere, Myhrvold suggests, why not use a "flexible, inflatable hot-air-balloon smokestack" 25 kilometers tall? The emissions from just two coal-fired plants might solve the problem, he says. He estimates that his solution would cost less than $100 million a year, including the cost of replacing balloons damaged by storms.

Not surprisingly, climate scientists are not ready to sign off on such a scheme. Some problems are obvious. No one has ever tried to build a 25-kilometer smokestack, for one thing. Moreover, scientists don't understand atmospheric chemistry well enough to be sure what would happen; far from alleviating climate change, shooting tons of sulfates into the stratosphere could have disastrous consequences. The chemistry is too complex for us to be certain, and climate models aren't powerful enough to tell the whole story.

"We know Pinatubo cooled the earth, but that's not the question," Schrag says. "Average temperature is not the only issue." You've also got to account for regional variations in temperature and effects on precipitation, he explains—the very things that climate models are notoriously bad at accounting for. Prinn concurs: "If we lower levels of sunlight, we are unsure of the exact response of the climate system to doing that, for the same reason that we don't know exactly how the climate will respond to a particular level of greenhouse gases." He adds, "That's the big issue. How can you engineer a system you don't fully understand?"

The actual effects of Mount Pinatubo were, in fact, complex. Climate models at the time predicted that by decreasing the amount of sunlight hitting the surface of the earth, the haze of sulfates produced in such an eruption would reduce evaporation, which in turn would lower the amount of precipitation worldwide. Rainfall did decrease—but by much more than scientists had expected. "The year following Mount Pinatubo had by far the lowest amount of rainfall on record," says Kevin Trenberth, a senior scientist at the National Center for Atmospheric Research in Boulder, CO. "In fact, it was 50 percent lower than the previous low of any year." The effects, however, weren't uniform; in some places, precipitation actually increased. A human-engineered sulfate haze could have similarly unpredictable results, scientists warn.

Even in a best-case scenario, where side effects are small and manageable, cooling the planet by deflecting sunlight would not reduce the carbon dioxide in the atmosphere, and elevated levels of that gas have consequences beyond raising the temperature. One is that the ocean absorbs more carbon dioxide and becomes more acidic as a result. That harms shellfish and some forms of plankton, a key source of food for fish and whales. The fishing industry could be devastated. What's more, carbon dioxide levels will continue to rise if we don't address them directly, so any sunlight-reducing technology would have to be continually ratcheted up to compensate for their warming effects.

And if the geoengineering had to stop—say, for environmental or economic reasons—the higher levels of greenhouse gases would cause an abrupt warm-up. "Even if the geoengineering worked perfectly," says Raymond Pierrehumbert, a professor of geophysical sciences at the University of Chicago, "you're still in the situation where the whole planet is just one global war or depression away from being hit with maybe a hundred years' worth of global warming in under a decade, which is certainly catastrophic. Geoengineering, if it were carried out, would put the earth in an extremely precarious state."

Smarter Sulfates

Figuring out the consequences of various geoengineering plans and developing strategies to make them safer and more effective will take years, or even decades, of research. "For every dollar we spend figuring out how to actually do geoengineering," says Schrag, "we need to be spending 10 dollars learning what the impacts will be."

To begin with, scientists aren't even sure that sulfates delivered over the course of decades, rather than in one short volcanic blast, will work to cool the planet down. One key question is how microscopic particles interact in the stratosphere. It's possible that sulfate particles added repeatedly to the same area over time would clump together. If that happened, the particles could start to interact with longer-wave radiation than just the wavelengths of electromagnetic energy in visible light. This would trap some of the heat that naturally escapes into space, causing a net heating effect rather than a cooling effect. Or the larger particles could fall out of the sky before they had a chance to deflect the sun's heat. To study such phenomena, David Keith, the director of the Energy and Environmental Systems Group at the University of Calgary, envisions experiments in which a plane would spray a gas at low vapor pressure over an area of 100 square kilometers. The gas would condense into particles in the stratosphere, and the plane would fly back through the particle cloud to take measurements. Systematically altering the size of the particles, the quantity of particles in a given area, the timing of their release, and other variables could reveal key details about their microscale interactions.

Yet even if the behavior of sulfate particles can be understood and managed, it's far from clear how injecting them into the stratosphere would affect vast, complex climate systems. So far, most models have been crude; only recently, for example, did they start taking into account the movement of ice and ocean currents. Sulfates would cool the planet during the day, but they'd make no difference when the sun isn't shining. As a result, nights would probably be warmer relative to days, but scientists have done little to model this effect and study how it could affect ecosystems. "Similarly, you could affect the seasons," Schrag says: the sulfates would lower temperatures less during the winter (when there's less daylight) and more during the summer. And scientists have done little to understand how stratospheric circulation patterns would change with the addition of sulfates, or precisely how any of these things could affect where and when we might experience droughts, floods, and other disasters.

If scientists could learn more about the effects of sulfates in the stratosphere, it could raise the intriguing possibility of "smart" geoengineering, Schrag says. Volcanic eruptions are crude tools, releasing a lot of sulfur in the course of a few days, and all from one location. But geoengineers could choose exactly where to send sulfates into the stratosphere, as well as when and how fast.

"So far we're thinking about a very simplistic thing," Schrag says. "We're talking about injecting stuff in the stratosphere in a uniform way." The effects that have been predicted so far, however, aren't evenly distributed. Changes in evaporation, for example, could be devastating if they caused droughts on land, but if less rain falls over the ocean, it's not such a big deal. By taking advantage of stratospheric circulation patterns and seasonal variations in weather, it might be possible to limit the most damaging consequences. "You can pulse injections," he says. "You could build smart systems that might cancel out some of those negative effects."

Rather than intentionally polluting the stratosphere, a different and potentially less risky approach to geoengineering is to pull carbon dioxide out of the air. But the necessary technology would be challenging to develop and put in place on large scale.

In his 10th-floor lab in the Manhattan neighborhood of Morningside Heights, Klaus Lackner, a professor of geophysics in the Department of Earth and Environmental Engineering at Columbia University, is experimenting with a material that chemically binds to carbon dioxide in the air and then, when doused in water, releases the gas in a concentrated form that can easily be captured. The work is at an early stage. Lackner's carbon-capture devices look like misshapen test-tube brushes; they have to be hand dipped in water, and it's hard to quickly seal them into the improvised chamber used to measure the carbon dioxide they release. But he envisions automated systems—millions of them, each the size of a small cabin—scattered over the countryside near geologic reservoirs that could store the gases they capture. A system based on this material, he calculates, could remove carbon dioxide from the air a thousand times as fast as trees do now. Others at Columbia are working on ways to exploit the fact that peridotite rock reacts with carbon dioxide to form magnesium carbonate and other minerals, removing the greenhouse gas from the atmosphere. The researchers hope to speed up these natural reactions.

It's far from clear that these ideas for capturing carbon will be practical. Some may even require so much energy that they create a net increase in carbon dioxide. "But even if it takes us a hundred years to learn how to do it," Pierrehumbert says, "it's still useful, because CO_2 naturally takes a thousand years to get out of the atmosphere."

The Seeds of War

Several existing geoengineering schemes, though, could be attempted relatively cheaply and easily. And even if no one knows whether they would be safe or effective, that doesn't mean they won't be tried.

David Victor, the director of the Laboratory on International Law and Regulation at the University of California, San Diego, sees two scenarios in which it might happen. First, "the desperate Hail Mary pass": "A country quite vulnerable to changing climate is desperate to alter outcomes and sees that efforts to cut emissions are not bearing fruit. Crude geoengineering schemes could be very inexpensive, and thus this option might even be available to a Trinidad or Bangladesh—the former rich in gas exports and quite vulnerable, and the latter poor but large enough that it might do something seen as essential for survival." And second, "the Soviet-style arrogant engineering scenario": "A country run by engineers and not overly exposed to public opinion or to dissenting voices undertakes geoengineering as a national mission—much like massive building of poorly designed nuclear reactors, river diversion projects, resettlement of populations, and other national missions that are hard to pursue when the public is informed, responsive, and in power." In either case, a single country acting alone could influence the climate of the entire world.

How would the world react? In extreme cases, Victor says, it could lead to war. Some countries might object to cooling the earth, especially if higher temperatures have brought them advantages such as longer growing seasons and milder winters. And if geoengineering decreases rainfall, countries that have experienced droughts due to global warming could suffer even more.

No current international laws or agreements would clearly prevent a country from unilaterally starting a geoengineering project. And too little is known now for a governing body such as the United Nations to establish sound regulations—regulations that might in any case be ignored by a country set on trying to save itself from a climate disaster. Victor says the best hope is for leading scientists around the world to collaborate on establishing as clearly as possible what dangers could be involved in geoengineering and how, if at all, it might be used. Through open international research, he says, we can "increase the odds—not to 100 percent—that responsible norms would emerge."

Ready or Not

In 2006, Paul Crutzen, the Dutch scientist who won the Nobel Prize in chemistry for his discoveries about the depletion of the stratospheric ozone layer, wrote an essay in the journal *Climatic Change* in which he declared that efforts to reduce greenhouse-gas emissions "have been grossly unsuccessful." He called for increased research into the "feasibility and environmental consequences of climate engineering," even though he acknowledged that injecting sulfates into the stratosphere could damage the ozone layer and cause large, unpredictable side effects. Despite these dangers, he said, climatic engineering could ultimately be "the only option available to rapidly reduce temperature rises."

At the time, Crutzen's essay was controversial, and many scientists called it irresponsible. But since then it has served to bring geoengineering into the open, says David Keith, who started studying the subject in 1989. After a scientist of Crutzen's credentials, who understood the stratosphere as well as anyone, came out in favor of studying sulfate injection as a way to cool the earth, many other scientists were willing to start talking about it.

Among the most recent converts is David Battisti, a professor of atmospheric sciences at the University of Washington. One problem in particular worries him. Studies of heat waves show that crop yields drop off sharply when temperatures rise 3°C to 4°C above normal—the temperatures that MIT's Prinn predicts we might reach even with strict emissions controls. Speaking at a geoengineering symposium at MIT this fall, Battisti said, "By the end of the century, just due to temperature alone, we're looking at a 30 to 40 percent reduction in [crop] yields, while in the next 50 years demand for food is expected to more than double."

Battisti is well aware of the uncertainties that surround geoengineering. According to research he's conducted recently, the first computer models that tried to show how shading the earth would affect climate were off by 2°C to 3°C in predictions of regional temperature change and by as much as 40 percent in predictions of regional rainfall. But with a billion people already malnourished, and billions more who could go hungry if global warming disrupts agriculture, Battisti has reluctantly conceded that we may need to consider "a climate-engineering patch." Better data and better models will help clarify the effects of geoengineering. "Give us 30 or 40 years and we'll be there," he said at the MIT symposium. "But in 30 to 40 years, at the level we're increasing CO_2, we're going to need this, whether we're ready or not."

James R. Fleming

NO

The Climate Engineers

Beyond the security checkpoint at the National Aeronautics and Space Administration's Ames Research Center at the southern end of San Francisco Bay, a small group gathered in November for a conference on the innocuous topic of "managing solar radiation." The real subject was much bigger: how to save the planet from the effects of global warming. There was little talk among the two dozen scientists and other specialists about carbon taxes, alternative energy sources, or the other usual remedies. Many of the scientists were impatient with such schemes. Some were simply contemptuous of calls for international cooperation and the policies and lifestyle changes needed to curb greenhouse-gas emissions; others had concluded that the world's politicians and bureaucrats are not up to the job of agreeing on such reforms or that global warming will come more rapidly, and with more catastrophic consequences, than many models predict. Now, they believe, it is time to consider radical measures: a technological quick fix for global warming.

"Mitigation is not happening and is not going to happen," physicist Lowell Wood declared at the NASA conference. Wood, the star of the gathering, spent four decades at the University of California's Lawrence Livermore National Laboratory, where he served as one of the Pentagon's chief weapon designers and threat analysts. (He reportedly enjoys the "Dr. Evil" nickname bestowed by his critics.) The time has come, he said, for "an intelligent elimination of undesired heat from the biosphere by technical ways and means," which, he asserted, could be achieved for a tiny fraction of the cost of "the bureaucratic suppression of CO_2." His engineering approach, he boasted, would provide "instant climatic gratification."

Wood advanced several ideas to "fix" the earth's climate, including building up Arctic sea ice to make it function like a planetary air conditioner to "suck heat in from the mid-latitude heat bath." A "surprisingly practical" way of achieving this, he said, would be to use large artillery pieces to shoot as much as a million tons of highly reflective sulfate aerosols or specially engineered nanoparticles into the Arctic stratosphere to deflect the sun's rays. Delivering up to a million tons of material via artillery would require a constant bombardment—basically declaring war on the stratosphere. Alternatively, a fleet of B-747 "crop dusters" could deliver the particles by flying continuously around the Arctic Circle. Or a 25-kilometer-long sky hose could be tethered to a military superblimp high above the planet's surface to pump reflective particles into the atmosphere.

Far-fetched as Wood's ideas may sound, his weren't the only Rube Goldberg proposals aired at the meeting. Even as they joked about a NASA staffer's apology for her inability to control the temperature in the meeting room, others detailed their own schemes for manipulating earth's climate. Astronomer J. Roger Angel suggested placing a huge fleet of mirrors in orbit to divert incoming solar radiation, at a cost of "only" several trillion dollars. Atmospheric scientist John Latham and engineer Stephen Salter hawked their idea of making marine clouds thicker and more reflective by whipping ocean water into a froth with giant pumps and eggbeaters. Most frightening was the science-fiction writer and astrophysicist Gregory Benford's announcement that he wanted to "cut through red tape and demonstrate what could be done" by finding private sponsors for his plan to inject diatomaceous earth—the chalk-like substance used in filtration systems and cat litter—into the Arctic stratosphere. He, like his fellow geoengineers, was largely silent on the possible unintended consequences of his plan.

The inherent unknowability of what would happen if we tried to tinker with the immensely complex planetary climate system is one reason why climate engineering has until recently been spoken of only sotto voce in the scientific community. Many researchers recognize that even the most brilliant scientists have a history of blindness to the wider ramifications of their work. Imagine, for example, that Wood's scheme to thicken the Arctic icecap did somehow become possible. While most of the world may want to maintain or increase polar sea ice, Russia and some other nations have historically desired an ice-free Arctic ocean, which would liberate shipping and open potentially vast oil and mineral deposits for exploitation. And an engineered Arctic ice sheet would likely produce shorter growing seasons and harsher winters in Alaska, Siberia, Greenland, and elsewhere, and could generate super winter storms in the midlatitudes. Yet Wood calls his brainstorm a plan for "global climate stabilization," and hopes to create a sort of "planetary thermostat" to regulate the global climate.

Who would control such a "thermostat," making life-altering decisions for the planet's billions? What is to prevent other nations from undertaking unilateral climate modification? The United States has no monopoly on such dreams. In November 2005, for example, Yuri Izrael, head of the Moscow-based Institute of Global Climate and Ecology Studies, wrote to Russian president Vladimir Putin to make the case for immediately burning massive amounts of sulfur in the stratosphere to lower the earth's temperature "a degree or two"—a correction greater than the total warming since pre-industrial times.

There is, moreover, a troubling motif of militarization in the history of weather and climate control. Military leaders in the United States and other countries have pondered the possibilities of weaponized weather manipulation for decades. Lowell Wood himself embodies the overlap of civilian and military interests. Now affiliated with the Hoover Institution, a think tank at Stanford University, Wood was a protégé of the late Edward Teller, the weapons scientist who was credited with developing the hydrogen bomb and was the architect of the Reagan-era Star Wars missile defense system (which Wood worked on, too). Like Wood, Teller was known for his advocacy of controversial military and technological solutions to complex problems, including the chimerical "peaceful uses of nuclear weapons." Teller's plan to excavate an artificial harbor in Alaska using thermonuclear explosives actually came close to receiving government approval. Before his death in 2003, Teller was advocating a climate control scheme similar to what Wood proposed.

Despite the large, unanswered questions about the implications of playing God with the elements, climate engineering is now being widely discussed in the scientific community and is taken seriously within the U.S. government. The Bush administration has recommended the addition of this "important strategy" to an upcoming report of the Intergovernmental Panel on Climate Change, the UN-sponsored organization whose February study seemed to persuade even the Bush White House to take global warming more seriously. And climate engineering's advocates are not confined to the small group that met in California. Last year, for example, Paul J. Crutzen, an atmospheric chemist and Nobel laureate, proposed a scheme similar to Wood's, and there is a long paper trail of climate and weather modification studies by the Pentagon and other government agencies.

As the sole historian at the NASA conference, I may have been alone in my appreciation of the irony that we were meeting on the site of an old U.S. Navy airfield literally in the shadow of the huge hangar that once housed the ill-starred Navy dirigible U.S.S. *Macon.* The 785-foot-long *Macon,* a technological wonder of its time, capable of cruising at 87 miles per hour and launching five Navy biplanes, lies at the bottom of the Pacific Ocean, brought down in 1935 by strong winds. The Navy's entire rigid-airship program went down with it. Coming on the heels of the crash of its sister ship, the *Akron,* the *Macon's* destruction showed that the design of these technological marvels was fundamentally flawed. The hangar, built by the Navy in 1932, is now both a historic site and a Superfund site, since it has been discovered that its "galbestos" siding is leaching PCBs into the drains. As I reflected on the fate of the Navy dirigible program, the geoengineers around the table were confidently and enthusiastically promoting techniques of climate intervention that were more than several steps beyond what might be called state of the art, with implications not simply for a handful of airship crewmen but for every one of the 6.5 billion inhabitants of the planet.

Ultimate control of the weather and climate excites some of our wildest fantasies and our greatest fears. It is the stuff of age-old myths. Throughout history, we mortals have tried to protect ourselves against harsh weather. But weather *control* was reserved for the ancient sky gods. Now the power has seemingly devolved to modern Titans. We are undoubtedly facing an uncertain future. With rising temperatures, increasing emissions of greenhouse gases, and a growing world population, we may be on the verge of a worldwide climate crisis. What shall we do? Doing nothing or too little is clearly wrong, but so is doing too much.

Largely unaware of the long and checkered history of weather and climate control and the political and ethical challenges it poses, or somehow considering themselves exempt, the new Titans see themselves as heroic pioneers, the first generation capable of alleviating or averting natural disasters. They are largely oblivious to the history of the charlatans and sincere but deluded scientists and engineers who preceded them. If we fail to heed the lessons of that history, and fail to bring its perspectives to bear in thinking about public policy, we risk repeating the mistakes of the past, in a game with much higher stakes.

Three stories (there are many more) capture the recurring pathologies of weather and climate control schemes. The first involves 19th-century proposals by the U.S. government's first meteorologist and other "pluviculturalists" to make artificial rain and relieve drought conditions in the American West. The second begins in 1946 with promising discoveries in cloud seeding that rapidly devolved into exaggerated claims and attempts by cold warriors to weaponize the technique in the jungles of Vietnam. And then there is the tale of how computer modeling raised hopes for perfect forecasting and ultimate control of weather and climate—hopes that continue to inform and encourage present-day planetary engineers. . . .

Weather warfare took a macro-pathological turn between 1967 and '72 in the jungles over North and South Vietnam, Laos, and Cambodia. Using technology developed at the naval weapons testing center at China Lake, California, to seed clouds by means of silver iodide flares, the military conducted secret operations intended, among other goals, to "reduce trafficability" along portions of the Ho Chi Minh Trail, which Hanoi used to move men and materiel to South Vietnam. Operating out of Udorn Air Base, Thailand, without the knowledge of the Thai government or almost anyone else, but with the full and enthusiastic support of presidents Lyndon B. Johnson and Richard M. Nixon, the Air Weather Service flew more than 2,600 cloud seeding sorties and expended 47,000 silver iodide flares over a period of approximately five years at an annual cost of some $3.6 million. The covert operation had several names, including "POPEYE" and "Intermediary-Compatriot."

In March 1971, nationally syndicated columnist Jack Anderson broke the story about Air Force rainmakers in Southeast Asia in *The Washington Post*, a story confirmed several months later with the leaking of the Pentagon Papers and splashed on the front page of *The New York Times* in 1972 by Seymour Hersh. By 1973, despite stonewalling by Nixon administration officials, the U.S. Senate had adopted a resolution calling for an international treaty "prohibiting the use of any environmental or geophysical modification activity as a weapon of war." The following year, Senator Claiborne Pell (D.-R.I.), referring to the field as a "Pandora's box," published the transcript of a formerly top-secret briefing by the Defense Department on the topic of weather warfare. Eventually, it was revealed that the CIA had tried rainmaking in South Vietnam as early as 1963 in an attempt to break up the protests of Buddhist monks, and that cloud seeding was probably used in Cuba to disrupt the sugarcane harvest. Similar technology had been employed, yet proved ineffective, in drought relief efforts in India and Pakistan, the Philippines, Panama, Portugal, and Okinawa. All of the programs were conducted under military sponsorship and had the direct involvement of the White House.

Operation POPEYE, made public as it was at the end of the Nixon era, was dubbed the "Watergate of weather warfare." Some defended the use of environmental weapons, arguing that they were more "humane" than nuclear weapons. Others suggested that inducing rainfall to reduce trafficability was preferable to dropping napalm. As one wag put it, "Make mud, not war." At a congressional briefing in 1974, military officials downplayed the impact of Operation POPEYE, since the most that could be claimed were 10 percent increases in local rainfall, and even that result was "unverifiable." Philip Handler, president of the National Academy of Sciences, represented the mainstream of scientific opinion when he observed, "It is grotesquely immoral that scientific understanding and technological capabilities developed for human welfare to protect the public health, enhance agricultural productivity, and minimize the natural violence of large storms should be so distorted as to become weapons of war."

At a time when the United States was already weakened by the Watergate crisis, the Soviet Union caused considerable embarrassment to the Ford administration by bringing the issue of weather modification as a weapon of war to the attention of the United Nations. The UN Convention on the Prohibition of Military or Any Other Hostile Use of Environmental Modification Techniques (ENMOD) was eventually ratified by nearly 70 nations, including the United States. Ironically, it entered into force in 1978, when the Lao People's Democratic Republic, where the American military had used weather modification technology in war only six years earlier, became the 20th signatory.

The language of the ENMOD Convention may become relevant to future weather and climate engineering, especially if such efforts are conducted unilaterally or if harm befalls a nation or region. The convention targets those techniques having "widespread, longlasting or severe effects as the means of destruction, damage, or injury to any other State Party." It uses the term "environmental modification" to mean "any technique for changing—through the deliberate manipulation of natural processes—the dynamics, composition, or structure of the Earth, including its biota, lithosphere, hydrosphere, and atmosphere, or of outer space."

A vision of perfect forecasting ultimately leading to weather and climate control was present at the birth of modern computing, well before the GE cloud seeding experiments. In 1945 Vladimir Zworykin, an RCA engineer noted for his early work in television technology, promoted the idea that electronic computers could be used to process and analyze vast amounts of meteorological data, issue timely and highly accurate forecasts, study the sensitivity of weather systems to alterations of surface conditions and energy inputs, and eventually intervene in and control the weather and climate. He wrote:

> The eventual goal to be attained is the international organization of means to study weather phenomena as global phenomena and to channel the world's weather, as far as possible, in such a way as to

> minimize the damage from catastrophic disturbances, and otherwise to benefit the world to the greatest extent by improved climatic conditions where possible.

Zworykin imagined that a perfectly accurate machine forecast combined with a paramilitary rapid deployment force able literally to pour oil on troubled ocean waters or even set fires or detonate bombs might someday provide the capacity to disrupt storms before they formed, deflect them from populated areas, and otherwise control the weather.

John von Neumann, the multi-talented mathematician extraordinaire at the Institute for Advanced Study in Princeton, New Jersey, endorsed Zworykin's view, writing to him, "I agree with you completely. . . . This would provide a basis for scientific approach[es] to influencing the weather." Using computer-generated predictions, von Neumann wrote, weather and climate systems "could be controlled, or at least directed, by the release of perfectly practical amounts of energy" or by "altering the absorption and reflection properties of the ground or the sea or the atmosphere." It was a project that neatly fit von Neumann's overall philosophy: "All stable processes we shall predict. All unstable processes we shall control." Zworykin's proposal was also endorsed by the noted oceanographer Athelstan Spilhaus, then a U.S. Army major, who ended his letter of November 6, 1945, with these words: "In weather control meteorology has a new goal worthy of its greatest efforts."

In a 1962 speech to meteorologists, "On the Possibilities of Weather Control," Harry Wexler, the MIT-trained head of meteorological research at the U.S. Weather Bureau, reported on his analysis of early computer climate models and additional possibilities opened up by the space age. Reminding his audience that humankind was modifying the weather and climate "whether we know it or not" by changing the composition of the earth's atmosphere, Wexler demonstrated how the United States or the Soviet Union, perhaps with hostile intent, could alter the earth's climate in a number of ways. Either nation could cool it by several degrees using a dust ring launched into orbit, for example, or warm it using ice crystals lofted into the polar atmosphere by the explosion of hydrogen bombs. And while most practicing atmospheric chemists today believe that the discovery of ozone-destroying reactions dates to the early 1970s, Wexler sketched out a scenario for destroying the ozone layer using chlorine or bromine in his 1962 speech.

"The subject of weather and climate control is now becoming respectable to talk about," Wexler claimed, apparently hoping to reduce the prospects of a geophysical arms race. He cited Soviet premier Nikita Khrushchev's mention of weather control in an address to the Supreme Soviet and a 1961 speech to the United Nations by John F. Kennedy in which the president proposed "cooperative efforts between all nations in weather prediction and eventually in weather control." Wexler was actually the source of Kennedy's suggestions, and had worked on them behind the scenes with the President's Science Advisory Committee and the State Department. But if weather control's "respectability" was not in question, its attainability—even using computers, satellites, and 100-megaton bombs—certainly was.

In 1965, the President's Science Advisory Committee warned in a report called *Restoring the Quality of Our Environment* that increases in atmospheric carbon dioxide due to the burning of fossil fuels would modify the earth's heat balance to such an extent that harmful changes in climate could occur. This report is now widely cited as the first official statement on "global warming." But the committee also recommended geoengineering options. "The possibilities of deliberately bringing about countervailing climatic changes . . . need to be thoroughly explored," it said. As an illustration, it pointed out that, in a warming world, the earth's solar reflectivity could be increased by dispersing buoyant reflective particles over large areas of the tropical sea at an annual cost, not considered excessive, of about $500 million. This technology might also inhibit hurricane formation. No one thought to consider the side effects of particles washing up on tropical beaches or choking marine life, or the negative consequences of redirecting hurricanes, much less other effects beyond our imagination. And no one thought to ask if the local inhabitants would be in favor of such schemes. The committee also speculated about modifying high-altitude cirrus clouds to counteract the effects of increasing atmospheric carbon dioxide. It failed to mention the most obvious option: reducing fossil fuel use.

After the embarrassment of the 1978 ENMOD Convention, federal funding for weather modification research and development dried up, although freelance rainmakers continued to ply their trade in the American West with state and local funding. Until recently, a 1991 National Academy of Sciences report, *Policy Implications of Greenhouse Warming,* was the only serious document in decades to advocate climate control. But the level of urgency and the number of proposals have increased dramatically since the turn of the new century.

In September 2001, the U.S. Climate Change Technology Program quietly held an invitational conference, "Response Options to Rapid or Severe Climate Change." Sponsored by a White House that was officially skeptical about global warming, the meeting gave new status to the control fantasies of the climate engineers. According to one participant, "If they had broadcast that meeting live to people in Europe, there would have been riots." . . .

[The] National Research Council issued a study, *Critical Issues in Weather Modification Research,* in 2003. It cited looming social and environmental challenges such as water shortages and drought, property damage and loss of life from severe storms, and the threat of "inadvertent" climate change as justifications for investing in major new national and international programs in weather modification research. Although the NRC study included an acknowledgment that there is "no convincing scientific proof of the efficacy of intentional weather modification efforts," its authors nonetheless argued that there should be "a renewed commitment" to research in the field of intentional and unintentional weather modification.

The absence of such proof after decades of efforts has not deterred governments here and abroad from a variety of ill-advised or simply fanciful undertakings. . . .

With great fanfare, atmospheric chemist Paul J. Crutzen, winner of a 1995 Nobel Prize for his work on the chemistry of ozone depletion, recently proposed to cool the earth by injecting reflective aerosols or other substances into the tropical stratosphere using balloons or artillery. He estimated that more than five million metric tons of sulfur per year would be needed to do the job, at an annual cost of more than $125 billion. The effect would emulate the 1991 eruption of Mount Pinatubo in the Philippines, which covered the earth with a cloud of sulfuric acid and other sulfates and caused a drop in the planet's average temperature of about 0.5°C for roughly two years. Unfortunately, Mount Pinatubo may also have contributed to the largest ozone hole ever measured. The volcanic eruption was also blamed for causing cool, wet summers, shortening the growing season, and exacerbating Mississippi River flooding and the ongoing drought in the Sahel region of Africa.

Overall, the cooling caused by Mount Pinatubo's eruption temporarily suppressed the greenhouse warming effect and was stronger than the influence of the El Niño event that occurred at the same time. Crutzen merely noted that if a Mount Pinatubo-scale eruption were emulated every year or two, undesired side effects and ozone losses should not be "as large," but some whitening of the sky and colorful sunsets and sunrises would occur. His "interesting alternative" method would be to release soot particles to create minor "nuclear winter" conditions.

Crutzen later said that he had only reluctantly proposed his planetary "shade," mostly to "startle" political leaders enough to spur them to more serious efforts to curb greenhouse-gas emissions. But he may well have produced the opposite effect. The appeal of a quick and seemingly painless technological "fix" for the global climate dilemma should not be underestimated. The more practical such dreams appear, the less likely the world's citizens and political leaders are to take on the difficult and painful task of changing the destiny that global climate models foretell.

These issues are not new. In 1956, F. W. Reichelderfer, then chief of the U.S. Weather Bureau, delivered an address to the National Academy of Sciences, "Importance of New Concepts in Meteorology." Reacting to the widespread theorizing and speculation on the possibilities of weather and climate control at the time, he pointed out that the crucial issue was "practicability" rather than "possibility." In 1956 it was possible to modify a cloud with dry ice or silver iodide, yet it was impossible to predict what the cloud might do after seeding and impracticable to claim any sense of control over the weather. This is still true today. Yet thanks to remarkable advances in science and technology, from satellite sensors to enormously sophisticated global climate models, the fantasies of the weather and climate engineers have only grown. Now it is possible to tinker with scenarios in computer climate models—manipulating the solar inputs, for example, to demonstrate that artificially increased solar reflectivity will generate a cooling trend in the model.

But this is a far cry from conducting a practical global field experiment or operational program with proper data collection and analysis; full accounting for possible liabilities, unintended consequences, and litigation; and the necessary international support and approval. Lowell Wood blithely declares that if his proposal to turn the polar icecap into a planetary air conditioner were implemented and didn't work, the process could be halted after a few years. He doesn't mention what harm such a failure could cause in the meantime.

There are signs among the geoengineers of an overconfidence in technology as a solution of first resort. Many appear to possess a too-literal belief in progress that produces an anything-is-possible mentality, abetted by a basic misunderstanding of the nature of today's climate models. The global climate system is a "massive, staggering beast," as oceanographer Wallace Broecker describes it, with no simple set of controlling parameters. We are more than a long way from understanding how it works, much less the precise prediction and practical "control" of global climate.

Assume, for just a moment, that climate control were technically possible. Who would be given the authority to manage it? Who would have the wisdom to dispense drought, severe winters, or the effects of storms to some so that the rest of the planet could prosper? At what cost, economically, aesthetically, and in our moral relationship to nature, would we manipulate the climate?

These questions are never seriously contemplated by the climate wizards who dream of mastery over nature. If, as history shows, fantasies of weather and climate control have chiefly served commercial and military interests, why should we expect the future to be different? . . .

When Roger Angel was asked at the NASA meeting last November how he intended to get the massive amount of material required for his space mirrors into orbit, he dryly suggested a modern cannon of the kind originally proposed for the Strategic Defense Initiative: a giant electric rail gun firing a ton or so of material into space roughly every five minutes. Asked where such a device might be located, he suggested a high mountaintop on the Equator.

I was immediately reminded of Jules Verne's 1889 novel *The Purchase of the North Pole*. For two cents per acre, a group of American investors gains rights to the vast and incredibly lucrative coal and mineral deposits under the North Pole. To mine the region, they propose to melt the polar ice. Initially the project captures the public imagination, as the backers promise that their scheme will improve the climate everywhere by reducing extremes of cold and heat, making the earth a terrestrial heaven. But when it is revealed that the investors are retired Civil War artillerymen who intend to change the inclination of the earth's axis by building and firing the world's largest cannon, public enthusiasm gives way to fears that tidal waves generated by the explosion will kill millions. In secrecy and haste, the protagonists proceed with their plan, building the cannon on Mount Kilimanjaro. The plot fails only when an error in calculation renders the massive shot ineffective. Verne concludes, "The world's inhabitants could thus sleep in peace." Perhaps he spoke too soon.

EXPLORING THE ISSUE

Is It Time to Think Seriously About "Climate Engineering"?

Critical Thinking and Reflection

1. Are we likely to solve the global warming problem without geoengineering?
2. If we cannot, is it ethical to modify global climate deliberately, and thereby affect the world's ecosystems?
3. How could climate engineering lead to war?
4. How does belief in progress lead people to think that anything is possible?

Is There Common Ground?

There is a consensus that the hazards of global warming are so great that humanity must do something to ward them off. When geoengineering was first proposed, the general reaction was, "These guys are out of their minds." In the years since, that reaction has shifted to, "Well, maybe, if everything else fails."

1. Does it seem likely that the world will succeed in reducing greenhouse gas emissions enough to prevent global warming? (To start your research, Google on "congress climate change.")
2. What are the benefits of combining geoengineering with efforts to cut back emissions of greenhouse gases?
3. What kind of geoengineering seems likely to work best in the long term?

Internet References . . .

The Future of Nuclear Power

When an "MIT faculty group decided to study the future of nuclear power because of a belief that this technology is an important option for the United States," the result was this report.

http://web.mit.edu/nuclearpower/

Nuclear Energy

The U.S. Department of Energy's Office of Nuclear Energy leads U.S. efforts to develop new nuclear energy generation technologies; to develop advanced, proliferation-resistant nuclear fuel technologies that maximize energy from nuclear fuel; and to maintain and enhance the national nuclear technology infrastructure.

http://www.ne.doe.gov/

The La Hague Nuclear Reprocessing Plant

The AREVA NC La Hague site, located on the western tip of the Cotentin Peninsula in Normandy, France, reprocesses spent power reactor fuel to recycle reusable energy materials—uranium and plutonium—and to condition the waste into suitable final form. Virtual tours are no longer offered, due to security concerns.

http://www.lahague.areva-nc.com/scripts/areva-nc/publigen/content/templates/Show.asp?P=13&L=EN

UNIT 3

Nuclear Power

Nuclear power has drawbacks, but many people, including environmentalists, are concluding that it has fewer drawbacks than fossil fuels and may be essential to our energy future. However, nuclear power does not exist by itself, for it generates worrisome wastes whose handling poses additional problems. Understanding the issues in this unit of the book is essential to understanding the associated debates.

ISSUE 12

Is It Time to Revive Nuclear Power?

YES: Allison MacFarlane, from "Nuclear Power: A Panacea for Future Energy Needs?" *Environment* (March/April 2010)

NO: Kristin Shrader-Frechette, from "Five Myths About Nuclear Energy," *America* (June 23–30, 2008)

Learning Outcomes

After studying this issue, students will be able to:

- Compare the hazards of nuclear power and global warming.
- Explain how nuclear power avoids the release of greenhouse gases.
- Explain why it will be difficult to shift from fossil fuels to nuclear power rapidly.
- Describe the obstacles that must be overcome before nuclear power can be more widely used.

ISSUE SUMMARY

YES: Allison MacFarlane argues that although nuclear power poses serious problems to be overcome, it "offers a potential avenue to significantly mitigate carbon dioxide emissions while still providing baseload power required in today's world." However, it will take many years to build the necessary number of new nuclear power plants.

NO: Professor Kristin Shrader-Frechette argues that nuclear power is one of the most impractical and risky of energy sources. Renewable energy sources such as wind and solar are a sounder choice.

The technology of releasing for human use the energy that holds the atom together did not get off to an auspicious start. Its first significant application was military, and the deaths associated with the Hiroshima and Nagasaki explosions have ever since tainted the technology with negative associations. It did not help that for the ensuing half-century, millions of people grew up under the threat of nuclear Armageddon. But almost from the beginning, nuclear physicists and engineers wanted to put nuclear energy to more peaceful uses, largely in the form of power plants. Touted in the 1950s as an astoundingly cheap source of electricity, nuclear power soon proved to be more expensive than conventional sources, largely because safety concerns caused delays in the approval process and prompted elaborate built-in precautions. Safety measures have worked well when needed—Three Mile Island, often cited as a horrific example of what can go wrong, released very little radioactive material to the environment. The Chernobyl disaster occurred when safety measures were ignored. In both cases, human error was more to blame than the technology itself. The related issue of nuclear waste has also raised fears and proved to add expense to the technology.

It is clear that two factors—fear and expense—impede the wide adoption of nuclear power. If both could somehow be alleviated, it might become possible to gain the benefits of the technology. Among those benefits are that nuclear power does not burn oil, coal, or any other fuel, does not emit air pollution and thus contribute to smog and haze, does not depend on foreign sources of fuel and thus weaken national independence, and does not emit carbon. Avoiding the use of fossil fuels is an important benefit; see Robert L. Hirsch, Roger H. Bezdek, and Robert M. Wendling, "Peaking Oil Production: Sooner Rather than Later?" *Issues in Science and Technology* (Spring 2005). But avoiding carbon dioxide emissions may be more important at a time when society is concerned about global warming, and this is the benefit that prompted James Lovelock, creator of the Gaia Hypothesis and hero to environmentalists everywhere, to say, "If we had nuclear power we wouldn't be in this mess now, and whose fault was it? It was [the anti-nuclear environmentalists']." See his autobiography, *Homage to Gaia: The Life of an Independent Scientist* (Oxford University Press, 2001).

Others have also seen this point. The OECD's Nuclear Energy Agency ("Nuclear Power and Climate Change" (Paris, France, 1998); www.nea.fr/html/ndd/climate/climate.pdf) found that a greatly expanded deployment of nuclear power to combat global warming was both technically and economically feasible. Robert C. Morris published *The Environmental Case for Nuclear Power: Economic, Medical, and Political Considerations* (Paragon House) in 2000. "The time seems right to reconsider the future of nuclear power," say James A. Lake, Ralph G. Bennett, and John F. Kotek in "Next-Generation Nuclear Power," *Scientific American* (January 2002). Stewart Brand, for long a leading environmentalist, predicts in "Environmental Heresies," *Technology Review* (May 2005), that nuclear power will soon be seen as the "green" energy technology. David Talbot, "Nuclear Powers Up," *Technology Review* (September 2005), notes, "While the waste problem remains unsolved, current trends favor a nuclear renaissance. Energy needs are growing. Conventional energy sources will eventually dry up. The atmosphere is getting dirtier." Peter Schwartz and Spencer Reiss, "Nuclear Now!" *Wired* (February 2005), argue that nuclear power is the one practical answer to global warming and coming shortages of fossil fuels. Iain Murray, "Nuclear Power? Yes, Please," *National Review* (June 16, 2008), argues that the world's experience with nuclear power has shown it to be both safe and reliable.

Costs can be contained, and if one is concerned about global warming, the case for nuclear power is unassailable.

Robert Evans, "Nuclear Power: Back in the Game," *Power Engineering* (October 2005), reports that a number of power companies are considering new nuclear power plants. See also Eliot Marshall, "Is the Friendly Atom Poised for a Comeback?" and Daniel Clery, "Nuclear Industry Dares to Dream of a New Dawn," *Science* (August 19, 2005). Nuclear momentum is growing, says Charles Petit, "Nuclear Power: Risking a Comeback," *National Geographic* (April 2006), thanks in part to new technologies. Karen Charman, "Brave Nuclear World? (Part I)" *World Watch* (May/June 2006), objects that producing nuclear fuel uses huge amounts of electricity derived from fossil fuels, so going nuclear can hardly prevent all releases of carbon dioxide (although using electricity derived from nuclear power would reduce the problem). She also notes that "Although no comprehensive and integrated study comparing the collateral and external costs of energy sources globally has been done, all currently available energy sources have them. . . . Burning coal—the single largest source of air pollution in the United States—causes global warming, acid rain, soot, smog, and other toxic air emissions and generates waste ash, sludge, and toxic chemicals. Landscapes and ecosystems are completely destroyed by mountaintop removal mining, while underground mining imposes high fatality, injury, and sickness rates. Even wind energy kills birds, can be noisy, and, some people complain, blights landscapes."

Stephen Ansolabehere et al., "The Future of Nuclear Power," *An Interdisciplinary MIT Study* (MIT, 2003), note that in 2000 there were 352 nuclear power plants in the developed world as a whole, and a mere 15 in developing nations, and that even a very large increase in the number of nuclear power plants—from 1000 to 1500—will not stop all releases of carbon dioxide. In fact, if carbon emissions double by 2050 as expected, from 6500 to 13,000 million metric tons per year, the 1800 million metric tons not emitted because of nuclear power will seem relatively insignificant. Nevertheless, say John M. Deutch and Ernest J. Moniz, "The Nuclear Option," *Scientific American* (September 2006), such a cut in carbon emissions would be "significant." Christine Laurent, in "Beating Global Warming with Nuclear Power?" *UNESCO Courier* (February 2001), notes, "For several years, the nuclear energy industry has attempted to cloak itself in different ecological robes. Its credo: nuclear energy is a formidable asset in the battle against global warming because it emits very small amounts of greenhouse gases. This stance, first presented in the late 1980s when the extent of the phenomenon was still the subject of controversy, is now at the heart of policy debates over how to avoid droughts, downpours and floods." Laurent adds that it makes more sense to focus on reducing carbon emissions by reducing energy consumption.

Even though President Obama declared support for "a new generation of safe, clean nuclear power plants" in his January 27, 2010, State of the Union speech and the Department of Energy soon proposed massive loan guarantees for the industry (see Pam Russell and Pam Hunter, "Nuclear Resurgence Poised for Liftoff," *ENR: Engineering News-Record* (March 1, 2010), the debate over the future of nuclear power is likely to remain vigorous for some time to come. But as Richard A. Meserve says in a *Science* editorial ("Global Warming and Nuclear Power," *Science* (January 23, 2004)), "For those who are serious about confronting global warming, nuclear power should be seen as part of the solution. Although it is unlikely that many environmental groups will become enthusiastic proponents of nuclear power, the harsh reality is that any serious program to address global warming cannot afford to jettison any technology prematurely. . . . The stakes are large, and the scientific and educational community should seek to ensure that the public understands the critical link between nuclear power and climate change." Paul Lorenzini, "A Second Look at Nuclear Power," *Issues in Science and Technology* (Spring 2005), argues that the goal must be energy "sufficiency for the foreseeable future with minimal environmental impact." Nuclear power can be part of the answer, but making it happen requires that we shed ideological biases. "It means ceasing to deceive ourselves about what might be possible." Charles Forsberg, "The Real Path to Green Energy: Hybrid Nuclear-Renewable Power," *Bulletin of the Atomic Scientists* (November/December 2009), suggests that the best use of nuclear power will be to provide energy for biofuel refineries and as backup for solar and wind power.

Alvin M. Weinberg, former director of the Oak Ridge National Laboratory, notes in "New Life for Nuclear Power," *Issues in Science and Technology* (Summer 2003), that to make a serious dent in carbon emissions would require perhaps four times as many reactors as suggested in the MIT study. The accompanying safety and security problems would be challenging. If the challenges can be met, says John J. Taylor, retired vice president for nuclear power at the Electric Power Research Institute, in "The Nuclear Power Bargain," *Issues in Science and Technology* (Spring 2004), there are a great many potential benefits. Are new reactor technologies needed? Richard K. Lester, "New Nukes," *Issues in Science and Technology* (Summer 2006), says that better centralized waste storage is what is needed, at least in the short term, despite the Obama Administration's declaration that Yucca Mountain, the only U.S. storage site under development, will no longer be supported. On the other hand, some new technologies are available already; see Carol Matlack, "A High-End Bet on Nuclear Power," *BusinessWeek* (March 15, 2010).

Environmental groups such as Friends of the Earth are adamantly opposed, but there are signs that some environmentalists do not agree; see William M. Welch, "Some Rethinking Nuke Opposition," *USA Today* (March 23, 2007). Judith Lewis, "The Nuclear Option," *Mother Jones* (May/June 2008), concludes that "When rising seas flood our coasts, the idea of producing electricity from the most terrifying force ever harnessed may not seem so frightening—or expensive—at all."

In the YES selection, Allison MacFarlane argues that although nuclear power poses serious problems to be overcome, it "offers a potential avenue to significantly mitigate carbon dioxide emissions while still providing baseload power required in today's world." However, it will take many years to build the necessary number of new nuclear power plants. Professor Kristin Shrader-Frechette argues that nuclear power is one of the most impractical and risky of energy sources. Renewable energy sources such as wind and solar are a sounder choice.

YES 

Allison MacFarlane

Nuclear Power: A Panacea for Future Energy Needs?

Each week seems to bring further evidence that the Earth is warming at a faster rate than previously estimated. Pressure is building to replace power sources that emit carbon dioxide with those that do not. It is in this "climate" that nuclear energy is getting a second look. Once relegated to the junk heap after the Three Mile Island and Chernobyl disasters brought the dangers associated with nuclear power to everyone's attention, nuclear power may now be undergoing a "renaissance," as the nuclear industry likes to say. Environmentalists such as Stewart Brand, originator of the Whole Earth Catalog, and Patrick Moore of Greenpeace have started pushing nuclear power as a ready solution to the problem of electricity production without carbon emissions.

Over the past few years, more than 40 countries that currently do not have nuclear power have expressed interest in acquiring it to address their future energy needs, and a few are making significant progress. "Every country has the right to make use of nuclear power, as well as the responsibility to do it in accordance with the highest standards of safety, security, and non-proliferation," stated Mohamed ElBaradei, then–Director General of the International Atomic Energy Agency (IAEA).

All this enthusiasm for nuclear power must be tempered by plain reality. Construction of new nuclear power reactors is still one of the most capital-intensive ventures compared with other energy sources such as coal and natural gas. High-level nuclear waste from the nuclear industry waits in above-ground temporary storage until a final solution is implemented. And the connection between nuclear power and nuclear weapons continues to be perceived as a threat by the international community. Given the pushes and pulls for nuclear power, what is the likelihood that the future will entail a vast expansion of nuclear power on a global scale, and if so, what are its implications, and can this expansion be considered "sustainable" in a development sense?

How Does Nuclear Perform Now?

Globally, there are 436 reactors with a generating capacity of 372 GWe (gigawatts electric) located in 31 countries. [In] 2006, nuclear power generated 15 percent of the world's electricity. Currently, the IAEA estimates that

52 reactors are under construction, whereas the World Nuclear Association lists 71 reactors that will come online between 2009 and 2015. Of the 52 listed by the IAEA, . . . over 85 percent of them are located in Asia and Eastern Europe, and they would add 46 GWe to the existing generating capacity. Of the 52 under construction, 12 of them began construction before 1990, and it is not clear that these projects will ever be completed. At the same time, existing reactors are aging. The average age of the operating fleet in 2007 was 23 years. Most reactors were originally designed for a 40-year life span, but some countries, such as the United States, have recently been granting 20-year life extensions to their fleet.

Carbon Dioxide Emissions Reductions

Nuclear reactors do not emit carbon dioxide to produce electricity because their fuel is uranium-based. This is not to say that nuclear power is emission free—it is not. Carbon dioxide is emitted during the lifecycle of nuclear power production, particularly during the uranium mining, milling, and fuel fabrication processes, and during the construction of new nuclear plants. Nonetheless, nuclear power displaces large volumes of carbon dioxide in comparison with fossil fuel plants.

For instance, in 2006, global nuclear power provided 2594 TWh of electricity. If that electricity was produced by coal and natural gas plants combined, it would have added 3904 million tons of carbon-dioxide equivalent to the atmosphere. Compared to annual emissions of 29,195 million tons of CO_2 equivalent from fossil fuel burning, nuclear power saved about 13 percent.

Certainly if nuclear power grew its capacity, it could contribute significantly to carbon dioxide emissions reductions. Scholars have suggested that 1000 GWe nuclear capacity (in addition to the existing capacity) could reduce potential carbon dioxide emissions growth by 15–25 percent by 2050. Of course, adding 1000 GWe is equivalent to building 1000 new large-scale 1000-MWe plants over the next 40 years, which requires that 25 plants be built per year, an ambitious schedule, but doable. Such plans are not on anyone's drawing board at the moment, though.

Past and Present Safety Issues

Since the Chernobyl accident in 1986, there have been no catastrophic failures at nuclear power plants. That's not to say that there have not been problems, but the overall global safety record since 1986 has been good. This is, in large part, due to vigilant oversight of the nuclear industry and redundant design in safety features.

Issues may arise from aging plants as equipment fails. A case in point is the Davis-Besse reactor in Ohio, which developed an undetected hole in the reactor pressure vessel head, leaving only 0.95 cm of stainless steel to protect against a loss of coolant accident. The hole, which was almost 13 cm in diameter and 17 cm deep, was caused by boric acid corrosion from the borated cooling water in the reactor. The U.S. Nuclear Regulatory Commission's (NRC)

Inspector General found that the NRC itself did not demand more rapid action to address the potentially dangerous issue because it wanted to "lessen the financial impact" on the utility plant operator.

The best way to continue to ensure reactor safety is to follow the advice of the Kemeney Commission, which investigated the 1979 Three Mile Island accident in the United States. They recommended that constant vigilance was required to ensure reactor safety; complacency about the safety of nuclear power would lead to accidents.

Current Waste Management Strategies

With over 50 years of power reactor operation, by 2007 the world had accumulated almost 365,000 m^3 of high-level nuclear waste, including spent nuclear fuel and reprocessing wastes; over 3 million m^3 of long-lived low- and intermediate-level waste; and over 2 million m^3 of short-lived low- and intermediate-level waste. Though many countries have low-level waste disposal facilities and a few have intermediate-level facilities, none have opened a high-level nuclear waste disposal facility. Most spent nuclear fuel remains in cooling pools and dry storage casks at power reactors; some has been reprocessed, and the resulting high-level waste has been vitrified and sits in storage facilities.

Most countries with significant nuclear power generation have decided that high-level nuclear waste should be disposed of in mined geologic repositories. Four countries have been focusing on a single site within their borders. In the United States, the Yucca Mountain, Nevada, site is currently undergoing licensing review for repository construction by the Nuclear Regulatory Commission, although the current Obama Administration has said it will not support the Yucca Mountain site. . . .

Currently there are two management pathways for handling the spent nuclear fuel generated by power reactors. One is the "open cycle," in which spent fuel is considered high-level waste and will be directly disposed of in a repository. The United States, Sweden, Finland, Canada, and others follow this pathway. The second is the "closed cycle," as practiced by France, the United Kingdom, Japan, Russia, and India, in which spent nuclear fuel is reprocessed, the plutonium and uranium extracted, and the remaining waste turned into glass logs destined for a repository. The separated plutonium is then made into new nuclear fuel known as MOX (mixed oxide), but as of yet, it is not reprocessed a second time. The extracted uranium tends to be considered a waste product, as it is too expensive to clean up for reuse.

Reprocessing does reduce the volume of high-level waste by a factor of 4 to 10 times. But this volume reduction does not imply a corresponding capacity reduction in a geologic repository, because volume is not the relevant unit of measure for capacity; heat production and radionuclide composition of the waste are. In addition, reprocessing generates large volumes of low- and intermediate-level wastes, some of which require their own repository. As a result, reprocessing as it is currently practiced has a limited impact on repository size. And no matter which cycle is used, open or closed, a repository will be required in the end.

Issues for the Future

For developed countries, nuclear power can significantly offset carbon dioxide emissions, whereas for developing countries, nuclear power has the potential to electrify the country as well as provide desalination services to address dwindling water supplies. An expansion of nuclear power will not occur without overcoming significant hurdles, though. Whether nuclear power will be sustainable is another issue. The UK Sustainable Development Commission recently concluded that nuclear power was not necessary to achieve a low-carbon future. Some issues associated with nuclear power are unique when compared with other electricity sources, especially in terms of safety and security.

Safety

To operate safely, nuclear power requires a large, active, and well-established infrastructure. The industry needs a trained workforce for plant construction and operation, and also for the regulatory system. In places like the United States, the existing nuclear engineering workforce is aging, and questions remain as to how rapidly it could be replaced. The construction workforce for nuclear power plants requires specialized skills, and the current number of these workers is small. Moreover, there are supply chain issues globally for the nuclear industry. Most significant is the limited capacity for heavy forgings for the reactor pressure vessels. At the moment, a single Japanese plant is capable of making these, and its availability is limited because of commitments to other industries.

For countries with no nuclear plants, an indigenous workforce would have to obtain training abroad, they would need to hire foreign workers, or they would have to rely on the reactor suppliers for a workforce. The same will be true for a regulatory infrastructure. Countries with no existing reactors will need to establish a nuclear infrastructure, again by developing it from the bottom up or by hiring/adapting one from abroad. For example, the United Arab Emirates has hired a former U.S. Nuclear Regulatory Commission official to be the Director-General of its Federal Authority for Nuclear Regulation.

Another key issue for emerging nuclear energy countries is that of liability in the case of nuclear accident for adequate protection of the victims and predictability for nuclear suppliers and insurers. A catastrophic nuclear accident could result in compensation costs that would run into the hundreds of billions of dollars. There is no single global liability regime, and currently 236 of the 436 operating reactors are not covered by liability conventions. Those countries that are party to liability conventions must assure that the legal liability is the responsibility of the reactor operator, and that there are monetary and temporal limits on liability claims.

Cost

Costs associated with the construction of new nuclear reactors may pose the greatest roadblock to the global expansion of nuclear power. . . .

Historically, reactor construction experiences delays and cost escalation. Costs of constructing U.S. nuclear power plants exceed original estimates by 200 percent to 400 percent on average. Experience in the U.S. also shows that increased construction times lead to greatly increased costs from accumulating interest costs. And construction time has not improved that much. Average construction time globally, in most recent experience between 2001–2005, was 6.8 years; at best it was five years, on average.

In the United States, nuclear reactors may cost even more per unit. Florida Power and Light, the utility poised to build two new reactors at its Turkey Point site, recently released figures of $5,780/kW to $8,071/kW, for a total of $12.1 billion to $24.3 billion, depending on the type of plants built. Costs such as these will be prohibitive for many developing countries and even for most utility companies in developed countries. Indeed, the U.S. nuclear industry is seeking further loan guarantees from the federal government over the original $18 billion laid out in the Energy Policy Act of 2005.

All of the reactor designs currently available for construction ("generation III") are large reactors, ranging in size between 600 MWe and 1700 MWe, with most averaging around 1000 MWe. The reason for this is the economies of scale associated with large reactors. Costs associated with reactor design, licensing, regulation, and operation are independent of reactor size, so larger reactors tend to cost less per kilowatt than smaller reactors. These large reactors require large, sophisticated electrical grid networks, something that many developing countries lack. Developing countries seeking nuclear power for electricity production or desalination would be better served by smaller reactors, but none is currently available, though new designs are being put forth.

Capital costs can be decreased by shortening construction time, building standardized designs (in the United States, all existing plants have unique designs), using factory construction of plant components, relying on local resources for labor and construction materials, and building more than one reactor at a site, which allows for costsharing in licensing. For example, Westinghouse's new AP-1000, under construction in Sanmen, China, uses a modular design in which factory prefabricated components can be assembled at the reactor site and put in place using a large crane. Smaller reactors might be helpful in reducing capital costs, because the industry might be able get the economic advantages of producing many units, thus overcoming the problems associated with site-built reactors. Moreover, for small, modular reactors, it may be possible to do the licensing at the manufacturing plant instead of the individual site, greatly reducing costs.

Security

One of the main security concerns about nuclear power programs is their connection to nuclear weapons development. This concern is the reason for international agreements, such as the Nuclear Nonproliferation Treaty (NPT), which guarantees that countries that do not have nuclear weapons are allowed nuclear energy technology, and for institutions such as the IAEA, whose main charge is to monitor nuclear energy programs. The general problem is that the fuel for nuclear power plants is also the same fuel for nuclear weapons: highly enriched uranium or plutonium.

Running a nuclear power reactor poses little proliferation threat. The problem comes either at the "front-end" of the nuclear fuel cycle, associated with fuel production, or the "back-end," associated with the reprocessing of spent nuclear fuel. . . .

One additional security issue requires comment. Nuclear power plants could become terrorist targets in the future. Terrorists might desire to cause a loss of coolant accident of some sort. Particularly vulnerable in this situation are older plants with near-full or full cooling pools in which the fuel is placed in high density arrangements. These pools are located in different areas depending on the reactor design, though some are located several stories above ground. Were the pool to lose coolant, a fire could ensue, causing the release of huge amounts of radioactivity (more than Chernobyl). Fortunately, unlike dealing with proliferation issues associated with the fuel cycle, there is a relatively straightforward solution to this problem. Older fuel can be moved from the cooling pools into dry cask storage on site. Dry casks are passively cooled and therefore much safer if attacked. One of the only reasons reactors have not gone to low-density fuel arrangement in pools is cost.

Environmental/Health Impacts

Life cycle wastes from nuclear power are relatively small in volume in comparison to other energy sources such as fossil fuels. Nuclear power has low emissions from reactor operation, and overall waste production is low, especially in the open fuel cycle. More wastes are generated by reprocessing spent nuclear fuel, though again, in comparison with fossil fuels, they are small. . . .

The bulk of the volume of wastes associated with nuclear power originates from mining and processing the uranium ore for fabrication into fuel. The largest and most problematic wastes associated with mining are uranium mill tailings, which are the rock residue from the extraction of uranium from the ore. The problem is that all the uranium daughter products, in particular radium (Ra-226) and its daughter radon (Rn-222), from millions to billions of years of decay, remain with the residual rock. These pose risks from windblown dust exposure, radon emissions, and leaching into groundwater. . . .

High-level nuclear wastes will increase if nuclear power expands. At current rates of production, an additional 1000 GWe will produce an extra 20,000–30,000 metric tons of spent fuel annually. This amount is approximately one-third to one-half the currently statutory capacity for the proposed Yucca Mountain repository in the United States. Thus, an expanded nuclear world would require more or larger repositories than are envisaged today.

Prospects for the Future

Nuclear power offers a potential avenue to significantly mitigate carbon dioxide emissions while still providing baseload power required in today's world. But nuclear power cannot provide significant baseload power over the next 15 years or so, because it will take many years before a considerable number of new plants are licensed and built. The high capital costs of the plants and the risks of investing in this technology will likely enforce a slow start to large-scale nuclear build. New policies that institute a price on carbon dioxide emissions will certainly improve nuclear power's prospects. In some ways, though, nuclear power suffers from a chicken-and-egg situation: more manufacturing experience may reduce prices, but to achieve the necessary experience, many new plants must first be paid for and built.

New reactors in emerging nuclear energy countries will take even longer because of the large infrastructure required and the need for the international community to respond in an organized fashion to the potential proliferation threats posed by nuclear energy. As a result, nuclear power may begin to provide significant carbon emissions relief in the longer term, starting 20 or 30 years from now.

Unlike most other energy sources, nuclear power does have "doomsday" scenarios. History may repeat itself. Both the accidents at Three Mile Island and Chernobyl had a chilling effect on the global nuclear industry. In the United States, no new nuclear plants have been ordered since the Three Mile Island accident. In Sweden in the 1980s, the country voted to end its use of nuclear power (though they have more recently reversed this decision). If a similar accident occurs, nuclear power's future will grow dim. Similarly, if a country breaks out and develops nuclear weapons using its nuclear energy technology, expansion of nuclear power to non-nuclear energy countries will stop. Countries with nuclear energy technology will be loathe to provide it to emerging nuclear energy nations. Unlike fossil fuels and renewable resources, nuclear energy's success rests on the performance and behavior of all of those involved in producing nuclear power, wherever they are. And this is a somewhat tenuous situation on which to base an industry.

Kristin Shrader-Frechette

Five Myths About Nuclear Energy

Atomic energy is among the most impractical and risky of available fuel sources. Private financiers are reluctant to invest in it, and both experts and the public have questions about the likelihood of safely storing lethal radioactive wastes for the required million years. Reactors also provide irresistible targets for terrorists seeking to inflict deep and lasting damage on the United States. The government's own data show that U.S. nuclear reactors have more than a one-in-five lifetime probability of core melt, and a nuclear accident could kill 140,000 people, contaminate an area the size of Pennsylvania, and destroy our homes and health.

In addition to being risky, nuclear power is unable to meet our current or future energy needs. Because of safety requirements and the length of time it takes to construct a nuclear-power facility, the government says that by the year 2050 atomic energy could supply, at best, 20 percent of U.S. electricity needs; yet by 2020, wind and solar panels could supply at least 32 percent of U.S. electricity, at about half the cost of nuclear power. Nevertheless, in the last two years, the current U.S. administration has given the bulk of taxpayer energy subsidies—a total of $20 billion—to atomic power. Why? Some officials say nuclear energy is clean, inexpensive, needed to address global climate change, unlikely to increase the risk of nuclear proliferation and safe.

On all five counts they are wrong. Renewable energy sources are cleaner, cheaper, better able to address climate change and proliferation risks, and safer. The government's own data show that wind energy now costs less than half of nuclear power; that wind can supply far more energy, more quickly, than nuclear power; and that by 2015, solar panels will be economically competitive with all other conventional energy technologies. The administration's case for nuclear power rests on at least five myths. Debunking these myths is necessary if the United States is to abandon its current dangerous energy course.

Myth 1. Nuclear Energy Is Clean

The myth of clean atomic power arises partly because some sources, like a pro-nuclear energy analysis published in 2003 by several professors at the Massachusetts Institute of Technology, call atomic power a "carbon-free source" of energy. On its Web site, the U.S. Department of Energy, which is also a proponent of nuclear energy, calls atomic power "emissions free." At best, these claims are half-truths because they "trim the data" on emissions.

While nuclear reactors themselves do not release greenhouse gases, reactors are only part of the nine-stage nuclear fuel cycle. This cycle includes mining uranium ore, milling it to extract uranium, converting the uranium to gas, enriching it, fabricating fuel pellets, generating power, reprocessing spent fuel, storing spent fuel at the reactor and transporting the waste to a permanent storage facility. Because most of these nine stages are heavily dependent on fossil fuels, nuclear power thus generates at least 33 grams of carbon-equivalent emissions for each kilowatt-hour of electricity that is produced. (To provide uniform calculations of greenhouse emissions, the various effects of the different greenhouse gases typically are converted to carbon-equivalent emissions.) Per kilowatt-hour, atomic energy produces only one-seventh the greenhouse emissions of coal, but twice as much as wind and slightly more than solar panels.

Nuclear power is even less clean when compared with energy-efficiency measures, such as using compact-fluorescent bulbs and increasing home insulation. Whether in medicine or energy policy, preventing a problem is usually cheaper than curing or solving it, and energy efficiency is the most cost-effective way to solve the problem of reducing greenhouse gases. Department of Energy data show that one dollar invested in energy-efficiency programs displaces about six times more carbon emissions than the same amount invested in nuclear power. Government figures also show that energy-efficiency programs save $40 for every dollar invested in them. This is why the government says it could immediately and cost-effectively cut U.S. electricity consumption by 20 percent to 45 percent, using only existing strategies, like time-of-use electricity pricing. (Higher prices for electricity used during daily peak-consumption times—roughly between 8 a.m. and 8 p.m.—encourage consumers to shift their time of energy use. New power plants are typically needed to handle only peak electricity demand.)

Myth 2. Nuclear Energy Is Inexpensive

Achieving greater energy efficiency, however, also requires ending the lopsided system of taxpayer nuclear subsidies that encourage the myth of inexpensive electricity from atomic power. Since 1949, the U.S. government has provided about $165 billion in subsidies to nuclear energy, about $5 billion to solar and wind together, and even less to energy-efficiency programs. All government efficiency programs—to encourage use of fuel-efficient cars, for example, or to provide financial assistance so that low-income citizens can insulate their homes—currently receive only a small percentage of federal energy monies.

After energy-efficiency programs, wind is the most cost-effective way both to generate electricity and to reduce greenhouse emissions. It costs about half as much as atomic power. The only nearly finished nuclear plant in the West, now being built in Finland by the French company Areva, will gener-

WHAT DOES THE CHURCH SAY?

Though neither the Vatican nor the U.S. bishops have made a statement on nuclear power, the church has outlined the ethical case for renewable energy. In *Centesimus Annus* Pope John Paul II wrote that just as Pope Leo XIII in 1891 had to confront "primitive capitalism" in order to defend workers' rights, he himself had to confront the "new capitalism" in order to defend collective goods like the environment. Pope Benedict XVI warned that pollutants "make the lives of the poor especially unbearable." In their 2001 statement *Global Climate Change,* the U.S. Catholic bishops repeated his point: climate change will "disproportionately affect the poor, the vulnerable, and generations yet unborn."

The bishops also warn that "misguided responses to climate change will likely place even greater burdens on already desperately poor peoples." Instead they urge "energy conservation and the development of alternate renewable and clean-energy resources." They argue that renewable energy promotes care for creation and the common good, lessens pollution that disproportionately harms the poor and vulnerable, avoids threats to future generations and reduces nuclear-proliferation risks.

---

ate electricity costing 11 cents per kilowatt-hour. Yet the U.S. government's Lawrence Berkeley National Laboratory calculated actual costs of new wind plants, over the last seven years, at 3.4 cents per kilowatt-hour. Although some groups say nuclear energy is inexpensive, their misleading claims rely on trimming the data on cost. The 2003 M.I.T. study, for instance, included neither the costs of reprocessing nuclear material, nor the full interest costs on nuclear-facility construction capital, nor the total costs of waste storage. Once these omissions—from the entire nine-stage nuclear fuel cycle—are included, nuclear costs are about 11 cents per kilowatt-hour.

The cost-effectiveness of wind power explains why in 2006 utility companies worldwide added 10 times more wind-generated, than nuclear, electricity capacity. It also explains why small-scale sources of renewable energy, like wind and solar, received $56 billion in global private investments in 2006, while nuclear energy received nothing. It explains why wind supplies 20 percent of Denmark's electricity. It explains why, each year for the last several years, Germany, Spain and India have each, alone, added more wind capacity than all countries in the world, taken together, have added in nuclear capacity.

In the United States, wind supplies up to 8 percent of electricity in some Midwestern states. The case of Louis Brooks is instructive. Utilities pay him $500 a month for allowing 78 wind turbines on his Texas ranch, and he can still use virtually all the land for farming and grazing. Wind's cost-effectiveness also explains why in 2007 wind received $9 billion in U.S. private investments, while nuclear energy received zero. U.S. wind energy has been growing by nearly 3,000 megawatts each year, annually producing new electricity equivalent to what three new nuclear reactors could generate. Meanwhile, no new U.S. atomic-power reactors have been ordered since 1974.

Should the United States continue to heavily subsidize nuclear technology? Or, as the distinguished physicist Amory Lovins put it, is the nuclear industry dying of an "incurable attack of market forces"? Standard and Poor's, the credit- and investment-rating company, downgrades the rating of any utility that wants a nuclear plant. It claims that even subsidies are unlikely to make nuclear investment wise. *Forbes* magazine recently called nuclear investment "the largest managerial disaster in business history," something pursued only by the "blind" or the "biased."

Myth 3. Nuclear Energy Is Necessary to Address Climate Change

Government, industry and university studies, like those recently from Princeton, agree that wind turbines and solar panels already exist at an industrial scale and could supply one-third of U.S. electricity needs by 2020, and the vast majority of U.S. electricity by 2050—not just the 20 percent of electricity possible from nuclear energy by 2050. The D.O.E. says wind from only three states (Kansas, North Dakota and Texas) could supply all U.S. electricity needs, and 20 states could supply nearly triple those needs. By 2015, according to the D.O.E., solar panels will be competitive with all conventional energy technologies and will cost 5 to 10 cents per kilowatt hour. Shell Oil and other fossil-fuel companies agree. They are investing heavily in wind and solar.

From an economic perspective, atomic power is inefficient at addressing climate change because dollars used for more expensive, higher-emissions nuclear energy cannot be used for cheaper, lower-emissions renewable energy. Atomic power is also not sustainable. Because of dwindling uranium supplies, by the year 2050 reactors would be forced to use low-grade uranium ore whose greenhouse emissions would roughly equal those of natural gas. Besides, because the United States imports nearly all its uranium, pursuing nuclear power continues the dangerous pattern of dependency on foreign sources to meet domestic energy needs.

Myth 4. Nuclear Energy Will Not Increase Weapons Proliferation

Pursuing nuclear power also perpetuates the myth that increasing atomic energy, and thus increasing uranium enrichment and spent-fuel reprocessing, will increase neither terrorism nor proliferation of nuclear weapons. This myth has been rejected by both the International Atomic Energy Agency and the U.S. Office of Technology Assessment. More nuclear plants means more weapons materials, which means more targets, which means a higher risk of terrorism and proliferation. The government admits that Al Qaeda already has targeted U.S. reactors, none of which can withstand attack by a large airplane. Such an attack, warns the U.S. National Academy of Sciences, could cause fatalities as far away as 500 miles and destruction 10 times worse than that caused by the nuclear accident at Chernobyl in 1986.

Nuclear energy actually increases the risks of weapons proliferation because the same technology used for civilian atomic power can be used for weapons, as the cases of India, Iran, Iraq, North Korea and Pakistan illustrate. As the Swedish Nobel Prize winner Hannes Alven put it, "The military atom and the civilian atom are Siamese twins." Yet if the world stopped building nuclear-power plants, bomb ingredients would be harder to acquire, more conspicuous and more costly politically, if nations were caught trying to obtain them. Their motives for seeking nuclear materials would be unmasked as military, not civilian.

Myth 5. Nuclear Energy Is Safe

Proponents of nuclear energy, like Patrick Moore, cofounder of Greenpeace, and the former Argonne National Laboratory adviser Steve Berry, say that new reactors will be safer than current ones—"meltdown proof." Such safety claims also are myths. Even the 2003 M.I.T. energy study predicted that tripling civilian nuclear reactors would lead to about four core-melt accidents. The government's Sandia National Laboratory calculates that a nuclear accident could cause casualties similar to those at Hiroshima or Nagasaki: 140,000 deaths. If nuclear plants are as safe as their proponents claim, why do utilities need the U.S. Price-Anderson Act, which guarantees utilities protection against 98 percent of nuclear-accident liability and transfers these risks to the public? All U.S. utilities refused to generate atomic power until the government established this liability limit. Why do utilities, but not taxpayers, need this nuclear-liability protection?

Another problem is that high-level radioactive waste must be secured "in perpetuity," as the U.S. National Academy of Sciences puts it. Yet the D.O.E. has already admitted that if nuclear waste is stored at Nevada's Yucca Mountain, as has been proposed, future generations could not meet existing radiation standards. As a result, the current U.S. administration's proposal is to allow future releases of radioactive wastes, stored at Yucca Mountain, provided they annually cause no more than one person—out of every 70 persons exposed to them—to contract fatal cancer. These cancer risks are high partly because Yucca Mountain is so geologically unstable. Nuclear waste facilities could be breached by volcanic or seismic activity. Within 50 miles of Yucca Mountain, more than 600 seismic events, of magnitude greater than two on the Richter scale, have occurred since 1976. In 1992, only 12 miles from the site, an earthquake (5.6 on the Richter scale) damaged D.O.E. buildings. Within 31 miles of the site, eight volcanic eruptions have occurred in the last million years. These facts suggest that Alvin Weinberg was right. Four decades ago, the then-director of the government's Oak Ridge National Laboratory warned that nuclear waste required society to make a Faustian bargain with the devil. In exchange for current military and energy benefits from atomic power, this generation must sell the safety of future generations.

Yet the D.O.E. predicts harm even in this generation. The department says that if 70,000 tons of the existing U.S. waste were shipped to Yucca

Mountain, the transfer would require 24 years of dozens of daily rail or truck shipments. Assuming low accident rates and discounting the possibility of terrorist attacks on these lethal shipments, the D.O.E. says this radioactive-waste transport likely would lead to 50 to 310 shipment accidents. According to the D.O.E., each of these accidents could contaminate 42 square miles, and each could require a 462-day cleanup that would cost $620 million, not counting medical expenses. Can hundreds of thousands of mostly unguarded shipments of lethal materials be kept safe? The states do not think so, and they have banned Yucca Mountain transport within their borders. A better alternative is onsite storage at reactors, where the material can be secured from terrorist attack in "hardened" bunkers.

Where Do We Go From Here?

If atomic energy is really so risky and expensive, why did the United States begin it and heavily subsidize it? As U.S. Atomic Energy Agency documents reveal, the United States began to develop nuclear power for the same reason many other nations have done so. It wanted weapons-grade nuclear materials for its military program. But the United States now has more than enough weapons materials. What explains the continuing subsidies? Certainly not the market. *The Economist* (7/7/05) recently noted that for decades, bankers in New York and London have refused loans to nuclear industries. Warning that nuclear costs, dangers and waste storage make atomic power "extremely risky," *The Economist* claimed that the industry is now asking taxpayers to do what the market will not do: invest in nuclear energy. How did *The Economist* explain the uneconomical $20 billion U.S. nuclear subsidies for 2005–7? It pointed to campaign contributions from the nuclear industry.

Despite the problems with atomic power, society needs around-the-clock electricity. Can we rely on intermittent wind until solar power is cost-effective in 2015? Even the Department of Energy says yes. Wind now can supply up to 20 percent of electricity, using the current electricity grid as backup, just as nuclear plants do when they are shut down for refueling, maintenance and leaks. Wind can supply up to 100 percent of electricity needs by using "distributed" turbines spread over a wide geographic region—because the wind always blows somewhere, especially offshore.

Many renewable energy sources are safe and inexpensive, and they inflict almost no damage on people or the environment. Why is the current U.S. administration instead giving virtually all of its support to a riskier, more costly nuclear alternative?

EXPLORING THE ISSUE

Is It Time to Revive Nuclear Power?

Critical Thinking and Reflection

1. Which is more dangerous? Nuclear power or global warming? (See *The Anatomy of a Silent Crisis,* Global Humanitarian Forum Geneva (2009), www.ghfgeneva.org/LinkClick.aspx?fileticket=pg6PNIoVEoA%3d.)
2. How are nuclear wastes handled today?
3. Pretend that your campus is a nuclear power plant. Now devise an evacuation plan to protect the residents of neighboring communities in case of a serious accident.
4. What are the advantages of a "closed cycle" approach to handling spent nuclear fuel?

Is There Common Ground?

There is no argument that today's society needs vast amounts of energy to function. The debate largely centers on how to meet the need (which is projected to grow greatly in the future). The greatest agreement may well be in the solutions people refuse to consider. As an exercise, consider the following major uses of energy:

Global trade

Commuting

Air-conditioning

Central heating

1. In what ways could we change our lifestyles to reduce energy use in each of these areas?
2. Would reducing population help?

ISSUE 13

Should the United States Stop Planning for Permanent Nuclear Waste Disposal at Yucca Mountain?

YES: U.S. Department of Energy (DOE), from "Motion to Withdraw," filed before the Nuclear Regulatory Commission (March 2, 2010)

NO: Luther J. Carter, Lake H. Barrett, and Kenneth C. Rogers, from "Nuclear Waste Disposal: Showdown at Yucca Mountain," *Issues in Science and Technology* (Fall 2010)

Learning Outcomes

After studying this issue, students will be able to:

- Explain why the Obama administration chose to end support for the Yucca Mountain nuclear waste repository.
- Explain the political difficulties involved in finding an acceptable location for a nuclear waste repository.
- Describe the physical requirements a nuclear waste repository must satisfy.
- Explain how various incentives can affect acceptance of a nuclear waste repository.

ISSUE SUMMARY

YES: The U.S. Department of Energy (DOE) moves to withdraw its application for a license to operate a permanent repository for spent nuclear fuel and high-level radioactive waste at Yucca Mountain, Nevada, calling Yucca Mountain "not a workable option" and saying that it has no plans ever to refile the application.

NO: Luther J. Carter, Lake H. Barrett, and Kenneth C. Rogers argue that the decision to withdraw the application for a nuclear waste repository at Yucca Mountain was motivated by politics rather than by evidence. If successful, it will impede future efforts to use nuclear power to combat global warming.

Nuclear waste is generated when uranium and plutonium atoms are split to make energy in nuclear power plants, when uranium and plutonium are purified to make nuclear weapons, when nuclear wastes are reprocessed, and when radioactive isotopes useful in medical diagnosis and treatment are made and used. These wastes are radioactive, meaning that as they break down they emit radiation of several kinds. Those that break down fastest are most radioactive; they are said to have a short half-life (the time needed for half the material to break down). Uranium-238, the most common isotope of uranium, has a half-life of 4.5 billion years and is not very radioactive at all. Plutonium-239 (bomb material) has a half-life of 24,000 years and is radioactive enough to be hazardous to humans.

According to the U.S. Department of Energy, high-level waste includes spent reactor fuel and waste from weapons production. Transuranic waste includes clothing, equipment, and other materials contaminated with plutonium and other radioactive materials, some of which has been buried in the Waste Isolation Pilot Plant (WIPP) salt cavern in New Mexico (WIPP started receiving transuranic waste in 1999). It too was surrounded by controversy, as summarized by Chris Hayhurst in "WIPP Lash: Doubts Linger about a Controversial Underground Nuclear Waste Storage Site," *E Magazine* (January–February 1998). Its Web site is at www.wipp.energy.gov/. Low and mixed-level waste includes waste from hospitals and research labs, remnants of decommissioned nuclear plants, and air filters. The high-level waste is the most hazardous and poses the most severe disposal problems. In general, experts say, such materials must be kept away from people and other living things, with no possibility of contaminating air, water (including ground water), or soil for 10 half-lives.

For a good summary of the nuclear waste problem and the disposal controversy, see Michael E. Long, "Half Life: The Lethal Legacy of America's Nuclear Waste," *National Geographic* (July 2002). Gary Taubes, in "Whose Nuclear Waste?" *Technology Review* (January/February 2002), argues that a whole new approach may be necessary. One such approach is an interim, above-ground storage facility for commercial nuclear waste at Yucca Mountain. This has been urged as a way to create commitment to continue developing the Yucca Mountain site and to meet government responsibilities to deal with commercial waste. Steven Ashley, in "Divide and Vitrify," *Scientific American* (June 2002), describes work on potential methods of separating the most hazardous components of nuclear waste. One such approach is to expose nuclear waste to neutrons from particle accelerators or special nuclear reactors and thereby greatly hasten the process of radioactive decay.

The Nuclear Age began in the 1940s. As nuclear waste accumulated, there also developed a sense of urgency about finding a place to put it where it would not threaten humans or ecosystems for a quarter million years or more. In 1982, the Nuclear Waste Policy Act called for locating candidate disposal sites for high-level wastes and choosing one by 1998. Since no state chosen as a candidate site was happy about being chosen and many sites were for various reasons less than ideal, the schedule proved impossible to meet. In 1987, Congress attempted to settle the matter by designating Yucca Mountain, Nevada, as the one site to be intensively studied and developed. It would be opened for use in 2010. Risk assessment expert D. Warner North wrote in "Unresolved Problems of Radioactive Waste: Motivation for a New Paradigm," *Physics Today* (June 1997) that the technical and political problems related to nuclear waste disposal remained formidable and a new approach was needed. Luther J. Carter and Thomas H. Pigford wrote in "Getting Yucca Mountain Right," *Bulletin of the Atomic Scientists* (March/April 1998) that those formidable problems could be defeated, given technical and congressional attention, and the Yucca Mountain strategy was both sensible and realistic. However, problems have continued to plague the project, as summarized by Chuck McCutcheon, "High-Level Acrimony in Nuclear Storage Standoff," *Congressional Quarterly Weekly Report* (September 25, 1999), and Sean Paige, "The Fight at the End of the Tunnel," *Insight on the News* (November 15, 1999). Jon Christensen, in "Nuclear Roulette," *Mother Jones* (September/October 2001), argues that one of the most basic problems is that estimates of Yucca Mountain's long-term safety are based on probabilistic computer models that are too uncertain to trust. Per F. Peterson, William E. Kastenberg, and Michael Corradini, "Nuclear Waste and the Distant Future," *Issues in Science and Technology* (Summer 2006), argue that the risks of waste disposal have been sensibly addressed by the EPA and we should be focusing more attention on other risks (such as those of global warming).

Even those who favor using Yucca Mountain for high-level nuclear waste disposal admit that in time the site is bound to leak. The intensity of the radioactivity emitted by the waste will decline rapidly as short-half-life materials decay, and by 2300 AD, when the site is expected to be sealed, that intensity will be less than 5 percent of the initial level. After that, however, radiation intensity will decline much more slowly. The nickel-alloy containers for the waste are expected to last at least 10,000 years, but they will not last forever. The U.S. Department of Energy's computer simulations predict that the radiation released to the environment will rise rapidly after about 100,000 years, with a peak annual dose after 400,000 years that is about double the natural background exposure. Whether the site can be protected for any significant fraction of such time periods arouses considerable skepticism among those who point out that 10,000 years is about the same length of time as has passed since humans built their first cities, and 400,000 years is about twice as long as modern *Homo sapiens* has existed.

Despite the controversy, in February 2002, U.S. Secretary of Energy Spencer Abraham recommended to the president that the nation go ahead with development of the Yucca Mountain site. His report argues that a disposal site is necessary, that Yucca Mountain has been thoroughly studied, and that moving ahead with the site best serves "our energy future, our national security, our economy, our environment, and safety." Objections to the site are not serious enough to stop the project. However, that decision hardly settled the matter. Gar Smith, "A Gift to Terrorists?" *Earth Island Journal* (Winter 2002–2003), argues that transporting nuclear waste to Yucca Mountain will expose millions of Americans to risks from accidents and terrorists. Senator Hillary Clinton said in testimony before the Senate Environment and Public Works Committee on October 31, 2007, that she thought it was time to scrap both the work done so far and the controversy and start over. In February 2008, Senator James Inhofe introduced a bill intended to speed up the licensing process for Yucca Mountain. Journalist Chuck Muth, "Nevada Kids, They Glow in the Dark," *Las Vegas Business Press* (January 7, 2008), says that waste deposited at Yucca Mountain could be very valuable if the nation chooses to reprocess waste and that Nevada could wind up being "the nuclear research and reprocessing capital of the world."

On June 3, 2008, the U.S. Department of Energy (DOE) submitted a license application to the U.S. Nuclear Regulatory Commission, seeking authorization to construct a deep geologic repository for disposal of high-level radioactive waste at Yucca Mountain, Nevada; see www.nrc.gov/waste/hlw-disposal.html. Early in 2010, the DOE, following up on a commitment made by President Obama, moved to withdraw its application. William Beaver, "The Demise of Yucca Mountain," *Independent Review* (Spring 2010), questions the wisdom of the decision, citing future need for nuclear power. Sean Davies, "End of the Road for Yucca Mountain," *Engineering & Technology* (May 14, 2010), argues that we can now begin to look for truly workable alternatives. James M. Hylko and Robert Peltier, "The U.S. Spent Nuclear Fuel Policy: Road to Nowhere," *Power* (May 2010), favor nuclear fuel reprocessing. Matthew L. Wald, "What Now for Nuclear Waste?" *Scientific American* (August 2009), suggests that for now it may be best to do nothing, in part because Yucca Mountain may reenter the picture.

The YES selection is DOE's motion before the Nuclear Regulatory Commission to withdraw its application, calling Yucca Mountain "not a workable option" and saying that it has no plans ever to refile the application. In the NO selection, Luther J. Carter, Lake H. Barrett, and Kenneth C. Rogers argue that the decision to withdraw the application for a nuclear waste repository at Yucca Mountain was motivated by politics rather than by evidence. If successful, it will impede future efforts to use nuclear power to combat global warming.

U.S. Department of Energy

U.S. Department of Energy's Motion to Withdraw

The United States Department of Energy ("DOE") hereby moves . . . to withdraw its pending license application for a permanent geologic repository at Yucca Mountain, Nevada. DOE asks the Board to dismiss its application with prejudice and to impose no additional terms of withdrawal.

While DOE reaffirms its obligation to take possession and dispose of the nation's spent nuclear fuel and high-level nuclear waste, the Secretary of Energy has decided that a geologic repository at Yucca Mountain is not a workable option for long-term disposition of these materials. Additionally, at the direction of the President, the Secretary has established the Blue Ribbon Commission on America's Nuclear Future, which will conduct a comprehensive review and consider alternatives for such disposition. And Congress has already appropriated $5 million for the Blue Ribbon Commission to evaluate and recommend such "alternatives." . . . In accord with those decisions, and to avoid further expenditure of funds on a licensing proceeding for a project that is being terminated, DOE has decided to discontinue the pending application in this docket, and hereby moves to withdraw that application with prejudice.

Under the Nuclear Waste Policy Act of 1982, as amended, 42 U.S.C. §§ 10101 *et seq.* ("NWPA"), this licensing proceeding must be conducted "in accordance with the laws applicable to such applications. . . ." Those laws necessarily include the NRC's regulations governing license applications. . . .

Thus, applicable Commission regulations empower this Board to regulate the terms and conditions of withdrawal. . . . Any terms imposed for withdrawal must bear a rational relationship to the conduct and legal harm at issue. . . . And the record must support any findings concerning the conduct and harm in question to impose a term. . . .

The Board Should Grant Dismissal With Prejudice

In this instance, the Board should prescribe only one term of withdrawal—that the pending application for a permanent geologic repository at the Yucca Mountain site shall be dismissed with prejudice.[1]

That action will provide finality in ending the Yucca Mountain project for a permanent geologic repository and will enable the Blue Ribbon Commission,

From U.S. Nuclear Regulatory Commission, March 2, 2010, Docket No. 63-001.

as established by the Department and funded by Congress, to focus on alternative methods of meeting the federal government's obligation to take high-level waste and spent nuclear fuel. It is the Secretary of Energy's judgment that scientific and engineering knowledge on issues relevant to disposition of high-level waste and spent nuclear fuel has advanced dramatically over the twenty years since the Yucca Mountain project was initiated. . . . Future proposals for the disposition of such materials should thus be based on a comprehensive and dismissal. The statute simply requires that the Secretary "shall submit . . . an application for a construction authorization." . . . It neither directs nor circumscribes the Secretary's actions on the application after that submission.

Indeed, far from imposing special limitations on DOE after the submission, the NWPA expressly requires that the application be considered "in accordance with the laws applicable to such applications." . . . Those laws include 10 C.F.R. § 2.107, which, as this Board has recognized, authorizes withdrawals on terms the Board prescribes. Congress, when it enacted the NWPA in 1982, could have dictated that special rules applied to this proceeding to prevent withdrawal motions, or could have prescribed duties by DOE with respect to prosecution of the application after filing, but it chose not to do so.

Nor does the structure of the NWPA somehow override the plain textual indication in the statute that ordinary NRC rules govern here or dictate that the Secretary must continue with an application he has decided is contrary to the public interest. The NWPA does not prescribe a step-by-step process that leads inexorably to the opening of a repository at Yucca Mountain. Indeed, even if the NRC granted the pending application today, the Secretary would not have the authority to create an operational repository. That would require further action by DOE, other agencies, and Congress itself, yet none of those actions is either mandated or even mentioned by the NWPA. The NWPA does not require the Secretary to undertake the actions necessary to obtain the license to receive and possess materials that would be necessary to open a repository. . . . Rather, the NWPA refers only to the need for a "construction authorization" . . . —and even there, as discussed, it mandates only the submission of an application. To open a facility, moreover, the Department would be required to obtain water rights, rights of way from the Bureau of Land Management for utilities and access roads, and Clean Water Act § 404 permits for repository construction, as well as all the state and federal approvals necessary for an approximately 300-mile rail line, among many other things. None of those actions is mandated by the NWPA. At least as important, as the prior Administration stressed, *Congress* would need to take further action not contained in the NWPA before any such repository could be opened. In short, there are many acts between the filing of the application and the actual use of the repository that the NWPA does not require.

Where, even if the NRC granted the pending application, Congress has not authorized the Secretary to make the Yucca Mountain site operational, or even mandated that he take the many required steps to make it operational, it would be bizarre to read the statute to impose a non-discretionary duty to continue with any particular intermediate step (here, prosecuting the application), absent clear statutory language mandating that result. More generally, it has not been the NRC's practice to require any litigant to maintain a license application that the litigant does not wish to pursue. That deference to an applicant's decisions should apply more strongly where a government official has decided not to pursue a license application because he believes that other courses would better serve the public interest.

Finally, the fact that Congress has approved Yucca Mountain as the site of a repository, *see* Pub. L. No. 107-200, 116 Stat. 735 (2002) ("there hereby is approved the site at Yucca Mountain, Nevada, for a repository, with respect to which a notice of disapproval was submitted by the Governor of the State of Nevada on April 8, 2002"), means, in the D.C. Circuit's words, simply that the Secretary is "permitted" to seek authority to open such a site and that challenges to the prior process to select that site are moot. . . . It does *not* require the Secretary to continue with an application proceeding if the Secretary decides that action is contrary to the public interest. . . . That conclusion is even more strongly compelled now, in light of Congress's recent decision to provide funding to a Blue Ribbon Commission, whose explicit purpose is to propose "alternatives" for the disposal of high-level waste and spent nuclear fuel.

Even if there were any ambiguity on these points, the Secretary's interpretation of the NWPA would be entitled to deference. . . . Simply put, the text of the NWPA does not specify actions the Secretary can or must take once the application is filed. Accordingly, while some may disagree with the wisdom of the Secretary's underlying policy decision, the Secretary may fill this statutory "gap." The Secretary's interpretation is a reasonable one that should be given great weight and sustained. . . .

No Conditions Are Necessary as to the Licensing Support Network

Finally, there is no reason to impose conditions relating to the Licensing Support Network ("LSN") as a term of withdrawal. As DOE's prior filings with this Board explain, DOE will, at a minimum, maintain the LSN throughout this proceeding, including any appeals, and then archive the LSN materials in accordance with the Federal Records Act and other relevant law. . . . Thus, DOE will retain the full LSN functionality throughout this proceeding, including appeal, and then follow well established legal requirements that already govern DOE's obligations regarding these documents. DOE is also considering whether sound public and fiscal policy, and the goal of preserving the knowledge gained both inside and outside of this proceeding, suggest going even further than those legal requirements. There is thus no need for this Board to impose additional conditions concerning the preservation of records.

DOE counsel has communicated with counsel for the other parties commencing on February 24, 2010, in an effort to resolve any issues raised by them prior to filing this Motion. . . . The State of Nevada and the State of California have stated that they agree with the relief requested here. The Nuclear Regulatory Commission Staff has stated that it takes no position at this time. The Nuclear Energy Institute has stated that it does not consent to the relief requested and will file its position in a response. All other parties that have responded have stated that they reserve their positions until they see the final text of the motion.

Note

1. DOE seeks this form of dismissal because it does not intend ever to refile an application to construct a permanent geologic repository for spent nuclear fuel and high-level radioactive waste at Yucca Mountain.

Luther J. Carter, Lake H. Barrett, and Kenneth C. Rogers

Nuclear Waste Disposal: Showdown at Yucca Mountain

If the nation is to seriously confront a growing inventory of highly radioactive waste, a key step is to determine the merits of its geologic repository project at Yucca Mountain in Nevada. A board of the U.S. Nuclear Regulatory Commission (NRC) has for nearly two years been conducting an open and transparent licensing proceeding to accomplish exactly that. Moreover, in its forceful ruling of June 29, 2010, the board rejected as contrary to law a motion by Secretary of Energy Steven Chu to withdraw the licensing application and shut the proceeding down. Yet the administration's attempt to abandon Yucca Mountain continues and in our view poses a significant risk of a major setback for public acceptance of nuclear energy.

The licensing application was filed by the Bush administration under the Nuclear Waste Policy Act (NWPA) of 1982, and the proceeding itself began in October 2008. The NRC staff has almost completed its safety evaluation of repository performance for many tens of thousands of years. With this report in hand, the licensing board (acting for the commission) could begin hearing and adjudicating scores of critical contentions by the state of Nevada and other opposing parties. If the case for licensing is convincing, the granting of a construction license could come in 2012. But the licensing board is a creature of the NRC, and if the commission should order the proceeding terminated in keeping with Secretary Chu's motion, the board must comply.

The attempt by the current administration to withdraw the licensing application and abandon Yucca Mountain follows a commitment made by Barack Obama in early 2008 during the competitive scramble for Nevada delegates to the Democratic National Convention. Hillary Clinton, then the hands-on favorite for the nomination, had long sided with Nevada in its opposition to a repository at Yucca Mountain. Not to be outdone, Senator Obama declared his own categorical opposition to the project. Earlier this year, when President Obama, acting through Secretary Chu, moved to withdraw the licensing application, no scientific justification or showing of alternatives was offered. The project was simply dismissed as "not a workable option."

To cover Obama's political debt to Nevada, repository licensing would be terminated without congressional review and approval despite the fact that this vital project was sanctioned by Congress in elaborate detail and handsomely funded by a fee imposed on tens of millions of consumers of electricity produced by nuclear reactors. The licensing proceeding marks the culmination of a 25-year site investigation that has cost over $7 billion for the Nevada project itself and over $10 billion for the larger national screening of repository sites from which the Yucca Mountain site was chosen.

What's At Stake

To summarily kill the project would cap with still another failure a half-century of frustrated endeavors to site, license, and construct a geologic repository. The roughly 64,000 metric tons of spent reactor fuel that await permanent geologic disposal are now in temporary storage at 120 operating and shut-down commercial nuclear power reactors in 36 states. In addition, there are the thousands of containers of highly radioactive waste arising from the cleanup of nuclear weapons production sites in Washington, South Carolina, and Idaho.

Now pending before the U.S. Circuit Court of Appeals for the District of Columbia are lawsuits brought by Washington, South Carolina, the National Association of Regulatory Utility Commissioners, and several other plaintiffs to stop the licensing withdrawal. Most tellingly, the plaintiffs allege violations of the NWPA of 1982, with its detailed prescriptions for repository site selection, approval, and construction licensing. But also in play is the Administrative Procedure Act, under which agency decisions can be voided as "arbitrary and capricious" and an abuse of discretion.

In its refusal to accede to the Department of Energy's (DOE's) motion to withdraw the licensing, the licensing board questioned why the Congress, in enacting the NWPA, would have set out an elaborate sequence of steps and procedures for the selection and approval of a repository site if in the end the Secretary of Energy could undo everything by withdrawing the licensing application. "Unless Congress directs otherwise, DOE may not single-handedly derail the legislatively mandated decision-making process," the board said.

The Court of Appeals initially called for arguments in the pending litigation to begin this September but has now decided to first await an outcome at the NRC.

Coupled with the attempted withdrawal of the licensing application is a self-evident violation of the Federal Advisory Committee Act of 1972, which is intended to keep advisory committees from being "inappropriately influenced by the appointing authority or any special interest." According to its charter, the Blue Ribbon Commission on America's Nuclear Future (BRC), which Secretary Chu unveiled early this year, is to conduct a "comprehensive review of policies for managing the back end of the nuclear fuel cycle, including all alternatives for the storage, processing, and disposal of civilian and defense used nuclear fuel [and] high-level waste. . . ." Left unstated, to say the least, was the fact that the commission was created in substantial part to show that Yucca Mountain was not being abandoned without identifying a full suite of waste management options—but with no intention to have the repository project serve as a baseline for this review.

In March 2009, Secretary Chu and Nevada's Senator Harry Reid, the Senate's Democratic Majority Leader and a relentless foe of Yucca Mountain, struck a deal wherein Reid would drop his proposed legislation for a blue ribbon commission that Congress would appoint in favor of a commission that the Secretary of Energy would choose. In a press conference announcing the formation of the BRC on January 29, 2010, and later at their first formal meeting, commission members were told by Secretary Chu and White House aide Carol Browner that Yucca Mountain is past history and is not among the waste management options to be considered.

A Blue Ribbon Agenda

The BRC's eminent co-chair, Lee Hamilton, the former Indiana congressman who served as vice chairman of the 9/11 commission, has made the general point that his study group's "recommendations will be ours and ours alone." Indeed, whatever the motivations of those who created it, the BRC is an independent advisory body chartered to provide a comprehensive review of waste management alternatives, and it cannot reasonably and honorably exclude Yucca Mountain from that review. The intellectual gyrations at play with respect to Yucca Mountain may be especially disturbing to those commission members well versed in nuclear energy issues, such as Richard Meserve (a former chair of the NRC), Per Peterson (chair of nuclear engineering at the University of California, Berkeley), and Phil Sharp (head of Resources for the Future and formerly a congressman from Indiana).

In turning its back on Yucca Mountain, the commission would put itself at high risk of failing to produce a report of significant policy impact and of coming across as little more than a fig leaf of respectability for the president's decision to abandon the repository. We don't think it will do that. This body could in fact prove itself enormously useful, not least by an insistence on recognizing and protecting the integrity of the NRC as an independent regulatory agency.

The commission could also emphasize that solid public acceptance of nuclear energy, together with the continued storage of large amounts of spent fuel in temporary surface facilities, may well turn on a credible promise of a geologic repository becoming available within the next few decades. This we see as a fundamental political reality that is accorded too little weight by the utility industry, the Secretary of Energy, and the NRC itself.

The utilities that are generating nuclear energy certainly want a repository, but they do not want their lack of one to stand in the way of public support and federal subsidies for a nuclear expansion. So from this contorted position they argue the safety and acceptability of surface storage of spent fuel for decades into the future while quite properly pressing the government to honor its long-past-due obligation to take custody of most of that fuel.

But the politically critical nexus between reactors and spent fuel disposal has been evident since 1976, when Californians approved a referendum that declared that no more nuclear plants could be built in the state until a means for permanent disposal of spent reactor fuel and high-level waste was achieved.

Waste Confidence

The NRC's successive "waste confidence" rule-makings during the past 25 years have been a milder response to the same issue. A lawsuit begun by the Natural Resources Defense Council in 1977 gave rise to the first such NRC rule-making in 1984. In that ruling, "reasonable assurance" was found on three critical points: that at least one mined geologic repository would be available by the years 2007–2009; that spent fuel from any reactor could go to geologic disposal within 30 years of the expiration of the reactor's operating license; and that during the interim, the spent fuel could be safely kept in surface storage facilities either at the reactor site or elsewhere.

These confidence findings were renewed in 1990, then again in 1999, but with the difference that the latter finding envisioned a geologic repository becoming available "within the first quarter of the twenty-first century." In September 2009, a new confidence proceeding was initiated wherein the NRC expressed reasonable assurance of having a repository within 50 to 60 years of the licensed life of existing reactors, which for some reactors may extend to the year 2060.

In plain English, what this meant was that the commission would be comfortable not having a repository until sometime well beyond the year 2100, when our great-great grandchildren may be left to worry about the disposal of nuclear waste arising from the generation of nuclear electricity from which we benefit today. The NRC, with two vacancies at the time, had but three members to consider this confidence finding and only one was willing to adopt it without receiving public comment on policy changes affecting Yucca Mountain. That one was the commission's new chair, Gregory B. Jaczko, formerly a senior aide and close associate of Senator Reid. President Bush appointed Jaczko to the commission in 2005 and reappointed him in 2008, and last year President Obama named him chairman.

Since then, the NRC has undergone major changes in membership, and whether there is among the five commissioners a legally qualified quorum of three to decide pending Yucca Mountain issues is being challenged. Of the two members who opposed issuance of a confidence finding last year, Commissioner Kristine L. Svinicki continues to serve but her former colleague Dale E. Klein has completed his term and departed.

Meanwhile, three new members—George E. Apostolakis, William D. Magwood IV, and William C. Ostendorff—have come aboard. At their Senate confirmation hearing in February, Senator Barbara Boxer of California asked each of the three this question on behalf of Senator Reid: "If confirmed, would you second guess the DOE decision to withdraw license application for Yucca Mountain from NRC review?" All three answered, no. In the pending litigation, Washington State and South Carolina, plus a few other parties, cite this exchange as compelling grounds why, by law, they should recuse themselves from any decision on the Yucca Mountain licensing issue.

Apostolakis, a professor of nuclear science and engineering at the Massachusetts Institute of Technology (MIT) and a member of the National Academy of Engineering, has in fact since recused himself. But his stated reason for doing so was not his response to Senator Boxer but the fact that he chaired the Sandia National Laboratory panel that reviewed the Yucca Mountain performance assessment and found it adequate to support submittal of a license application.

Commissioners Magwood and Ostendorff, on the other hand, have now refused to disqualify themselves, contending that Boxer's question was vaguely put and that they were at the time unaware that a White House decision to withdraw the licensing application would be coming up for NRC review. But the DOE had already filed a motion to stay the licensing board proceeding and announced that a motion to withdraw the licensing application would soon follow. Counsel for Washington et al., citing Supreme Court precedents, argue that whether a judge or regulatory official recuses himself should turn not on "the reality of bias or prejudice but its appearance" and on whether a "reasonable man, [knowing] all the circumstances, would harbor doubts about the judge's impartiality."

Of course, in principle there's nothing to keep Magwood and Ostendorff from deciding not to join their chairman, Gregory Jaczko, in overriding the licensing board. This would deny Jaczko a majority on the issue and leave in force the board's refusal to stop the licensing. But however that may be resolved by the commissioners, the matter of the new waste confidence finding is also pending. All five commissioners, including Magwood and Ostendorff, have issued position papers in which, despite differences in detail, there is broad agreement as to strategy. They have studiously avoided recognition of the elephant in the room, Yucca Mountain. The project's fate is either ignored or treated as by no means impeding a confidence finding.

The commissioners are counting on continued surface storage for up to 120 years or even much longer, and on having either a mined geologic repository or some other means of final disposal available "when necessary." The House report that accompanied the Nuclear Waste Policy Act almost 28 years ago noted that "an opiate of confidence" had led to a long trail of paper analyses and plans that had come to nothing. The record of frustration and failure that preceded that 1982 Act may well be extended right up to the present if the commissioners rubber-stamp the administration's withdrawal plans for Yucca Mountain or ignore the implications for waste confidence of the project's being abandoned at the very point of construction licensing.

Whatever happens at the NRC, the BRC must weigh in with its own judgments. A central fact to be recognized is that geologic storage or disposal of highly radioactive waste will not begin within this generation without a renewed commitment to Yucca Mountain. Apart from the continued surface storage of spent fuel, other waste management options that the commission is considering—spent fuel reprocessing, "recycling," and transmutation of dangerously radiotoxic species to more benign forms—have little to offer for the next half century or longer.

This is true for a mix of technical and financial reasons explained at length in studies done by experts at Harvard, MIT, and elsewhere. A primary reference is the National Research Council's Separations Technology and Transmutation Systems report of 1996. For the foreseeable future, waste management systems resting on such technologies would come at prohibitive cost and could not in any case eliminate all of the dangerously radioactive and long-lived wastes of concern. For final disposal of such waste, geologic containment is the only option, and Yucca Mountain is the one place where this might happen in the next few decades.

Redefining Yucca Mountain

The commission has an opportunity to broadly redefine the Yucca Mountain project to suggest how advantage might be taken of the repository's early potentialities and how uncertainties about its long-term performance might be reduced. Bear in mind that operation of the repository would come in two phases. There is, first, a pre-closure phase of up to several hundreds of years during which spent fuel and high-level waste would be emplaced retrievably. This is followed by a post-closure phase that begins when the repository is sealed.

Built in volcanic rock high above the water table and accessed by gently inclined ramps from the ridge slopes, a Yucca Mountain repository would be ideally situated to serve for monitored geologic storage of spent fuel, which ultimately could be retrieved if, say, fuel recycling should become economically attractive. Regrettably, in 1987, when the investigation of repository sites was narrowed to Yucca Mountain, the Congress, as a concession to Nevada, declared that no "monitored retrievable storage facility" could be built in that state. Here, Congress was, without doubt, referring to the kind of monitored retrievable surface storage facility that some sponsors of the NWPA of 1982 had deemed no less essential than a geologic repository and much more easily achieved.

But DOE officials did not believe that the NRC, under its licensing policies, would permit them to seek a license allowing retrievable emplacement of spent fuel and high-level waste early in the pre-closure phase while work continued on meeting the more stringent standards for permanent emplacement. They knew, too, that to propose such a two-phased strategy would arouse Senator Reid's wrath.

But the BRC could strongly advocate a two-phased approach to licensing, with vigorous pursuit of repository design alternatives to continue in parallel with the program of monitored retrievable geologic storage.

The National Research Council's Board on Radioactive Waste Management has long recommended that repository design be approached in a phased, stepwise manner that allows intensive testing and analysis and a flexible, adaptive response to the setbacks and surprises sure to come. This concept was most recently articulated in the board's 2003 report *One Step at a Time: The Staged Development of Geologic Repositories for High-Level Radioactive Waste.*

In sorting things out, the commission might note with emphasis that commercial spent fuel and defense high-level waste differ greatly in the degree of hazard posed. Because there is relatively little presence of plutonium and other actinides of long half-life in the defense wastes, the period of hazard for these wastes may be as short as 10,000 years, compared to up to a million years for spent fuel.

A Fair Deal for Nevada

As for Nevada's grievances, the commission doubtless will note that when the Congress, in its 1987 amendment to the NWPA, narrowed the search for a repository site to Yucca Mountain, this came as an abrupt departure from the procedure originally mandated to go to a single candidate site only after an in-depth, in-situ exploration of three candidates. But the volcanic tuff site at Yucca Mountain had emerged from the first round of studies as clearly superior to the other two candidates: the site in volcanic basalt at Hanford, Washington, and the one in deep bedded salt in Deaf Smith County, Texas. A more tentative or contingent congressional choice of Yucca Mountain would almost certainly have survived an impartial technical review, so in our view the hasty adoption of what soon came to be known as the "screw Nevada bill" was as unnecessary as it was politically provocative.

We think Nevada's cause for redress turns chiefly on regional fairness and equity, on having been fingered to take dangerously radioactive and long-lived nuclear waste that probably no other state would willingly accept. A major question for the BRC to consider is what compensation is due the state chosen for the nation's first repository for permanent disposal of spent fuel and high-level waste? The state could, for example, be given preference in the siting of various other new government-sponsored or -encouraged enterprises, civil or military, nuclear or non-nuclear, promising to bring Nevada more high-tech jobs and attract other business.

Even today, Nevada's Nye County (host to Yucca Mountain) and several other rural counties see a duly licensed repository project as a distinct economic asset and quite safe. Also, some of Nevada's more visible Republican politicians openly advocate the project, too, but on condition that the "nuclear dump" many Nevadans envision be made more acceptable by adding other nuclear-related industrial activities. Although Senator Reid surely has had the wind at his back in opposing the repository, the oft-repeated claim that Nevadans are overwhelmingly opposed to the repository is a canard that dies hard.

President Obama, at the Copenhagen climate change summit last December, announced a goal of reducing carbon emissions by 83% by the year 2050. In pondering the nation's nuclear future, the BRC must be aware that a nuclear contribution on a scale truly relevant to that hugely ambitious goal might entail a fivefold expansion of the present suite of 104 large reactors and a fivefold increase in the annual production of spent fuel from 2,000 to 10,000 metric tons. Surely this is not the time to abandon the only currently viable option for very long-term geologic retrievable storage of spent fuel, and possibly final disposal.

But also at stake is the reputation of the NRC as an independent, trustworthy overseer of the civil nuclear enterprise. The NRC has been dealt with abusively by the Obama administration and Senator Reid in the matter of Yucca Mountain. So now will the commissioners acquiesce in the policies of the senator and the White House, or will they reassert the NRC's dignity and independence by upholding their own Yucca Mountain licensing board? Also, will they see the speciousness of their pending waste confidence finding that would ignore the blatantly political undoing of a sophisticated technical endeavor to build the world's first geologic repository for highly radioactive waste? How the commissioners exercise their great trust will soon be apparent.

EXPLORING THE ISSUE

Should the United States Stop Planning for Permanent Nuclear Waste Disposal at Yucca Mountain?

Critical Thinking and Reflection

1. Why, according to the U.S. Department of Energy, is the DOE free to simply withdraw its license application and in essence abandon the effort to build the Yucca Mountain nuclear waste repository?
2. Why is it politically difficult to find an acceptable location for a nuclear waste repository?
3. It is generally accepted that any nuclear waste repository must be stable for a great many years. What kinds of locations might satisfy this requirement?
4. What kinds of incentives might help persuade people to accept a nearby nuclear waste repository?

Is There Common Ground?

Both sides in this debate agree that nuclear waste already exists and must be dealt with, nuclear power will be with us for many years and more wastes will be generated, and there is a need for either an acceptable location for a nuclear waste repository or an alternative method of dealing with nuclear wastes. One such method is covered in the next issue in this book. Look ahead and answer the following questions:

1. What is nuclear fuel reprocessing?
2. How does it reduce the quantity of nuclear waste?

ISSUE 14

Should the United States Reprocess Spent Nuclear Fuel?

YES: Phillip J. Finck, from Statement before the House Committee on Science, Energy Subcommittee, Hearing on Nuclear Fuel Reprocessing (June 16, 2005)

NO: Charles D. Ferguson, from "An Assessment of the Proliferation Risks of Spent Fuel Reprocessing and Alternative Nuclear Waste Management Strategies," testimony before the U.S. House of Representatives Committee on Science and Technology Hearing on Advancing Technology for Nuclear Fuel Recycling: What Should Our Research, Development and Demonstration Strategy Be? (June 17, 2009)

Learning Outcomes

After studying this issue, students will be able to:

- Explain why fear of nuclear proliferation has kept the United States from developing nuclear fuel processing facilities.
- Explain how nuclear fuel reprocessing can help deal with the nuclear waste problem.
- Explain how nuclear fuel reprocessing supports expended use of nuclear power.
- Explain why it will be difficult to find a publicly acceptable site for a nuclear fuel reprocessing plant.

ISSUE SUMMARY

YES: Phillip Finck argues that by reprocessing spent nuclear fuel, the United States can enable nuclear power to expand its contribution to the nation's energy needs while reducing carbon emissions, nuclear waste, and the need for waste repositories such as Yucca Mountain.

NO: Charles Ferguson, president of the Federation of American Scientists, argues that even though reprocessing can help reduce nuclear waste management problems, because as currently practiced it both poses a significant risk that weapons-grade material will fall into the wrong hands and raises the price of nuclear fuel (compared to the once-through fuel cycle), it should not be pursued at present. There is time for further research. Meanwhile, we should concentrate our efforts on safe storage of nuclear wastes.

As nuclear reactors operate, the nuclei of uranium-235 atoms split, releasing neutrons and nuclei of smaller atoms called fission products, which are themselves radioactive. Some of the neutrons are absorbed by uranium-238, which then becomes plutonium. The fission product atoms eventually accumulate to the point where the reactor fuel no longer releases as much energy as it used to. It is said to be "spent." At this point, the spent fuel is removed from the reactor and replaced with fresh fuel.

Once removed from the reactor, the spent fuel poses a problem. Currently, it is regarded as high-level nuclear waste, which must be stored on the site of the reactor, initially in a swimming pool–sized tank and later, once the radioactivity levels have subsided a bit, in "dry casks." Until 2009, there was a plan to dispose of the casks permanently in a subterranean repository being built at Yucca Mountain, Nevada, but the Obama administration has proposed ending funding for Yucca Mountain (see Issue 13 and Dan Charles, "A Lifetime of Work Gone to Waste?" *Science* (March 20, 2009)).

It is worth noting that spent fuel still contains useful components. Not all the uranium-235 has been burned up, and the plutonium created as fuel is burned can itself be used as fuel. When spent fuel is treated as waste, these components of the waste are discarded. Early in the Nuclear Age, it was seen that if these components could be recovered, the amount of waste to be disposed of could be reduced. The fuel supply could also be extended, and in fact, since plutonium is made from otherwise useless uranium-238, new fuel could be created. Reactors designed to maximize plutonium creation, known as "breeder" reactors because they "breed" fuel, were built and are still in use as power plants in Europe. In the United States, breeder reactors have been built and operated only by the Department of Defense, for plutonium extracted from spent fuel is required for making nuclear bombs. They have not seen civilian use in part because of fear that bomb-grade material could fall into the wrong hands.

The separation and recycling of unused fuel from spent fuel is known as reprocessing. In the United States, a reprocessing plant operated in West Valley, New York, from 1966 to 1972 (see "Plutonium Recovery from Spent Fuel Reprocessing by Nuclear Fuel Services at West Valley, New York, from 1966 to 1972" (DOE, 1996; www.osti.gov/opennet/forms.jsp?formurl=document/purecov/nfsrepo.html)). After the Nuclear Nonproliferation Treaty went into force in 1970, it became U.S. policy not to reprocess spent nuclear fuel and thereby to limit the availability of bomb-grade material. As a consequence, spent fuel was not recycled, it was regarded as high-level waste to be disposed of, and the waste continued to accumulate.

Despite the proposed termination of the Yucca Mountain nuclear waste disposal site, the nuclear waste disposal problem is real and it must be dealt with. If it is not, we may face the same kinds of problems created by the former Soviet Union, which disposed of some nuclear waste simply by dumping it at sea. For a summary of the nuclear waste problem and the disposal controversy, see Michael E. Long, "Half Life: The Lethal Legacy of America's Nuclear Waste," *National Geographic* (July 2002). The need for care in nuclear waste disposal is underlined by Tom Carpenter and Clare Gilbert, "Don't Breathe the Air," *Bulletin of the Atomic Scientists* (May/June 2004); they describe the Hanford Site in Hanford, Washington, where wastes from nuclear weapons production were stored in underground tanks. Leaks from the tanks have contaminated groundwater, and an extensive cleanup program is under way. But cleanup workers are being exposed to both radioactive materials and toxic chemicals, and they are falling ill. And in June 2004, the U.S. Senate voted to ease cleanup requirements. Per F. Peterson, William E. Kastenberg, and Michael Corradini, "Nuclear Waste and the Distant Future," *Issues in Science and Technology* (Summer 2006), argue that the risks of waste disposal have been sensibly addressed and we should be focusing more attention on other risks (such as those of global warming). Behnam Taebi and Jan Kloosterman, "To Recycle or Not to Recycle? An Intergenerational Approach to Nuclear Fuel Cycles," *Science & Engineering Ethics* (June 2008), argue that the question of whether to accept reprocessing comes down to choosing between risks for the present generation and risks for future generations.

In November 2005, President Bush directed the Department of Energy to start work toward a reprocessing plant; see Eli Kintisch, "Congress Tells DOE to Take Fresh Look at Recycling Spent Reactor Fuel," *Science* (December 2, 2005). By April 2008, Senator Pete Domenici of the U.S. Senate Energy and Natural Resources Committee was working on a bill that would set up the nation's first government-backed commercialized nuclear waste reprocessing facilities. Reprocessing spent nuclear fuel will be expensive, but the costs may not be great enough to make nuclear power unacceptable; see "The Economic Future of Nuclear Power," University of Chicago (August 2004) (www.ne.doe.gov/np2010/reports/NuclIndustryStudy-Summary.pdf).

In February 2006, the U.S. Department of Energy announced the Global Nuclear Energy Partnership (GNEP), to be operated by the United States, Russia, Great Britain, and France. It would lease nuclear fuel to other nations, reprocess spent fuel without generating material that could be diverted to making nuclear bombs, reduce the amount of waste that must be disposed of, and help meet future energy needs. See Stephanie Cooke, "Just Within Reach?" *Bulletin of the Atomic Scientists* (July/August 2006), and Jeff Johnson, "Reprocessing Key to Nuclear Plan," *Chemical & Engineering News* (June 18, 2007). Critics such as Karen Charman, "Brave Nuclear World, Parts I and II," *World Watch* (May/June and July/August 2006), insist that nuclear power is far too expensive and carries too serious risks of breakdown and exposure to wastes to rely upon, especially when cleaner, cheaper, and less dangerous alternatives exist. Early in 2009, the Department of Energy announced it was closing down the GNEP.

It is an unfortunate truth that the reprocessing of nuclear spent fuel does indeed increase the risks of nuclear proliferation. Both nations and terrorists itch to possess nuclear weapons, whose destructive potential makes present members of the "nuclear club" tremble. Can the risks be controlled? John Deutch, Arnold Kanter, Ernest Moniz, and Daniel Poneman, in "Making the World Safe for Nuclear Energy," *Survival* (Winter 2004–2005), argue that present nuclear nations could supply fuel and reprocess spent fuel for other nations; nations that refuse to participate would be seen as suspect and subject to international action. Nuclear physicist and Princeton professor Frank N. von Hippel, "Rethinking Nuclear Fuel Recycling," *Scientific American* (May 2008) argues that reprocessing nuclear spent fuel is expensive and emits lethal radiation. There is also a worrisome risk that the increased availability of bomb-grade nuclear materials will increase the risk of nuclear war and terrorism. Prudence demands that spent fuel be stored until the benefits of reprocessing exceed the risks (if they ever do). See also Rodney C. Ewing and Frank N. von Hippel, "Nuclear Waste Management in the United States—Starting Over," *Science* (July 10, 2009).

In the YES selection, Phillip J. Finck, Deputy Associate Laboratory Director, Applied Science and Technology and National Security, Argonne National Laboratory, argues that by reprocessing spent nuclear fuel the United States can enable nuclear power to expand its contribution to the nation's energy needs while reducing carbon emissions, nuclear waste, and the need for waste repositories such as Yucca Mountain. Charles D. Ferguson, president of the Federation of American Scientists, argues that even though reprocessing can help reduce nuclear waste management problems, because as currently practiced it both poses a significant risk that weapons-grade material will fall into the wrong hands and raises the price of nuclear fuel (compared to the once-through fuel cycle), it should not be pursued at present. There is time for further research. Meanwhile, we should concentrate our efforts on safe storage of nuclear wastes.

Phillip J. Finck

Statement before the House Committee on Science, Energy Subcommittee, Hearing on Nuclear Fuel Reprocessing

Summary

Management of spent nuclear fuel from commercial nuclear reactors can be addressed in a comprehensive, integrated manner to enable safe, emissions-free, nuclear electricity to make a sustained and growing contribution to the nation's energy needs. Legislation limits the capacity of the Yucca Mountain repository to 70,000 metric tons from commercial spent fuel and DOE defense-related waste. It is estimated that this amount will be accumulated by approximately 2010 at current generation rates for spent nuclear fuel. To preserve nuclear energy as a significant part of our future energy generating capability, new technologies can be implemented that allow greater use of the repository space at Yucca Mountain. By processing spent nuclear fuel and recycling the hazardous radioactive materials, we can reduce the waste disposal requirements enough to delay the need for a second repository until the next century, even in a nuclear energy growth scenario. Recent studies indicate that such a closed fuel cycle may require only minimal increases in nuclear electricity costs, and are not a major factor in the economic competitiveness of nuclear power (The University of Chicago study, "The Economic Future of Nuclear Power," August 2004). However, the benefits of a closed fuel cycle can not be measured by economics alone; resource optimization and waste minimization are also important benefits. Moving forward in 2007 with an engineering-scale demonstration of an integrated system of proliferation-resistant, advanced separations and transmutation technologies would be an excellent first step in demonstrating all of the necessary technologies for a sustainable future for nuclear energy.

Nuclear Waste and Sustainability

World energy demand is increasing at a rapid pace. In order to satisfy the demand and protect the environment for future generations, energy sources must evolve from the current dominance of fossil fuels to a more balanced,

U.S. House of Representatives, June 16, 2005.

sustainable approach. This new approach must be based on abundant, clean, and economical energy sources. Furthermore, because of the growing worldwide demand and competition for energy, the United States vitally needs to establish energy sources that allow for energy independence.

Nuclear energy is a carbon-free, secure, and reliable energy source for today and for the future. In addition to electricity production, nuclear energy has the promise to become a critical resource for process heat in the production of transportation fuels, such as hydrogen and synthetic fuels, and desalinated water. New nuclear plants are imperative to meet these vital needs.

To ensure a sustainable future for nuclear energy, several requirements must be met. These include safety and efficiency, proliferation resistance, sound nuclear materials management, and minimal environmental impacts. While some of these requirements are already being satisfied, the United States needs to adopt a more comprehensive approach to nuclear waste management. The environmental benefits of resource optimization and waste minimization for nuclear power must be pursued with targeted research and development to develop a successful integrated system with minimal economic impact. Alternative nuclear fuel cycle options that employ separations, transmutation, and refined disposal (e.g., conservation of geologic repository space) must be contrasted with the current planned approach of direct disposal, taking into account the complete set of potential benefits and penalties. In many ways, this is not unlike the premium homeowners pay to recycle municipal waste.

The spent nuclear fuel situation in the United States can be put in perspective with a few numbers. Currently, the country's 103 commercial nuclear reactors produce more than 2000 metric tons of spent nuclear fuel per year (masses are measured in heavy metal content of the fuel, including uranium and heavier elements). The Yucca Mountain repository has a legislative capacity of 70,000 metric tons, including spent nuclear fuel and DOE defense-related wastes. By approximately 2010 the accumulated spent nuclear fuel generated by these reactors and the defense-related waste will meet this capacity, even before the repository starts accepting any spent nuclear fuel. The ultimate technical capacity of Yucca Mountain is expected to be around 120,000 metric tons, using the current understanding of the Yucca Mountain site geologic and hydrologic characteristics. This limit will be reached by including the spent fuel from current reactors operating over their lifetime. Assuming nuclear growth at a rate of 1.8% per year after 2010, the 120,000 metric ton capacity will be reached around 2030. At that projected nuclear growth rate, the U.S. will need up to nine Yucca Mountain-type repositories by the end of this century. Until Yucca Mountain starts accepting waste, spent nuclear fuel must be stored in temporary facilities, either storage pools or above ground storage casks.

Today, many consider repository space a scarce resource that should be managed as such. While disposal costs in a geologic repository are currently quite affordable for U.S. electric utilities, accounting for only a few percent of the total cost of electricity, the availability of U.S. repository space will likely remain limited.

Only three options are available for the disposal of accumulating spent nuclear fuel:

- Build more ultimate disposal sites like Yucca Mountain.
- Use interim storage technologies as a temporary solution.
- Develop and implement advanced fuel cycles, consisting of separation technologies that separate the constituents of spent nuclear fuel into elemental streams, and transmutation technologies that destroy selected elements and greatly reduce repository needs.

A responsible approach to using nuclear power must always consider its whole life cycle, including final disposal. We consider that temporary solutions, while useful as a stockpile management tool, can never be considered as ultimate solutions. It seems prudent that the U.S. always have at least one set of technologies available to avoid expanding geologic disposal sites.

Spent Nuclear Fuel

The composition of spent nuclear fuel poses specific problems that make its ultimate disposal challenging. Fresh nuclear fuel is composed of uranium dioxide (about 96% U238, and 4% U235). During irradiation, most of the U235 is fissioned, and a small fraction of the U238 is transmuted into heavier elements (known as "transuranics"). The spent nuclear fuel contains about 93% uranium (mostly U238), about 1% plutonium, less than 1% minor actinides (neptunium, americium, and curium), and 5% fission products. Uranium, if separated from the other elements, is relatively benign, and could be disposed of as low-level waste or stored for later use. Some of the other elements raise significant concerns:

- The fissile isotopes of plutonium, americium, and neptunium are potentially usable in weapons and, therefore, raise proliferation concerns. Because spent nuclear fuel is protected from theft for about one hundred years by its intense radioactivity, it is difficult to separate these isotopes without remote handling facilities.
- Three isotopes, which are linked through a decay process (Pu241, Am241, and Np237), are the major contributors to the estimated dose for releases from the repository, typically occurring between 100,000 and 1 million years, and also to the long-term heat generation that limits the amount of waste that can be placed in the repository.
- Certain fission products (cesium, strontium) are major contributors to the repository's shortterm heat load, but their effects can be mitigated by providing better ventilation to the repository or by providing a cooling-off period before placing them in the repository.
- Other fission products (Tc99 and I129) also contribute to the estimated dose.

The time scales required to mitigate these concerns are daunting: several of the isotopes of concern will not decay to safe levels for hundreds of thousands of years. Thus, the solutions to long-term disposal of spent nuclear fuel are limited to three options: the search for a geologic environment that will remain stable for that period; the search for waste forms that can contain these elements for that period; or the destruction of these isotopes. These three options underlie the major fuel cycle strategies that are currently being developed and deployed in the U.S. and other countries.

Options for Disposing of Spent Nuclear Fuel

Three options are being considered for disposing of spent nuclear fuel: the once-through cycle is the U.S. reference; limited recycle has been implemented in France and elsewhere and is being deployed in Japan; and full recycle (also known as the closed fuel cycle) is being researched in the U.S., France, Japan, and elsewhere.

1. Once-through Fuel Cycle

This is the U.S. reference option where spent nuclear fuel is sent to the geologic repository that must contain the constituents of the spent nuclear fuel for hundreds of thousands of years. Several countries have programs to develop these repositories, with the U.S. having the most advanced program. This approach is considered safe, provided suitable repository locations and space can be found. It should be noted that other ultimate disposal options have been researched (e.g., deep sea disposal; boreholes and disposal in the sun) and abandoned. The challenges of long-term geologic disposal of spent nuclear fuel are well recognized, and are related to the uncertainty about both the long-term behavior of spent nuclear fuel and the geologic media in which it is placed.

2. Limited Recycle

Limited recycle options are commercially available in France, Japan, and the United Kingdom. They use the PUREX process, which separates uranium and plutonium, and directs the remaining transuranics to vitrified waste, along with all the fission products. The uranium is stored for eventual reuse. The plutonium is used to fabricate mixed-oxide fuel that can be used in conventional reactors. Spent mixed-oxide fuel is currently not reprocessed, though the feasibility of mixed-oxide reprocessing has been demonstrated. It is typically stored or eventually sent to a geologic repository for disposal. Note that a reactor partially loaded with mixed-oxide fuel can destroy as much plutonium as it creates. Nevertheless, this approach always results in increased production of americium, a key contributor to the heat generation in a repository. This approach has two significant advantages:

- It can help manage the accumulation of plutonium.
- It can help significantly reduce the volume of spent nuclear fuel (the French examples indicate that volume decreases by a factor of 4).

Several disadvantages have been noted:

- It results in a small economic penalty by increasing the net cost of electricity a few percent.
- The separation of pure plutonium in the PUREX process is considered by some to be a proliferation risk; when mixed-oxide use is insufficient, this material is stored for future use as fuel.
- This process does not significantly improve the use of the repository space (the improvement is around 10%, as compared to a factor of 100 for closed fuel cycles).
- This process does not significantly improve the use of natural uranium (the improvement is around 15%, as compared to a factor of 100 for closed fuel cycles).

3. Full Recycle (The Closed Fuel Cycle)

Full recycle approaches are being researched in France, Japan, and the United States. This approach typically comprises three successive steps: an advanced separations step based on the UREX+ technology that mitigates the perceived disadvantages of PUREX, partial recycle in conventional reactors, and closure of the fuel cycle in fast reactors.

The first step, UREX+ technology, allows for the separations and subsequent management of highly pure product streams. These streams are:

- Uranium, which can be stored for future use or disposed of as low-level waste.
- A mixture of plutonium and neptunium, which is intended for partial recycle in conventional reactors followed by recycle in fast reactors.
- Separated fission products intended for short-term storage, possibly for transmutation, and for long-term storage in specialized waste forms.
- The minor actinides (americium and curium) for transmutation in fast reactors.

The UREX+ approach has several advantages:

- It produces minimal liquid waste forms, and eliminates the issue of the "waste tank farms."
- Through advanced monitoring, simulation and modeling, it provides significant opportunities to detect misuse and diversion of weapons-usable materials.
- It provides the opportunity for significant cost reduction.
- Finally and most importantly, it provides the critical first step in managing all hazardous elements present in the spent nuclear fuel.

The second step—partial recycle in conventional reactors—can expand the opportunities offered by the conventional mixed-oxide approach. In particular, it is expected that with significant R&D effort, new fuel forms can be developed that burn up to 50% of the plutonium and neptunium present in spent nuclear fuel. (Note that some studies also suggest that it might be possible to recycle fuel in these reactors many times—i.e., reprocess and recycle the irradiated advanced fuel—and further destroy plutonium and neptunium; other studies also suggest possibilities for transmuting americium in these reactors. Nevertheless, the practicality of these schemes is not yet established and requires additional scientific and engineering research.) The advantage of the second step is that it reduces the overall cost of the closed fuel cycle by burning plutonium in conventional reactors, thereby reducing the number of fast reactors needed to complete the transmutation mission of minimizing hazardous waste. This step can be entirely bypassed, and all transmutation performed in advanced fast reactors, if recycle in conventional reactors is judged to be undesirable.

The third step, closure of the fuel cycle using fast reactors to transmute the fuel constituents into much less hazardous elements, and pyroprocessing technologies to recycle the fast reactor fuel, constitutes the ultimate step in reaching sustainable nuclear energy. This process will effectively destroy the transuranic elements, resulting in waste forms that contain only a very small fraction of the transuranics (less than 1%) and all fission products. These technologies are being developed at Argonne National Laboratory and Idaho National Laboratory, with parallel development in Japan, France, and Russia.

The full recycle approach has significant benefits:

- It can effectively increase use of repository space by a factor of more than 100.
- It can effectively increase the use of natural uranium by a factor of 100.
- It eliminates the uncontrolled buildup of all isotopes that are a proliferation risk.
- The fast reactors and the processing plant can be deployed in small co-located facilities that minimize the risk of material diversion during transportation.
- The fast reactor does not require the use of very pure weapons usable materials, thus increasing their proliferation resistance.
- It finally can usher the way towards full sustainability to prepare for a time when uranium supplies will become increasingly difficult to ensure.
- These processes would have limited economic impact; the increase in the cost of electricity would be less than 10% (ref: OECD).
- Assuming that demonstrations of these processes are started by 2007, commercial operations are possible starting in 2025; this will require adequate funding for demonstrating the separations, recycle, and reactor technologies.
- The systems can be designed and implemented to ensure that the mass of accumulated spent nuclear fuel in the U.S. would always remain below 100,000 metric tons—less than the technical capacity of Yucca Mountain—thus delaying, or even avoiding, the need for a second repository in the U.S.

Conclusion

A well engineered recycling program for spent nuclear fuel will provide the United States with a long-term, affordable, carbon-free energy source with low environmental impact. This new paradigm for nuclear power will allow us to manage nuclear waste and reduce proliferation risks while creating a sustainable energy supply. It is possible that the cost of recycling will be slightly higher than direct disposal of spent nuclear fuel, but the nation will only need one geologic repository for the ultimate disposal of the residual waste.

Charles D. Ferguson

An Assessment of the Proliferation Risks of Spent Fuel Reprocessing and Alternative Nuclear Waste Management Strategies

. . . U.S. leadership is essential for charting a constructive and cooperative international course to prevent nuclear proliferation. An essential aspect of that leadership involves U.S. policy on reprocessing spent nuclear fuel. The United States has sought to prevent the spread of reprocessing facilities to other countries and to encourage countries with existing stockpiles of separated plutonium from reprocessing facilities to draw down those stockpiles. The previous administration launched the Global Nuclear Energy Partnership (GNEP), which proposed offering complete nuclear fuel services, including provision of fuel and waste management, from fuel service states to client states in order to discourage the latter group from enriching uranium or reprocessing spent nuclear fuel—activities that would contribute to giving these countries latent nuclear weapons programs. The current administration and the Congress seek to determine the best course for U.S. nuclear energy policy with the focus of this hearing on recycling or reprocessing of spent fuel and nuclear waste management strategies.

Here at the start, I give a brief summary of the testimony's salient points:

- Reprocessing of the type currently practiced in a handful of countries poses a significant proliferation threat because of the separation of plutonium from highly radioactive fission products. A thief, if he had access, could easily carry away separated plutonium. Fortunately, this reprocessing is confined to nuclear-armed states except for Japan. If this practice spreads to other non-nuclear-weapon states the consequences for national and international security could be dire. Presently, the vast majority of the 31 states with nuclear power programs do not have reprocessing plants.
- The types of reprocessing examined under GNEP do not appear to offer substantial proliferation-resistant benefits, according to research sponsored by the Department of Energy. However, more research is needed to determine what additional safeguards, if any, could provide

U.S. Senate, June 17, 2009.

greater assurances that reprocessing methods are not misused in weapons programs and whether it is possible to have assurances of timely detection of a diversion of a significant quantity of plutonium or other fissile material.

- Time is on the side of the United States. There is no need to rush toward development and deployment of recycling of spent nuclear fuel. Based on the foreseeable price for uranium and uranium enrichment services, this practice is presently far more expensive than the once-through uranium fuel cycle. Nonetheless, more research is needed to determine the costs and benefits of recycling techniques coupled with fast-neutron reactors or other types of reactor technologies. This cost versus benefit analysis would concentrate on the capability of these technologies to help alleviate the nuclear waste management challenge.
- In related research, there is a need to better understand the safeguards challenges in the use of fast reactors. Such reactors are dual-use in the sense that they can burn transuranic material and can breed new plutonium. In the former operation, they could provide a needed nuclear waste management benefit. In the latter operation, they can pose a serious proliferation threat.

Proliferation Risks

Reprocessing involves extraction of plutonium and/or other fissile materials from spent nuclear fuel in order to recycle these materials into new fuel for nuclear reactors. As discussed below, many reprocessing techniques are available for use. Regardless of the particular technique, fissile material is removed from all or almost all of the highly radioactive fission products, which provide a protective barrier against theft or diversion of plutonium in spent nuclear fuel. Plutonium-239 is the most prevalent fissile isotope of plutonium in spent nuclear fuel. The greater the concentration of this isotope the more weapons-usable is the plutonium mixture. Weapons-grade plutonium typically contains greater than 90 percent plutonium-239 whereas reactor-grade plutonium from commercial thermal-neutron reactors has usually less than 60 percent plutonium-239, depending on the characteristics of the reactor that produced the plutonium. The presence of non-plutonium-239 isotopes complicates production of nuclear weapons from the plutonium mixture, but the challenges are surmountable. According to an unclassified U.S. Department of Energy report, reactor-grade plutonium is weapons-usable.

The potential proliferation threats from reprocessing of spent nuclear fuel are twofold. First, a state operating a reprocessing plant could use that technology to divert weapons-usable fissile material into a nuclear weapons program or alternatively it could use the skills learned in operating that plant to build a clandestine reprocessing plant to extract fissile material. Second, a non-state actor such as a terrorist group could seize enough fissile material produced by a reprocessing facility in order to make an improvised nuclear device—a crude, but devastating, nuclear weapon. Such a non-state group may obtain help from insiders at the facility. While commercial reprocessing facilities have typically been well-guarded, some facilities such as those at Sellafield in the United Kingdom and Tokai-mura in Japan have not been able to account for several weapons' worth of plutonium. This lack of accountability does not mean that the fissile material was diverted into a state or non-state weapons program. The discrepancy was most likely due to plutonium caked on piping. But an insider could exploit such a discrepancy. For commercial bulk handing facilities, several tons of plutonium can be processed annually. Thus, if even one tenth of one percent of this material were accounted for, an insider could conceivably divert about one weapon's worth of plutonium every year.

Location matters when determining the proliferation risk of a reprocessing program. That is, a commercial reprocessing plant in a nuclear-armed state such as France, Russia, or the United Kingdom poses no risk of state diversion (but could pose a risk of non-state access) because this type of state, by definition, already has a weapons program. Notably, Japan is the only non-nuclear-armed state that has reprocessing facilities. Japan has applied the Additional Protocol to its International Atomic Energy Agency safeguards, but its large stockpile of reactor-grade plutonium could provide a significant breakout capability for a weapons program. (Chinese officials and analysts occasionally express concern about Japan's plutonium stockpile.) Since the Ford and Carter administrations, when the United States decided against reprocessing on proliferation and economic grounds, the United States has made stopping the spread of further reprocessing facilities especially to non-nuclear weapon states a top priority.

Another top priority of U.S. policy on reprocessing is to encourage countries with stockpiles of separated plutonium to draw down these stockpiles quickly. This drawdown can be done either through consuming the plutonium as fuel or surrounding it with highly radioactive fission products. Global stockpiles of civilian plutonium are growing and now at about 250 metric tons—equivalent to tens of thousands of nuclear bombs—are comparable to the global stockpile of military plutonium. More than 1,000 metric tons of plutonium is contained in spent nuclear fuel in about thirty countries.

While no country has used a commercial nuclear power program to make plutonium for nuclear weapons, certain countries have used research reactor programs to produce plutonium. India, notably, used a research reactor supplied by Canada to produce plutonium for its first nuclear explosive test in 1974. North Korea, similarly, has employed a research-type reactor to produce plutonium for its weapons program. Although nonproliferation efforts with Iran have focused on its uranium enrichment program, which could make fissile material for weapons, its construction of a heavy water research reactor, which when operational (perhaps early next decade) could produce at least one weapon's worth of plutonium annually, poses a latent proliferation threat. To date, Iran is not known to have constructed a reprocessing facility that would be needed to extract plutonium from this reactor's spent fuel. Further activities could take place in the Middle East and other regions. For instance, according to the U.S. government, Syria received assistance from North Korea in building a plutonium production reactor. In September 2007, Israel bombed this construction site.

The United States has been trying to balance the perceived need by many states in the Middle East for nuclear power plants versus restricting these states' access to enrichment and reprocessing technologies. Presently, as an outstanding example, the U.S.-UAE bilateral nuclear cooperation agreement is before the U.S. Congress. Proponents of this agreement tout the commitment made by the UAE to refrain from acquiring enrichment and reprocessing technologies and to rely on market mechanisms to purchase nuclear fuel. However, the last clause in the agreement appears to open the door for the UAE to engage in such activities in the future. . . .

Proliferation-Resistant Reprocessing

Can reprocessing be made more proliferation-resistant? "Proliferation resistance is that characteristic of a nuclear energy system that impedes the diversion or undeclared production of nuclear material or misuse of technology by the host state seeking to acquire nuclear weapons or other nuclear explosive devices." No nuclear energy system is proliferation proof because nuclear technologies are dual-use. Enrichment and reprocessing can be used either for peaceful or military purposes. However, through a defense-in-depth approach, greater proliferation-resistance may be achieved. Both intrinsic features (for example, physical and engineering characteristics of a nuclear technology) and extrinsic features (for example, safeguards and physical barriers) complement each other to deter misuse of nuclear technologies and materials in weapons programs. The potential threats that proliferation-resistance tries to guard against are:

- "Concealed diversion of declared materials;
- Concealed misuse of declared facilities;
- Overt misuse of facilities or diversion of declared materials; and
- Clandestine declared facilities."

For each of these threats, a detailed proliferation pathway analysis can be done in order to measure the proliferation risk and to determine the needed, if any, additional safeguards. The U.S. Department of Energy has sponsored such analysis for proposed reprocessing techniques considered under GNEP. These techniques include UREX+, COEX, NUEX, and Pyroprocessing, and they have been compared to the PUREX technique, which is the commercially used method. PUREX separates plutonium and uranium from highly radioactive fission products. It is an aqueous separations process and thus generates sizable amounts of liquid radioactive waste. UREX+, COEX, and NUEX are also aqueous processes. UREX+ is a suite of chemical processes in which pure plutonium is not separated but different product streams can be produced depending on the reactor fuel requirements. COEX and NUEX are related processes. COEX co-extracts uranium and plutonium (and possibly neptunium) into one recycling stream; another stream contains pure uranium, which can be recycled; and a final stream contains fission products. NUEX separates into three streams: uranium, transuranics (including plutonium), and fission products.

Pyroprocessing uses electrorefining techniques to extract plutonium in combination with other transuranic elements, some of the rare earth fission products, and uranium. This fuel mixture would be intended for use in fast-neutron reactors, which have yet to be proven commercially viable.

Can these reprocessing techniques meet the highest proliferation-resistance standard of the "spent fuel standard" in which plutonium in its final form should be as hard to acquire, process, and use in weapons as is plutonium embedded in spent fuel? The brief answer is "no" because the act of separating most or all of the highly radioactive fission products makes the fuel product less protected than the intrinsic protection provided by spent fuel. In fact, Dr. E.D. Collins of Oak Ridge National Laboratory has shown that the radiation emission from these reprocessed products is 100 times less than the spent fuel standard. In other words, a thief could carry these products and not suffer a lethal radiation dose whereas the same thief would experience a lethal dose in less than one hour of exposure to plutonium surrounded by highly radioactive fission products. But these methods may still be worth pursuing depending on a detailed systems analysis factoring in security risks on site and during transportation, the final disposition of the material once it has been recycled as fuel, as well as the costs and benefits of nuclear waste management.

According to DOE's draft nonproliferation assessment of GNEP, "for a state with preexisting PUREX or equivalent capability (or more broadly the capability to design and operate a reprocessing plant of this complexity), there is minimal proliferation resistance to be found by [using the examined reprocessing techniques] considering the potential for diversion, misuse, and breakout scenarios." Moreover, the DOE assessment points out that these techniques pose additional safeguards challenges. For example, it is difficult to do an accurate accounting of the amount of plutonium in a bulk handling reprocessing facility that produces plutonium mixed with other transuranic elements. This challenge raises the probability of diversion of plutonium by insiders.

Another set of considerations is the choice of reactors to burn up the transuranic elements. The DOE draft assessment examined several choices including light water reactors, heavy water reactors, high temperature gas reactors, and fast-neutron reactors. Only the fast-neutron reactors offered the most benefits in terms of net consumption of transuranic material. This material would have to be recycled multiple times in fast reactors to consume almost all of it. This is called a full actinide recycle in contrast to a partial actinide recycle with the other reactor methods. The benefit from a waste management perspective is that the amount of time required for spent fuel's radiotoxicity to reduce to that of natural uranium goes from more than tens of thousands of years for partial actinide recycle to about 400 years for the full actinide recycle.

The challenge of the full actinide route, however, is that fast reactors can relatively easily be changed from a burner mode to a breeder mode. That is, these reactors can breed more plutonium by the insertion of uranium target material. The perceived need for breeder reactors has driven a few countries such as France, India, Japan, and Russia to develop reprocessing programs.

Alternative Nuclear Waste Management Programs of Other Nations

Have reprocessing programs, to date, helped certain nations solve their nuclear waste problems? The short answer is, "no." Before explicating that further, it is worth briefly examining why these countries began these programs. About fifty years ago, when the commercial nuclear industry was just starting, concerns were raised about the availability of enough natural uranium to fuel the thousands of reactors that were anticipated. Natural uranium contains 0.71 percent uranium-235, 99.28 percent uranium-238, and less than 0.1 percent uranium-234. Uranium-235 is the fissile isotope and thus is needed for sustaining a chain reaction. However, uranium-238 is a fertile isotope and can be used to breed plutonium-239, a fissile isotope that does not occur naturally. Thus, if uranium-238 can be transformed into plutonium-239, the available fissile material could be expanded by more than one hundred times, in principle. This observation motivated several countries, including the United States, to pursue reprocessing.

A related motivation was the desire for better energy security and thus less dependence on outside supplies of uranium. France and Japan, in particular, as countries with limited uranium resources, developed reprocessing plants in order to try to alleviate their dependency on external sources of uranium. They had invested in these plants before the realization that the world would not run out of uranium soon. By the late 1970s, two developments happened that alleviated the perceived pending shortfall. First, the pace of proposed nuclear power plant deployments dramatically slowed. There were plans at that time for more than 1,000 large reactors (of about 1,000 MWe power rating) by 2000, but even before the Three Mile Island accident in 1979, the number of reactor orders in the United States and other countries slacked off although France and Japan launched a reactor building boom in the 1970s that lasted through the 1980s. By 2000, there was only the equivalent of about 400 reactors of 1,000 MWe size. Second, uranium prospecting identified enough proven reserves to supply the present nuclear power demand for several decades to come.

Because there is plentiful uranium at relatively low prices and the cost of uranium enrichment has decreased, the cost of the once-through uranium cycle is significantly less than the cost of reprocessing. However, because fuel costs are a relatively small portion of the total costs of a nuclear power plant, reprocessing adds a relatively small amount to the total cost of electricity. In France, the added cost is almost six percent, and in Japan about ten percent. Nonetheless, in competitive utility markets in which consumers have choices, most countries have not chosen the reprocessing route because of the significantly greater fuel costs. France and Japan have adopted government policies in favor of reprocessing and also have sunk many billions of dollars into their reprocessing facilities. The French government owns and controls the electric utility Electricité de France (EDF) and the nuclear industry Areva. Despite this extensive government control, a 2000 French government study determined that if France stops reprocessing, it would save $4 to 5 billion over the remaining life of its reactor fleet. EDF assigns a negative value to recycled plutonium.

While France's La Hague plant is operating, Japan is still struggling to start up its Rokkasho plant, which is largely based on the French design. Thus, the costs of the Japanese plant keep climbing and will likely be more than $20 billion. While the Japanese government wants to fuel up to one-third of its more than 50 reactors with plutonium-based mixed oxide fuel, local governments tend to look unfavorably on this proposal.

Only a few other nations are involved with reprocessing. Russia and the United Kingdom operate commercial-scale facilities. China and India are interested in heading down this path. But the United Kingdom is moving toward imminent shut down of its reprocessing mainly due to lack of customers. Moreover, the clean up and decommissioning costs are projected to be many billions of dollars. Russia and France also lack enough customers to keep their reprocessing plants at full capacity. In early April, I visited the French La Hague plant and was told that it is only operating at about half capacity. France only uses mixed oxide fuel in 20 of its 58 light water reactors. Presently, less than 10 percent of the world's commercial nuclear power plants burn MOX fuel. As stated earlier, the demand for MOX fuel has not kept up with the stockpiled quantities of plutonium.

With respect to nuclear waste management, an important point is that reprocessing, as currently practiced, does little or nothing to alleviate this management problem. For example, France practices a once-through recycling in which plutonium is separated once, made into MOX fuel, and the spent fuel containing this MOX is not usually recycled . . . (although France has done some limited recycling of MOX spent fuel). The MOX spent fuel is stored pending the further development and commercialization of fast reactors. But France admits that this full deployment of a fleet of fast reactors is projected to take place at the earliest by mid-century. France will shut down later this year its only fast reactor, the prototype Phénix. Perhaps around 2020, France may have constructed another fast reactor, but the high costs of these reactors have been prohibitive. In effect, France has shifted its nuclear waste problem from the power plants to the reprocessing plant.

France's practice of transporting plutonium hundreds of miles from the La Hague to the MOX plant at Marcoule poses a security risk. While there has never been a theft of plutonium or a major accident during the hundreds of trips to date, each shipment contains many weapons' worth of plutonium. Thus, just one theft of a shipment could be an international disaster.

No country has yet to open a permanent repository. But the country with the most promising record of accomplishment in this area is Sweden. A couple of weeks ago, Sweden announced the selection of its repository site but admits that the earliest the site will accept spent fuel is 2023. Sweden had carefully evaluated three different sites and obtained widespread community and local government involvement in the decision making process. France touts the benefits of the volume reduction of recycling in which highly radioactive fission products are formed into a glass-like compound, which is now stored at an interim storage site. By weight percentage, spent fuel typically consists of 95.6% uranium (with most of that being uranium-238), 3% stable or short-lived radioactive fission products, 0.3% cesium and strontium (the primary sources of high-level radioactive waste over a few hundred years), 0.1% long-lived iodine and technetium, 0.1% long-lived actinides (heavy radioactive elements), and 0.9% plutonium. But the critical physical factor for a repository is the heat load. For the first several hundred years of a repository the most heat emitting elements are the highly radioactive fission products. The benefit of a fast reactor recycling program could be the reduction or near elimination of the longer-lived transuranic elements that are the major heat producing elements beyond several hundred years.

Other countries may venture into reprocessing. Therefore, it is imperative for the United States to reevaluate its policies and redouble its efforts to prevent the further spread of reprocessing plants to non-nuclear-weapon states. In particular, the Republic of Korea is facing a crisis in the overcrowded conditions in the spent fuel pools at its power plants. One option is to remove older spent fuel and place it in dry storage casks, but the ROK government believes this option may cost too much because of the precedent set by the exorbitantly high price paid for a low level waste disposal facility. Another option is for the ROK to reprocess spent fuel. While this will provide significant volume reduction in the waste, it will only defer the problem to storage of MOX spent fuel, similar to the problem faced by France. This option will run counter to the agreement the ROK signed with North Korea in the early 1990s for both states to prohibit reprocessing or enrichment on the Korean Peninsula. A related option is to ship spent fuel to La Hague, but a security question is whether to ship plutonium back to the ROK. France would require shipment of the high level waste back to the ROK. Thus, the ROK will need a high level waste disposal facility. The main reason I raise this ROK issue at length is that the ROK and the United States have recently begun talks on the renewal of their peaceful nuclear cooperation agreement, which will expire in 2014. The United States has consent rights on ROK spent fuel because either it was produced with U.S.-supplied fresh fuel or U.S.-origin reactor systems. The ROK is seeking to have future spent fuel not subject to such consent rights by purchasing fresh fuel from other suppliers and by developing reactor systems that do not have critical components that are U.S.-origin or derived from U.S.-origin systems. The bottom line is that the United States is steadily losing its leverage with the ROK and other countries because of declining U.S. leadership in nuclear power plant systems and nuclear waste management.

Concerning lessons the United States can learn from other countries' nuclear waste management experience, the first lesson is that a fair political and sound scientific process is essential for selecting a permanent repository. Sweden demonstrates the effectiveness of examining multiple sites and gaining buy-in from the public and local governments. The second lesson is that reprocessing, as currently practiced, does not substantially alleviate the nuclear waste management problem. However, more research is needed to determine the costs and benefits of fast reactors for reducing transuranic waste. Any type of reprocessing will require safe and secure waste repositories.

While the United States investigates the costs and benefits of various recycling proposals through a research program, it has an opportunity now to exercise leadership in two waste management areas. First, as envisioned in GNEP, the United States should offer fuel leasing services. As part of those services, it should offer to take back spent fuel from the client countries. (Russia is offering this service to Iran's Bushehr reactor.) This spent fuel does not necessarily have to be sent to the United States. It could be sent to a third party country or location that could earn money for the spent fuel storage rental service. Spent fuel can be safely and securely stored in dry storage casks for up to 100 years. Long before this time ends, a research program will most likely determine effective means of waste management. The spent fuel leasing could be coupled to the second area where the United States can play a leadership role. That is, the United States can offer technical expertise and political support in helping to establish regional spent fuel repositories. A regional storage system would be especially helpful for countries with smaller nuclear power programs.

Recommendations

- Continue to discourage separation of plutonium from spent nuclear fuel.
- Limit the spread of reprocessing technologies to non-nuclear weapon states.
- Draw down the massive stockpile of civilian plutonium.
- Support a research program to assess the costs and benefits of various reprocessing technologies with attention focused on proliferation-resistance, safeguards, and nuclear waste management. Compare the costs and benefits of reprocessing to enrichment, factoring in the proliferation risks of both technologies.
- Increase funding for safeguards research.
- Promote safe and secure storage of spent fuel until the time when reprocessing may become economically attractive.
- Evaluate multiple sites for permanent waste repositories based on political fairness and sound scientific assessments. Obtain buy-in from the public and local governments.
- Use secure interim spent fuel storage employing dry storage casks to relieve build up on spent fuel pools.
- Provide fuel leasing services that would include take back of spent fuel to either the fuel supplier state or a third party.
- Develop regional spent fuel storage facilities.
- Obtain better estimates on the remaining global reserves of uranium.
- Provide research support for developing more efficient nuclear power plants that would produce more electrical power per thermal power than today's fleet of reactors. Similarly, research more effective ways to make more efficient use of uranium fuel and reduce the amounts of plutonium-239 produced.

EXPLORING THE ISSUE

Should the United States Reprocess Spent Nuclear Fuel?

Critical Thinking and Reflection

1. Why has fear of nuclear proliferation inhibited nuclear fuel reprocessing?
2. What are the advantages of putting off the decision to reprocess spent nuclear fuel until the necessary technology has been more fully developed?
3. What are the disadvantages of putting off the decision to reprocess spent nuclear fuel until the necessary technology has been more fully developed.
4. Economic incentives have been used to persuade people to accept nearby undesirable facilities such as dumps, power plants, and factories. Would it be ethical to do the same for a nuclear fuel reprocessing plant?

Is There Common Ground?

Both sides agree that nuclear fuel reprocessing can help support the use of nuclear power and reduce the nuclear waste problem. The major difference lies in whether the technology can be implemented immediately or must undergo further research and development. Present and under-development technologies are described on the Web site of the World Nuclear Association. Visit it at www.world-nuclear.org/info/inf69.html and answer the following questions:

1. What is the purpose of nuclear fuel reprocessing?
2. What is the dominant technology today?
3. What is transmutation and how does it reduce the nuclear waste problem?

Internet References . . .

The National Renewable Energy Laboratory

The National Renewable Energy Laboratory is the nation's primary laboratory for renewable energy and energy efficiency research and development. Among other things, it works on wind power and biofuels.

http://www.nrel.gov/learning/re_basics.html

American Wind Energy Association

The American Wind Energy Association is the national trade association for wind power project developers, equipment suppliers, services providers, parts manufacturers, utilities, researchers, and advocates.

http://www.awea.org/

Army Corps of Engineers

The Army Corps of Engineers discusses its role in the management of the nation's water resources on this site.

http://www.vtn.iwr.usace.army.mil/

American Rivers

American Rivers advocates for the protection of rivers as vital to our health, safety, and quality of life. It addresses pollution cleanup, dam removal, and heritage activities.

http://www.americanrivers.org

The DOE Idaho National Laboratory Geothermal Program

The mission of the Idaho National Laboratory (INL) Geothermal Program is to work in partnership with U.S. industry to establish geothermal energy as an economically competitive contributor to the U.S. energy supply.

https://inlportal.inl.gov/portal/server.pt?open=512&objID=422&parentname=CommunityPage&parentid=14&mode=2

UNIT 4

Alternative Energy Sources

Fossil fuels and nuclear power are not the only options. Alternatives discussed here are wind and hydroelectric power, biofuels and hydrogen. Others include tidal power, geothermal power, and solar power. All require funding to develop and deploy, and all have impacts on the environment and people nearby. Some enthusiasts tout the promise of various kinds of free or infinite energy, such as zero point and vacuum energy, but physicists warn that thanks to the laws of thermodynamics, "There Ain't No Such Thing as a Free Lunch." We need energy, but we will always have to pay for it.

ISSUE 15

Is Renewable Energy Green?

YES: Andrea Larson, from "Growing U.S. Trade in Green Technology," testimony before the U.S. House Committee on Energy and Commerce, Subcommittee on Commerce, Trade and Consumer Protection (October 7, 2009)

NO: Jesse H. Ausubel, from "Renewable and Nuclear Heresies," *International Journal of Nuclear Governance, Economy and Ecology* (vol. 1, no. 3, 2007)

Learning Outcomes

After studying this issue, students will be able to:

- Explain what "green" means.
- Describe the environmental impacts of renewable energy technologies.
- Explain how renewable energy technologies benefit the economy.
- Explain why nuclear power may be preferable to renewable energy technologies.

ISSUE SUMMARY

YES: Andrea Larson argues that "green" technologies include, among other things, renewable energy technologies and these technologies are essential to future U.S. domestic economic growth and to international competitiveness.

NO: Jesse Ausubel argues that renewable energy technologies are not green, largely because when developed to a scale at which they might contribute meaningfully to society's energy requirements, they will cause serious environmental harm. He considers nuclear power a much "greener" way to meet society's energy needs.

"Green" has long been understood to mean environmentally friendly. A green energy technology is sustainable. It is not based on fossil fuels—which is not quite the same as saying it is based on renewable energy sources—so it does not add carbon dioxide or other greenhouse gases to the atmosphere. It does not pollute air or water. It does not diminish biodiversity or harm ecosystems.

Examples of green energy technologies include solar power, wind power, biomass power, hydropower, tidal power, geothermal power, and wave power. The first two—solar and wind—are the ones most frequently discussed in magazines and newspapers, perhaps because they can be installed in units small enough to fit in a family's backyard. John Gulland and Wendy Milne describe in "Choosing Renewable Energy," *Mother Earth News* (April/May 2008), the process—and difficulties—of converting their home to run mostly on solar and wind power, with a wood stove (biomass power) for heat. "[U]sing renewables," they write, "has increased our independence and sense of security, and lessened our carbon emissions . . . another benefit is in knowing that we are contributing, even in a small way, to the health and sustainability of the local community."

Green energy technologies can also be scaled up to provide much larger amounts of electricity. An Australian study says they could provide 60 percent of Australia's electricity by 2040 and reduce associated carbon dioxide emissions by 78 percent; at present, Australia has more than 50 wind farms with about a thousand wind turbines. In Iceland, green energy technologies account for three quarters of the installed electrical generation capacity. In California, the corresponding figure is one quarter; in Sweden, one third; and in Norway, one half. Wind alone meets a fifth of Denmark's needs. See Rachel Sullivan and Mary-Lou Considine, "Hastening Slowly in the Global Renewables Race," *Ecos* (April–May 2010).

In the United States, green energy technologies play much less of a role. Hydropower, at 7 percent (77,000 MW) of national electrical generating capacity, is an important contributor; unfortunately, potential growth is very limited. Wind power is growing rapidly (see *Vital Signs 2010,* Worldwatch (2010)) and the potential is huge (see Issue 16), but at the end of 2009, it provided only 34,863 MW of electrical generating capacity, about 3 percent of the total (and only about 0.2 percent of actual electricity produced). Solar power plays an even smaller role, with just over 2000 MW installed; it too is growing rapidly. Wind and solar thus promise continued growth in manufacturing of equipment, jobs for installation, and displacement of fossil fuels.

Unfortunately, wind and solar power take up a lot of land. Wind farms cover miles of high or windy terrain with towers and spinning blades. One major trend is to move them offshore, where they are out of sight to most people (and the wind is steadier). Solar electric power requires large expanses of solar panels or concentrators, preferably in locations where the sky is rarely cloudy, meaning arid lands or deserts. There is in fact a plan afoot to develop solar power in the Sahara desert to supply Europe with up to 15 percent of its electricity as early as 2015. See Ashley Seager, "Solar Power from Sahara a Step Closer," *The Guardian* (November 1, 2009; www.guardian.co.uk/business/2009/nov/01/solar-power-sahara-europe-desertec), and Daniel Clery, "Sending African Sunlight to Europe, Special Delivery," *Science* (August 13, 2010). It has been noted that if just 1 percent of the Sahara were covered with solar concentrators, it could meet the electrical needs of the entire world.

Power lines can be run across or under the Mediterranean Sea to bring electricity from the Sahara to Europe. Getting it to the rest of the world is a more difficult problem. It thus seems sensible to install green energy technologies closer to where the energy will be used. In Ontario, Canada, the provincial government has embarked on a major effort to increase the green component of its energy supply. As one component of the effort, the Hay Solar company is offering to provide farmers with free barns—whose roofs are covered with solar cells. Hay calculates that the barns will provide a large area for solar energy collection and—despite many more cloudy days in Ontario than in the Sahara—pay for themselves over 20 years, after which they belong to the farmers. See Chris Sorensen, "Absolute Power?" *Maclean's* (June 14, 2010). Rooftop solar power has been discussed in the United States as well; it has a number of advantages—it does not interfere with other uses of land, has minimal red tape, and is close to end users. Southern California Edison has proposed generating 500 MW of electricity by scattering solar power units on rooftops in the region (see David Anthony, "Where Will Solar Power Plants Be Built—Deserts or Rooftops?" *Greentechmedia* (February 22, 2010; www.greentechmedia.com/articles/read/where-will-solar-power-plants-be-built-deserts-or-rooftops/).

It thus seems clear that green energy technologies can be deployed without necessarily having huge environmental impacts. The real question is whether we can deploy *enough* green energy technology to replace fossil fuels, and one major obstacle is the sheer scale of the need for energy; see Richard A. Kerr, "Do We Have the Energy for the Next Transition?" *Science* (August 13, 2010). Many people contend that nuclear power is much better suited to meeting future needs; see Charles Forsberg, "The Real Path to Green Energy: Hybrid Nuclear-Renewable Power," *Bulletin of the Atomic Scientists* (November/December 2009), and John Bradley, "The Nuclear Revivalist," *Popular Science* (July 2010). Allan Sloan and Marilyn Adamo, "If You Believe in Magic, Green Energy Will Be Our Salvation," *Fortune* (July 26, 2010), add that part of the problem in shifting to green energy is that we must start from such a small base. Another part is that we use so much energy for transportation.

Testifying at the October 7, 2009, hearing on "Growing U.S. Trade in Green Technology" before the U.S. House Committee on Energy and Commerce, Subcommittee on Commerce, Trade and Consumer Protection, Mary Saunders, Acting Assistant Secretary for Manufacturing and Services, International Trade Administration, U.S. Department of Commerce, says she sees increasing demand for green energy technologies providing great export opportunities and adds, "Policies that support the early development and commercialization of green technologies are critical to the competitiveness of U.S. firms and improve their competitive edge in the global marketplace." If we do not support such technologies, other nations will claim the lead and the economic benefits.

The current rapid growth of wind and solar power supports, in the YES selection, professor Andrea Larson's argument that green technologies—including, among other things, renewable energy technologies such as wind and solar power—are essential to future U.S. domestic economic growth and to international competitiveness. However, in the NO selection, Jesse Ausubel argues that renewable energy technologies are not really green, largely because when developed to a scale at which they might contribute meaningfully to society's energy requirements, they will cause serious environmental harm. He considers nuclear power a much "greener" way to meet society's energy needs.

Andrea Larson

Growing U.S. Trade in Green Technology

Green technology and clean commerce are the future. Green technology has become, and will increasingly be, a major economic growth area for the U.S. and world trade. There is no reason the U.S. cannot be a world leader through export of clean technology and clean commerce innovation, and U.S. leadership should be a strategic goal.

Why? Because:

1. Investing in clean energy and clean materials is essential for intelligent economic development, human health protection, and ecosystem preservation.
2. U.S. leadership in clean energy and materials (green technology) creates jobs, stimulates innovation, drives exports, and differentiates U.S. technology, education, and skills in global markets.
3. The U.S. could have an advantage in world trade, but on the current path the U.S. will continue to fall behind.

Green Tech and Clean Commerce Is the Future

Population and economic development pressures are colliding with the ability of nature to deliver clean air, water, and soil. Yet the design of the industrial system that brought us to this point in history was based on assumptions of limitless resources and limitless capacity for natural system regeneration, even in the face of our waste streams. Responding to climate change and green tech opportunities are just the beginning of a major shift in this century for business. New design for business is imperative because the forces of change are accelerating.

It is not just the current economic downturn that confounds us. We face unacceptable income and opportunity disparities at home and poverty worldwide as global population grows from 6.5 to 9 billion in the next few decades. Worldwide over 2 billion people are moving rapidly into the middle class, and they will want all the opportunities and material wealth that the richest populations in western societies now view as normal. Today we concurrently face an economic downturn, a climate crisis, an energy security crisis, energy price volatility, new environmental health challenges, and ecological systems in dramatic decline.

U.S. House of Representatives, October 7, 2009.

If that were not enough, the U.S. also faces a competitiveness crisis as it loses ground to other countries that are already strategically committed to mobilizing state resources behind domestic businesses that will produce solutions to these problems. Other countries have mounted national efforts to reach clean commerce goals (e.g. renewable energy, domestic "green" companies, dramatic efficiencies, accelerating advances in PV solar design innovation, advancing clean public transportation, protecting consumers from toxic materials, and providing subsidies and incentives to advance their industries in global markets).

The larger picture shows capitalism as currently designed is at a crossroads. It must deliver on its promise of broad prosperity, yet its very design appears to undermine the ecological systems and healthy communities on which it depends. It needs an overhaul: clean energy and materials provide an answer. The U.S. should be leading this change, not following. . . .

Defining Green Technology

Green technology is one term of several used today to encompass a range of activity and innovation to simultaneously address economic development needs, health protection, and preservation of ecosystem services (e.g. the natural systems that provide us with clean air, water, soil, and food). Other terms include sustainability, clean commerce, cleantech, sustainable business, and sustainability innovation. The activities these terms reference challenge existing ways of designing and delivering not just energy, but the entire set of interdependent systems and supply chains that provide food, shelter, consumer products, and transportation modes.

We will use the abbreviation GT/CC throughout this testimony to refer to green tech and clean commerce, two terms that represent the ideas under discussion.

GT/CC refers to technology innovation, but also non-technical innovation, the latter represented by innovative supply chain management or innovative financing mechanisms to install urban PV solar installations that pay residents to sell excess electricity back to the grid. The non-technical innovative frontier must also be a focus for green tech and clean commerce innovation and U.S. competitiveness.

Furthermore, GT/CC is not just about energy. The fundamental basis of commerce and trade is energy AND materials. Both must be managed and designed to meet human needs and optimize ecological system functions. Thus green chemistry and green engineering practices are equally as important to green tech and clean commerce (GT/CC) as renewable energy technologies. PV solar systems that expose their production workers to toxins, are thrown away in landfills after use, then pollute water supplies, are not the solutions we need. "Fresh" vegetables and fruit grown with agricultural chemicals, processed, and transported thousands of miles and lacking fundamental nutrients that urban garden-grown food provides are not the solutions we need. More efficient lighting replacements that create mercury waste may save energy but are still poor designs. In other words, poorly thought out, so-called green technology improvements focused on today's hot topics (climate and energy are the focus today) are common. But a deeper design perspective is needed. First, a systems view is required. One that understands every "green" energy solution, in fact every energy AND product selection by a company or a consumer, reflects materials choices and embedded energy decisions that must be made visible, examined and evaluated for their life cycle implications. Fortunately this is now happening, led by innovative entrepreneurs. But it must be expanded and accelerated.

Nor is green technology just about efficiency. It is about that, but more importantly it is about innovation. Efficiency just allows us to do the same old things at lower cost and using less energy and fewer materials. A laudable improvement, but not the solution. Innovation creates fundamentally new solutions, preferably systems-oriented solutions that prevent and eliminate the problems we face now with climate alteration and unsafe products.

The concept that ties together innovation and both clean energy and materials is the notion of cradle to cradle design. Our current commercial practices extract raw materials, make products, generate waste streams that impact air and water, expose production workers, sell to consumers who use the products and throw them away, and leave the materials to decompose and contaminate our air and water from the landfill, incinerator or Third World country dumping destination. Think about how the costs and benefits are allocated in this linear system. This is called a cradle to grave product life cycle. The alternative is cradle to cradle design derived from systems thinking, that reduces or eliminates energy and material inputs, including toxicity *by design from the outset* to avoid employee, user/consumer, and ecosystem contamination. Under a cradle to cradle design, selected materials can be safely returned to the earth or maintained within closed recycling systems that use waste from one production and use process, as the feedstock for another.

The "greentech" issues or what I am calling the green technology and clean commerce issues (GT/CC) constitute a central challenge for governments. Providing ever growing volumes of products and services (under current design parameters) to support economic development also gives us pollution and costs that are externalized (and inequitably so) onto the population in one form or another (higher taxes for regulation, disease, and more expensive health insurance for chronic illnesses). Examples are air pollution (excessive concentrations of toxins in the air contributing to the asthma epidemic, among other respiratory problems), unsafe foods (linked to diabetes, obesity, and food contamination), excessive carbon dioxide concentration in the atmosphere (climate change and volatility), and water supply threats and shortages due to industrial contamination.

As world population rises to 9 billion in the next few decades and capitalism as currently designed stumbles in its promise of greater prosperity and results instead in wealth creation accompanied by income disparities, climate change, and waste streams increasingly tied to chronic human health challenges, a clean commerce solution is emerging. This is an alternative approach to business that we call green technology and clean commerce. This movement is obvious in the current emphasis on clean energy alternatives in response to climate change. . . .

U.S. Competitiveness

Transformation in the next decade to an alternative mindset about energy and materials is key to U.S. competitiveness and mandatory if global society is to handle the challenges of population growth, energy demands, and material throughput volumes required to provide prosperity for billions more people. We can choose to let others lead or we can mobilize and combine all the elements we have in this country to lead.

This discussion acknowledges that the U.S. has declared 25% renewable energy goals by 2025 with the February 2009 ARRA legislation. The clean technology stimulus accounts for about $66 billion, just ahead of China's stimulus investment. The important fact, nonetheless, is that we come to the table late. By way of example, according to the U.S. International Trade Commission, "Denmark, Germany, India, Japan, and Spain accounted for a combined 91 percent of global exports of wind-powered generating sets in 2008."

Globally, investments in GT/CC have been growing rapidly. For instance, new investments in sustainable energy increased between 25% and 73% annually from 2002 to 2007, until growth fell to only 5% in 2008 following the 2007–08 recession. Nonetheless, even in 2008, total investments in sustainable energy projects and companies reached $155 billion, with wind power representing the largest share at $51.8 billion. Meanwhile, the world's 12 major economic stimulus packages proposed to invest another $180 billion collectively in coming years. Also in 2008, sustainability-focused companies as identified by the Dow Jones Sustainability Index or Goldman Sachs SUSTAIN list outperformed their industries by 15% over a six-month period. Longer horizon analyses indicate companies screened for sustainability factors match or exceed the performance of conventional firms. These are companies that focus not only on renewable energy sources but also energy conservation, environmentally safer products, and improved corporate governance.

Despite being a leader in some areas, however, the U.S. was not an overall leader in GT/CC. From 2000 to 2008, venture capital investments in U.S.-based renewable energy companies increased from 0.6% of all VC investments to 11.84%, and in 2008, venture capital and private equity made new investments in energy efficiency and renewable energy worth $7.72 billion in North America and $3.05 billion in Europe. Moreover, the U.S. had the most GT/CC business incubators in 2008, with 56. The UK was next in incubators with 21, and 16 were in Germany. Yet Europe as whole was home to 46% of the global total of incubators, versus 40% for the U.S. Furthermore, North American investments in sustainable energy shrank 8% in 2008 to $30.1 billion, while in Europe they increased 2% to $49.7 billion. Many other major emerging economies also saw investments in their renewable energy sectors increase: Brazil's increased 76% to $10.8 billion (mainly due to ethanol), China's increased 18% to $15.6 billion, and India's increased 12% to $3.7 billion. Even in Spain investments reached $17.4 billion in 2008, or $430 per capita compared to North America's $57 per capita. For investments specifically in publically traded renewable energy and efficiency companies, Chinese companies led in 2008 with $2.8 billion, followed by Portugal ($2.6 billion), the U.S. ($2.1 billion), and Germany ($1.5 billion). In fact, in 2008, China became the world's largest manufacturer of photovoltaic panels, with 95% of them destined for export. This output means China may soon surpass both German and American manufacturers.

Indeed, China has recently made massive moves toward a CT/CC economy. For instance, China now has 60% of the total global capacity for solar thermal water heaters. Even such a relatively minor innovation saved 3 million tons of oil equivalent in 2006 according to the International Energy Agency. China is also nurturing and protecting its domestic wind power producers, reserving contracts for them and restricting foreign firms. The size of China's market for GT/CC creates significant opportunities for development of domestic innovators and mass producers. Nonetheless, China has a way to go: other countries have put themselves into leadership positions over the past two decades through a series of policies. Those world leaders have been Japan, Denmark, Spain, and Germany.

In 1996, Japan set a target by 2010 of using 3% (roughly 19 gigaliters oil equivalent) of primary energy supply from renewable sources excluding hydropower and geothermal energy. In 2008, the target was amended to represent an upper bound while 15.1 Gl was established as a lower bound. That goal plus grants for residential solar PV installations allowed Japan to lead the world in installed solar capacity from 1999 to 2005, which also allowed Japanese companies such as Sharp to gain an early manufacturing lead. Sharp and other Japanese companies remain competitive in the U.S. market to this day, even though Germany overtook Japan in installed capacity in 2006. In 2007, Japan established Renewable Portfolio Standards that required utilities to use renewable sources of electricity generation, to reach 16 TWh by 2014. The RPS also set prices for solar PV rates, and in December 2008, Japan allocated another $9 billion for solar subsidies, which is less than California's current solar subsidy program but reaches more eligible people. Japan continues to invest in solar research, including space-based solar energy.

Denmark began to shape its lead in GT/CC in 1976, when its Energy Research Program granted generous subsidies to renewable energies. Danish renewable energy companies turned heavily toward wind power, selling that technology domestically and abroad, especially in California. In 1989, new laws required utilities to buy electricity from renewable sources and co-generation plants, and a series of subsidies and other government support boosted GT/CC through the 1990s. By 2003, Denmark dominated the global market for wind-power generator sets, selling $966 million or 79.5% of the market. Denmark still gets a larger share of its energy from wind than any other country and sold $1.2 billion worth of generator sets in 2008, or 23.4% of the global market. Meanwhile, Danish Vestas controls 17.8% of the wind turbine market, putting Danish companies behind Germany and ahead of the U.S., Spain, and China in that field. In 2008, the Danish government's Agreement on Energy Policy sets goals of 20% of gross energy consumption from renewable sources by 2011, with incentives for de-centralized production, research, and other activity.

On the other side of Europe, Spain had a mere 979 GWh of renewable energy generation, almost all of it hydro-electric, in 1990. Yet in 2007, that same generation had risen 33-fold to 32,714 GWh, with wind accounting for about two-thirds of total. A series of steps similar to those in Japan and Denmark led to this rapid rise, which has ultimately left Spain a major force in the world's solar and wind energy markets. Spain's 1980 Law for the Conservation of Energy first established subsidies for renewable energy sources feeding into grid. In 1997, the Law of the Electricity Sector guaranteed grid access for renewable sources and later laws set prices as well as targets, such as 12% of energy from renewable sources by 2010. With this support, Spain ranked third globally in 2008 in installed wind capacity with 16.8 GW and controlled 8.8% of the market for wind generator sets and 14.9% for turbines. It has also been a leader in solar thermal plants, building Europe's first in 2007 and continuing to develop others.

Germany, finally, has achieved some of the broadest, most profound changes en route to a GT/CC economy. It reached its Kyoto Protocol emissions target of a 20% reduction of GHG emissions from 1990 levels in 2007, a year early. A series of policies has enabled this progress, such as the 1991 Feed-in Tariff Act that required utilities to purchase electricity from any supplier on the grid. Later laws, such as the 2000 Renewable Energies Act and its subsequent updates, have guaranteed prices for renewable energies and set broad environmental targets. Germany in 2009 set even more ambitious plans for reducing overall emissions and dependence on fossil fuels. . . .

In 2008 in Germany, revenue from construction of renewable energy facilities was 13.1 billion Euros (approximately $19.7 billion) and from operation was 15.7 billion Euros ($23.6 billion), representing approximately 278,000 jobs in all. The total revenue from these two activities increased 188% relative to 2003. Meanwhile, the German government's Market Incentive Program, through grants and other incentives, encourages renewable energies by direct funding, which attracts additional investment. From 2000 to 2008, 1.2 billion Euros of direct funding attracted an additional 8.6 billion Euros of outside investment, with government funding for renewable energy R&D directed mainly to solar and wind. The results have been a near quintupling of electricity generated from renewable sources since 1990. In contrast, U.S. government subsidies totaled $29 billion from 2002–2008 for renewable energies, more than half for corn ethanol, which paled in comparison to $72 billion in subsidies for fossil fuels.

What you see when reviewing different countries' strategies is policy variation customized to local conditions but built upon a consistent pattern of core features that includes protections to control consumer costs and mitigation for windfall profits to any players. Simplicity is important to keep public administration costs low and company and individual transaction costs minimal. Consistent policies, gradual amendments to update, and stable supports (whether direct investments or tax incentives) are essential to encourage equipment manufacturers to innovate and to mass produce. Clear and consistent signals also reassure investors that markets will be relatively predictable within adequate time frames for generating returns. In summary, successful government policies appear to include key stakeholders and set ambitious targets, and then address concerns about price-gouging and the factors that typically drive innovators and companies away: instability, uncertainty, and inconsistency.

The U.S. can catch up, but when other countries are working from 20 year-plus guaranteed grid access for renewable energy producers in Spain and Germany (starting in 1991 in Germany) and well-established Spanish Feed-In Tariffs (TIFs) that built on German and Danish examples established well over a decade ago, it suggests the magnitude of the catch up challenge. These countries jumped in early, learned and adapted, and can now act faster and more effectively to build their CT/CC going forward. For the huge and rapidly growing markets for GT/CC in India and China, the U.S. faces governments quickly moving to protect and support fledgling industries that will produce clean cars and public transportation technologies to address pollution impacts, clean energy production (to offset reliance on dirty coal), and the state of the art green components and systems to address the many development and pollution/health problems they know they must solve.

Final Thoughts

The economic growth paradigm and accompanying common knowledge that told us growth had to come first, followed only much later by investment in environmental and health protection (the path of western industrialized societies) will not be sufficient for India and China. I tell my MBA students that given the pace of innovation in those countries around clean commerce goals, the U.S. will be buying most of its clean technology solutions from Indian and Chinese companies in 10 years.

I would also suggest that the U.S.'s geopolitical decline, should it come to pass, will be reflected in our unwillingness to step up to the GT/CC challenge that current population, resource, pollution, and technology development conditions impose.

I am not an advocate of government regulation unless the private sector lacks the ability to provide for the public good. Unfortunately, companies trying to move toward GT/CC, while admirable, are in a race against the cumulative decisions of firms and individuals that continue to erode the commons that is our ultimate source of all wealth, social and financial.

We tend to think of the commons as natural systems (air, water, or land); we might want to consider adding our children's bodies to that collective commons. The Centers for Disease Control [and Prevention] extensive research on contaminants in human blood, immune, and reproductive systems suggest that this century long industrial experiment that clearly has had decisive negative influences on our ecological systems and atmosphere, is also at work on the human body and children's health. Are we surprised?

The last thing I want to see is unnecessary regulation. I work with private sector innovators and emphasize the amazing capacity of markets and entrepreneurial forces in society to create the changes we need to see. But this activity must be framed with enabling and supporting policy that sets the rules and provides consistent and intelligent guidance so that markets and human ingenuity can do the rest.

In addition, let us keep in mind, in the polarized and ideologically laced discussions that pass for policy debate, that there are no purists. State subsidies and consistent long-term government support for fossil fuels played a large part in giving us the energy and materials system we live with today. Subsidies, just in recent years alone, explain why GT/CC activities remain vulnerable and investment capital moves slowly.

Can the U.S. build a GT/CC strategy? Through insufficient investment and lack of policy leadership the U.S. continues to lose ground in its learning pace and its domestic experience to countries willing to back their companies with capital and create mutually reinforcing incentives to mobilize citizen behavior, corporate investment, education, and state decision making. While the hesitancy of the U.S. to create industrial policy to lead in GT/CC is historically understandable, other countries without our political and ideological history (and gridlock) have put policies in place. First we must get our own house in order. It is only then that we will have built the necessary platform for leadership in world trade.

The challenge is straightforward, if ambitious. Future prosperity depends on economic development solutions that address poverty and extreme disparities in income distribution while simultaneously delivering on job creation, skill development, and education for the future. Industrial and commercial activity that fails to actively support provision of clean, healthy products, and clean air, water, shelter, transport, and food, by definition undermines that prosperity. Fortunately the know-how and tools are now available in the form of GT/CC practices and innovation. . . .

Jesse H. Ausubel

Renewable and Nuclear Heresies

Introduction

Heretics maintain opinions at variance with those generally received. Putting heretics to death, hereticide, is common through history. In 1531 the Swiss Protestant heretic Huldreich Zwingli soldiering anonymously in battle against the Catholic cantons was speared in the thigh and then clubbed on the head. Mortally wounded, he was offered the services of a priest. His declination caused him to be recognised, whereupon he was killed and quartered, and his body parts mixed with dung and ceremonially burned. Recall that the first heresy against the Roman Church in Switzerland in 1522 was the eating of sausages during Lent, and the signal heresy was opposition to the baptism of children. As nuclear experts know deeply, humans are not rational in their beliefs, actions or reactions.

I will offer both renewable and nuclear heresies. I trust readers will not commit hereticide. Because culture defines heresies, readers coming from a nuclear tribe will probably applaud my renewable heresies and grumble about the nuclear. While my heresies may not rival favouring polygamy or sharing all worldly goods, they will disturb many. My main heresies are that renewable sources of energy are not green and that the nuclear industry should make a product beside electricity.

Decarbonisation

The dogma that gives me conviction to uphold heresies is decarbonisation, which I accept as the central measure of energy evolution. Consider our hydrocarbon fuels as blends of carbon and hydrogen, both of which burn to release energy. Molecules of the main so-called fossil fuels, coal, oil and natural gas, each have a typical ratio of carbon to hydrogen atoms. Methane, CH_4, is obviously 1 to 4. An oil such as kerosene is 1 to 2. A typical coal's ratio of C:H is about 2 to 1. Importantly, coal's precursor, wood, has an even more primitive C:H ratio, 10 to 1, once the moisture is removed. Carbon blackens miners' lungs, endangers urban air and threatens climate change. Hydrogen is as innocent as an element can be, ending combustion as water.

Suppose we placed all the hydrocarbon fuels humanity used each year since about 1800, when British colliers first mined thousands of tons of coal,

From *International Journal of Nuclear Governance, Economy and Ecology,* vol. 1, no. 3, 2007, pp. 229–235, 236–238, 240–242.

in a blender, mixed them, and plotted the yearly ratio of carbon to hydrogen. While the trend may waver for a decade or two, over the long term H gains in the mix at the expense of C, like cars replacing horses, colour TV substituting for black-and-white, or email gaining the market over hard copies sent through the post office. The consequent decarbonisation is the single most important fact from 30 years of energy studies.

When my colleagues Cesare Marchetti, Nebojsa Nakicenovic, Arnulf Grubler and I discovered decarbonisation in the 1980s, we were pleasantly surprised. When we first spoke of decarbonisation, few believed and many ridiculed the word. Everyone 'knew' the opposite to be true. Now prime ministers and presidents speak of decarbonisation. Neither Queen Victoria nor Abraham Lincoln decreed a policy of decarbonisation. Yet, the energy system pursued it. Human societies pursued decarbonisation for 170+ years before anyone noticed. By the way, another of my heresies is the belief that much of the time politicians pull on disconnected levers.

Returning to carbon, if world economic production or all energy rather than all hydrocarbons form the denominator, the world is also decarbonising, that is, using less carbon per dollar of output or kilowatt. Moreover, China and India as well as France and Japan decarbonise. The slopes are quite similar, though China and India lag by several decades, as they do in the diffusion of other technologies besides energy. Economically and technically, carbon seems fated to fade gradually over this century. By 2100 we will feel nostalgia for carbon as some do now for steam locomotives. Londoners have mythologised their great fogs, induced by coal as late as the 1950s, and Berliners already reminisce about the 'East Smell' of burnt lignite whose use collapsed after the fall of The Wall in 1989.

The explanation for the persistence of decarbonisation is simple and profound. The overall evolution of the energy system is driven by the increasing spatial density of energy consumption at the level of the end user, that is, the energy consumed per square metre, for example, in a city. Finally, fuels must conform to what the end user will accept, and constraints become more stringent as spatial density of consumption rises. Rich, dense cities accept happily only electricity and gases, now methane and later hydrogen. These fuels reach consumers easily through pervasive infrastructure grids, right to the burner tip in your kitchen.

My city, New York, by the way, already consumes in electricity alone on a July day about 15 watts per square metre averaged over its entire 820 square kilometres of land, including Central Park.

A few decades ago, some visionaries dreamed of an all-electric society. Today people convert about 35–40% of all primary fuel to electricity. The fraction will rise, but now even electricity enthusiasts (as I am) accept that finally not much more than half of all energy is likely to be electrified. Reasons include the impracticality of a generating system geared entirely to the instant consumption of energy and lack of amenability of many vehicles to reliance on electricity. Surrendering the vision of an all-electric society is a minor nuclear heresy.

Ultimately the behaviour of end-users drives the system. Happily, the system can thus be rational even when individuals are not. When end-users want electricity and hydrogen, over time the primary energy sources that can produce on the needed scale while meeting the ever more stringent constraints that attend growth in turn will win. Economies of scale are a juggernaut over the long run. Think, for better or worse, of Walmart stores.

Appropriately, the historical growth of world primary energy consumption over the past 150 years shows rises in long waves of 50–60 years, each time formed around the development of a more desirable source of energy that scaled up readily. Coal lifted the first wave, and oil the second. A new growth wave is underway, lifted by methane, now almost everyone's favourite fuel and a subject to which I will return later.

According to the historical trend in decarbonisation, large-scale production of carbon-free hydrogen should begin about the year 2020. So how will humanity keep lifting electricity production while also introducing more H_2 into the system to lift the average above the norm of methane? The obvious competitors are nuclear and the so-called renewables, the false and minor, yet popular, idols.

Renewable Heresies

Let's consider the renewable idols: hydro, biomass, wind and solar. As a Green, I care intensely about land-sparing, about leaving land for Nature. In fact, a Green credo is 'No new structures.' Or, in milder form, 'New structures or infrastructures should fit within the footprint of the old structures or infrastructures.' So, I will examine renewables primarily by their use of land.

In the USA and much of the rest of the world, including Canada, renewables mean dammed rivers. Almost 80% of so-called US renewable energy is hydro, and hydro generates about 60% of all Canada's electricity.

For the USA as a whole, the capacity of all existing hydropower plants is about 97,500 MWe, and their average production is about 37,500 MWe. The average power intensity—the watts divided by the land area of the USA—is 0.005 watts per square metre, that is, the approximate power that can be obtained from a huge tract of land that drains into a reservoir for a power station.

Imagine the entire province of Ontario, about 900,000 square km, collecting its entire 680,000 billion litres of rain, an average annual rainfall of about 0.8 m. Imagine collecting all that water, every drop, behind a dam of about 60 metres height. Doing so might inundate half the province, and thus win the support of the majority of Canadians, who resent the force of Ontario. This comprehensive 'Ontario Hydro' would produce about 11,000 MW or about four fifths the output of Canada's 25 nuclear power stations, or about 0.012 watts per square metre or more than twice the USA average. In my 'flood Ontario' scenario, a square kilometre would provide the electricity for about 12 Canadians.

This low density and the attending ecological and cultural headaches explain the trend in most of the world from dam building to dam removal. About 40% of Canada's immense total land area is effectively dammed for electrons already. The World Commission on Dams issued a report in November

2000 that essentially signalled the end of hydropower development globally. While the Chinese are constructing more dams, few foresee even ten thousand megawatts' further growth from hydropower.

Though electricity and hydrogen from hydro would decarbonise, the idol of hydro is itself dammed. Hydro is not green.

In the USA, after hydro's 80% comes biomass's 17% of renewables. Surprisingly, most of this biomass comes, not from backyard woodsmen or community paper drives, but from liquors in pulp mills burned to economise their own heat and power. In terms of decarbonisation, biomass of course retrogresses, with 10 Cs or more per H.

If one argues that biomass is carbon-neutral because photosynthesis in plants recycles the carbon, one must consider its other attributes, beginning with productivity of photosynthesis. Although farmers usually express this productivity in tons per hectare, in the energy industry the heat content of the trees, corn and hay instead quantify the energy productivity of the land. For example, the abundant and untended New England or New Brunswick forests produce firewood at the renewable rate of about 1200 watts (thermal) per hectare averaged around the year. The 0.12 watts per square metre of biomass is about ten times more powerful than rain, and excellent management can multiply the figure again ten times

Imagine, as energy analyst Howard Hayden has suggested, farmers use ample water, fertiliser, and pesticides to achieve 12,000 watts *thermal* per hectare (10,000 square metres). Imagine replacing a 1000 MWe nuclear power plant with a 90% capacity factor. During a year, the nuclear plant will produce about 7.9 billion kWh. To obtain the same electricity from a power plant that burns biomass at 30% heat-to-electricity efficiency, farmers would need about 250,000 hectares or 2500 square kilometres of land with very high productivity. Harvesting and collecting the biomass are not 100% efficient; some gets left in fields or otherwise lost.

Such losses mean that in round numbers a 1000 MWe nuclear plant equates to more than 2500 square kilometres of prime land. A typical Iowa county spans about 1000 square kilometres, so it would take at least two and a half counties to fire a station. A nuclear power plant consumes about ten hectares per unit or 40 hectares for a power park. Shifting entirely from baconburgers to kilowatts, Iowa's 55,000 square miles might yield 50,000 MWe. Prince Edward Island might produce about 2000 MWe.

The USA already consumes about ten and the world about 40 times the kilowatt hours that Iowa's biomass could generate. Prime land has better uses, like feeding the hungry. Ploughing marginal lands would require ten or 20 times the expanse and increase erosion. One hundred twenty square metres of New Brunswick or Manitoba might electrify one square metre of New York City.

Note also that pumping water and making fertiliser and pesticides also consume energy. If processors concentrate the corn or other biomass into alcohol or diesel, another step erodes efficiency. Ethanol production yields a tiny net of 0.05 watts per square metre.

As in hydro, in biomass the lack of economies of scale loom large. Because more biomass quickly hits the ceiling of watts per square metre, it can become more extensive but not cheaper. If not false, the idol of biomass is not sustainable on the scale needed and will not contribute to decarbonisation. Biomass may photosynthesise but it is not green.

Although, or because, wind provides only 0.2% of US electricity, the idol of wind evokes much worship. The basic fact of wind is that it provides about 1.2 watts per square metre or 12,000 watts per hectare of year-round average electric power. Consider, for example, the $212 million wind farm about 30 kilometres south of Lamar, CO, where 108 1.5 MWe wind turbines stand 80 metres tall, their blades sweeping to 115 metres. The wind farm spreads over 4800 hectares. At 30% capacity, peak power density is the typical 1.2 watts per square metre.

One problem is that two of the four wind speed regimes produce no power at all. Calm air means no power of course, and gales faster than 25 metres per second (about 90 kilometres per hour) mean shutting down lest the turbine blow apart. Perhaps three to ten times more compact than biomass, a wind farm occupying about 770 square kilometres could produce as much energy as one 1000 MWe nuclear plant. To meet 2005 US electricity demand of about four million MWhr with around-the-clock-wind would have required wind farms covering over 780,000 square kilometres, about Texas plus Louisiana, or about 1.2 times the area of Alberta. Canada's demand is about 10% of the USA and corresponds to about the area of New Brunswick.

For linear thinkers, a single file line of windmills has a power density of about 5 kilowatts per metre. If Christo could string windmills single file along Rocky Mountain ridges half way from Vancouver to Calgary, about 1200 km, the output would be about the same as one of the four Darlington CANDU units.

Rapidly exhausted economies of scale stop wind. One hundred windy square metres, a good size for a Manhattan apartment, can power a lamp or two, but not the clothes washer and dryer, microwave oven, plasma TVs or computers or dozens of other devices in the apartment, or the apartments above or below it. New York City would require every square metre of Connecticut to become a wind farm if the wind blew in Hartford as in Lamar. The idol of wind would decarbonise but will be minor.

Although negligible as a source of electric power today, photovoltaics also earn a traditional bow. Sadly, PVs remain stuck at about 10% efficiency, with no breakthroughs in 30 years. Today performance reaches about 5–6 watts per square metre. But no economies of scale inhere in PV systems. A 1000 MWe PV plant would require about 150 square kilometres plus land for storage and retrieval. Present USA electric consumption would require 150,000 square kilometres or a square almost 400 kilometres on each side. The PV industry now makes about 600 metres by 600 metres per year. About 600,000 times this amount would be needed to replace the 1000 MWe nuclear plant, but only a few square kilometres have ever been manufactured in total.

Viewed another way, to produce with solar cells the amount of energy generated in one litre of the core of a nuclear reactor requires one hectare of solar cells. To compete at making the millions of megawatts for the baseload of the world energy market, the cost and complication of solar collectors still need to shrink by orders of magnitude while efficiency soars.

Extrapolating the progress (or lack) in recent decades does not carry the solar and renewable system to market victory. Electrical batteries, crucial to many applications, weigh almost zero in the global energy market. Similarly, solar and renewable energy may attain marvellous niches, but seem puny for providing the base power for 8–10 billion people later this century.

While I have denominated power with land so far, solar and renewables, despite their sacrosanct status, cost the environment in other ways as well. The appropriate description for PVs comes from the song of the Rolling Stones, 'Paint It Black'. Painting large areas with efficient, thus black, absorbers evokes dark 19th century visions of the land. I prefer colourful desert to a 150,000 km^2 area painted black. Some of the efficient PVs contain nasty elements, such as cadmium. Wind farms irritate with low-frequency noise and thumps, blight landscapes, interfere with TV reception, and chop birds and bats. At the Altamont windfarm in California, the mills kill 40–60 golden eagles per year. Dams kill rivers.

Moreover, solar and renewables in every form require large and complex machinery to produce many megawatts. Berkeley engineer Per Petersen reports that for an average MWe a typical wind-energy system operating with a 6.5 metres-per-second average wind speed requires construction inputs of 460 metric tons of steel and 870 cubic metres of concrete. For comparison, the construction of existing 1970-vintage US nuclear power plants required 40 metric tons of steel and 190 cubic metres of concrete per average megawatt of electricity generating capacity. Wind's infrastructure takes five to ten times the steel and concrete as that of nuclear. Bridging the cloudy and dark as well as calm and gusty weather takes storage batteries and their heavy metals. Without vastly improved storage, the windmills and PVs are supernumeraries for the coal, methane and uranium plants that operate reliably round the clock day after day.

Since 1980 the US DOE alone has spent about $6 billion on solar, $2 billion on geothermal, $1 billion on wind and $3 billion on other renewables. The nonhydro renewable energy remains about 2% of US capacity, much of that the wood byproducts used to fuel the wood products industry. Cheerful self-delusion about new solar and renewables since 1970 has yet to produce a single quad of the more than 90 quadrillion Btu of the total energy the US now yearly consumes. In the 21 years from 1979 to 2000 the percentage of US energy from renewables actually fell from 8.5 to 7.3%. Environmentally harmless increments of solar and renewable megawatts look puny in a 20 or 30 million megawatt world, and even in today's 10 million megawatt world. If we want to scale up, then hydro, biomass, wind, and solar all gobble land from Nature. Let's stop sanctifying false and minor gods and heretically chant 'Renewables are not Green'.

Nuclear Heresies

How then can we meet more stringent consumer demands and stay on course for decarbonisation? The inevitable reply is nuclear energy. I should mention that I am not naïve about nuclear. Privileged to work with Soviet colleagues who participated in the Chernobyl clean-up, I saw the Dead Zone in 1990 with my own eyes. I visited the concrete sarcophagus encasing the blasted reactor with employees of the site management enterprise. But I will not offer heresies here about safety, waste disposal and proliferation, though important heresies exist, particularly about waste disposal.

Rather, my first nuclear heresy is that nuclear must ally with methane. Electric utilities that operate nuclear plants often embrace another source of power, for example, coal or hydro. Yet, importantly, the sheltering wing of methane will help nuclear grow again. The biggest fact of the energy system over the next 20 to 30 years will continue to be a massive expansion of the gas system, methane for the present, and many people may feel more comfortable with the addition of nuclear power plants if they know that methane, an attractive fuel in many ways, is taking the overall energy lead. To stay on track in decarbonisation, methane must and will prevail. Were I a businessman, I would want to ally with a winner, and methane will prosper in the market. . . .

Anyway, for business to continue as usual, by 2020 the reference point for the world's energy will be CH_4, methane. Still, energy's evolution should not end with methane. The completion of decarbonisation ultimately depends on the production and use of pure hydrogen. In the 1970s journalists called hydrogen the Tomorrow Fuel, and critics have worried that hydrogen will remain forever on the horizon, like fusion. For hydrogen tomorrow is now today. Hydrogen is a thriving young industry. World commercial production in 2002 exceeded 40 billion standard cubic feet per day, equal to 75,000 MW if converted to electricity, and USA production, which is about a third of the world's, multiplied tenfold between 1970 and 2003. Over 16,000 kilometres of pipeline transport H_2 gas for big users, with pipes at 100 atmospheres as long as 400 kilometres from Antwerp to Normandy. High pressure containers such as tube trailers distribute the liquid product to small and moderate users throughout the world. With production experience, the hydrogen price is falling.

The fundamental question then becomes, from where will large quantities of cheap hydrogen come? Methane and water will compete to provide the hydrogen feedstock, while methane and nuclear will compete to provide the energy needed to transform the feedstock.

Steam reforming of methane to produce hydrogen is already a venerable chemical process. Because methane abounds, in the near term steam reforming of methane, using heat from methane, will remain the preferred way to produce hydrogen. Moreover, because much of the demand for hydrogen is within the petrochemical industry, nepotism gives methane an edge. Increasingly, as new applications such as fuel cells demand hydrogen, nuclear's chance to compete as the transformer improves.

My next heresy is that the production of hydrogen will revolutionise the economics of nuclear power much more than standardising plants or building plants quicker. Firstly, hydrogen manufacture allows nuclear plants to address the half of energy demand that will not be met by electricity. Secondly, it gives nuclear power plants the chance to make valuable product 24 hours per day. Recall that a great problem the electric power industry faces is that, notwithstanding the talk of the '24/7 society', electric power demand remains asymmetrical. Users demand most electricity during the day. So, immense capital sits on its hands between about 9 O'clock at night and 6 or 7 O'clock in the morning. Turning that capital into an asset is incredibly valuable. Like the hotel and airline industries, the power industry would rather operate at 90% capacity than 60% capacity. The nuclear industry is limited to providing baseload electric power unless it reaches out to hydrogen to store and distribute its tireless energy.

While I stated earlier that methane and nuclear compete, they can also cooperate in the hydrogen market. Let's accept that in the near term steam reforming of methane will dominate hydrogen making. Nuclear power as well as methane can provide the energy for the reforming. Here let me share a big technological idea, methane-nuclear-hydrogen (MNH) complexes, first sketched by Cesare Marchetti. An enormous amount of methane travels through a few giant pipeline clusters, for example, from Russia through Ukraine and Slovakia. These methane trunk routes are attractive places to assemble MNH industrial complexes. Here, if one builds a few nuclear power plants and siphons off some of the methane, the nuclear plants could profitably manufacture large amounts of hydrogen that could be re-introduced into the pipelines, say up to 20% of the composition of the gas. This decarbonisation enhances the value of the gas. Meanwhile, the carbon separated from the methane becomes CO_2 to be injected into depleted oil and gas fields and profitably help with tertiary recovery. The hydrogen mixture could be distribute around Europe, or the world, getting users accustomed to the new level of decarbonisation.

Over the next 10–15 years, I will keep my eye on the places where much gas flows and see whether these regions initiate this next generation energy system. Alberta is an obvious locale, especially when methane from the Mackenzie Delta flows through it. The experience of working with hydrogen from methane will benefit the nuclear industry as it put nuclear plants at the nodes of the webs of hydrogen distribution, anticipating the shift from CH_4 to H_2O as a feedstock. The methane-nuclear-hydrogen complexes can be the nurseries for the next generation of the energy system.

The surprising longevity of nuclear power plants, observed by Alvin Weinberg, spurs me to look beyond the imminent methane era to complete decarbonisation. Nuclear energy's long-range potential is unique as an abundant, scalable source of electricity and for water-splitting while the cities sleep. . . .

By now, I have revealed my final heresy, that nuclear is green. An American now yearly emits about five tons of carbon per year or 14 kg per day. Globally each year humanity already produces carbon waste measuring about 15 cubic kilometres, a very large refuse bag. The volume of nuclear wastes is usually measured in litres. By weight, a 1000 MWe light water reactor that produces energy for one million typical homes produces approximately 1080 kg of fission products per year, 4 milligrams per person. A large apartment building housing 500 people would produce annually high-level radioactive waste equal to a small jar of aspirin tablets.

Over 500 years, in a fully nuclear world the high level radioactive wastes might amount to 700 million tonnes, less than the 800 million tons of coal

Americans burn in one year to produce half of USA electricity. Hayden calculates that all the reactors from 500 years of production of 100% of the world's energy could be stacked one high in an area of a little over 250 square kilometres, about the land area for a solar farm to provide 1000 MW of power. I recur to scale. Compact enough to grow, nuclear is green.

Conclusion

Let's return to the heart of energy evolution, decarbonisation. Because hydrogen is much better stuff for burning than carbon, the hydrocarbons form a clear hierarchy. Methane tops the ranking, with an energy density of about 55 megajoules per kilo, about twice that of black coal and three times that of wood.

The energy density of nuclear fuel is 10,000 or even 100,000 times as great as methane. While the full footprint of uranium mining might add a few hundred square kilometres, the dense heart of the atom still has much to offer. The extraordinary energy density of nuclear fuel allows compact systems of immense scale, and finally suits the ever higher spatial density of energy consumption at the level of the end user, logically matching energy consumption and production.

During the past 100 years motors have grown from 10 kilowatts to more than 1000 megawatts, scaling up an astonishing 100,000 times, while shrinking sharply in size and cost per kilowatt. A mere 1.5% per year growth of total energy demand during the 21st century, about two-thirds the rate since 1800, will multiply demand for primary energy to make the electricity and hydrogen from the 13 million MW years in 2002 to 50 million in 2100. If size and power, of individual machines or the total system, grow in tandem, use of materials and land and other resources becomes unacceptably costly. Technologies succeed when economies of scale form part of their conditions of evolution. Like computers, to grow larger, the energy system must now shrink in size and cost. Considered in watts per square metre, nuclear has astronomical advantages over its competitors.

You might well wonder whether we need DO anything. Decarbonisation appears automatic. At one level this is true. Yet, we also know that the trend of decarbonisation is the outcome of all the blood, sweat and tears of persistent workers, engineers, managers, investors, regulators and consumers. If people stop bleeding, sweating, and crying, the game producing decarbonisation could just stop. Without heretics, there are no schisms.

And energy offers ample room for heresies. I have mentioned several, some large like reading the Bible in one's own tongue, and some small like sausages on Lent:

- decarbonisation has proceeded for almost two centuries and without a policy for it
- renewables are not green . . .
- utilities should embrace nuclear together with methane
- nuclear plants must diversify to make hydrogen as well as electricity
- nuclear is green.

I hope readers will not toss offending documents I have written on a public bonfire or, worse yet, quarter and immolate me like the Swiss heretic Zwingli. Rather, keep in mind the meaning of the Greek word from which heresy derives. The word means to take for oneself or choose.

Received, widely held doctrines may be wise and right. But history, including the history of science, is littered with doctrines discarded as delusions. At present, my conviction is that our best energy doctrine is decarbonisation, and let us complete it within one hundred years or sooner. Wishful thinking holds that the way is by returning to a renewable Eden. Resisting wishful thinking requires courage. Even the courageous Zwingli wrote in the margin of his copy of St. Augustine's *City of God,* 'Ah God, if only Adam had eaten a pear'.

EXPLORING THE ISSUE

Is Renewable Energy Green?

Critical Thinking and Reflection

1. What does it mean to be "green"?
2. Which is the better reason for promoting the development of an energy technology—impact on the environment or impact on domestic economic growth?
3. How could society diminish its energy needs to make the problem of supplying enough energy more manageable?
4. If we choose to build huge arrays of solar power collectors, where should we put them—in deserts, on rooftops, or at sides of skyscrapers?

Is There Common Ground?

In the debate over whether renewable energy technologies are truly "green," both sides agree that we need to replace fossil fuels. Unfortunately, there is no easy way to do that, for every energy technology has environmental impacts. But solar and wind power are only two of several renewable energy technologies. Investigate geothermal power (Issue 19 in this book), wave power (www.energysavers.gov/renewable_energy/ocean/index.cfm/mytopic=50009), and ocean thermal energy conversion (OTEC; see http://www.nrel.gov/otec/what.html) and answer the following questions:

1. Which of these three energy technologies has the smallest "footprint" on the landscape?
2. In what ways are these three energy technologies likely to interfere with human activities?
3. How large a contribution to society's energy needs can these three energy technologies make?

ISSUE 16

Is Wind Enough?

YES: Xi Lu, Michael B. McElroy, and Juha Kiviluoma, from "Global Potential for Wind-Generated Electricity," *Proceedings of the National Academy of Sciences* (July 7, 2009)

NO: John Etherington, from *The Wind Farm Scam: An Ecologist's Evaluation* (Stacey International, 2009)

Learning Outcomes

After studying this issue, students will be able to:

- Describe the potential contribution of wind power to meeting society's energy needs.
- Describe the drawbacks of wind power.
- Explain why many say that wind power needs other sources of energy for backup.
- Describe an ideal location for a wind farm.
- List alternatives to wind power.

ISSUE SUMMARY

YES: Xi Lu, Michael McElroy, and Juha Kiviluoma argue that a network of land-based 2.5 MW wind turbines operating at as little as 25 percent of rated capacity would be more than enough to meet total current and anticipated future global demand for electricity. In the contiguous United States, the potential is enough to supply more than 16 times current consumption. Offshore turbines add to the potential.

NO: John Etherington argues that wind power has been vastly oversold. It cannot provide a predictable electrical supply, saves remarkably little on carbon emissions, is not cheap, and has a huge landscape footprint. We would be better off without it.

For centuries, windmills have exploited the pressure exerted by blowing wind to grind grain and pump water. On U.S. farms, before rural electrification programs brought utility power, windmills provided small amounts of electricity.

These windmills fell out of favor when the powerlines arrived, bringing larger amounts of electricity more reliably (even when the wind was not blowing). The sources of electricity became hydroelectric dams, coal- and oil-fired power plants, and even nuclear power plants.

The oil crisis of the 1970s brought new attention to wind and other renewable energy sources. Today, high oil prices and concern about global warming caused by the carbon dioxide emitted by the burning of fossil fuels are renewing that attention. Researchers are looking for ways to reduce carbon emissions (see Jay Apt, David W. Keith, and M. Granger Morgan, "Promoting Low-Carbon Electricity Production," *Issues in Science and Technology* (Spring 2007)). Low-carbon techniques include capturing carbon dioxide instead of releasing it, biofuels, nuclear power, hydroelectric power, wind power, and geothermal power, all of which play minor roles at present. At present, "the electric power industry is the single largest emitter of carbon dioxide in the United States, accounting for 40% of emissions of CO_2 in 2006, up from 36% in 1990 and 25% in 1970."

Wind power growth suffered with the economy in 2010, but until then growth was rapid. According to Janet L. Sawin, "Wind Power Increase in 2008 Exceeds 10-Year Average Growth Rate," *Vital Signs 2010* (Worldwatch Institute, 2010), global wind power capacity increased 27,051 MW in 2008, bringing the global total to 120,798 MW. "Wind now generates more than 1.5 percent of the world's electricity, up from 0.1 percent in 1997." The United States took first place for new installations, with U.S. capacity increasing by 50 percent. Projections for future growth are healthy, with one source expecting wind to supply as much as 6 percent of the global total of electricity by 2017. Yet, these advances are not without resistance. Wind turbines take up space and affect views. They provide fatal obstacles to birds and bats. Although recent work suggests that birds apparently learn to avoid windmills, bats may actually be attracted to them; once bats are close enough, the low-pressure vortices created by moving windmill blades may rupture blood vessels in their lungs, killing them. Since bats are less active when wind speeds are high, one solution may be to turn windmills on only when wind speeds are higher. See Catherine Brahic, "Wind Turbines Make Bat Lungs Explode," *New Scientist* (August 25, 2008) (www.newscientist.com/article/dn14593-wind-turbines-make-bat-lungs-explode.html?feedId=online-news_rss20).

Wind turbines also make noise, and some people find that noise objectionable (see Johann Tasker, "Wind Farm Noise Is Driving Us Out of Our House," *Farmers Weekly* (January 12, 2007)). Jon Boone, in "Wayward Wind," a speech given in the Township of Perry, near Silver Lake, Wyoming County, New York (June 19, 2006), agrees and argues that wind power is better for corporate tax avoidance than for providing environmentally friendly energy. It is at best a placebo for our energy dilemma. "The only environmentally responsible short-range solution to the problem of our dependence upon fossil fuels must combine effective conservation with much higher efficiency standards."

Yet, wind turbines do generate electricity without burning fossil fuels and emitting CO_2 (or other pollutants). The question is whether they offer enough advantages to be worth their price. Charles Komanoff, "Whither Wind?" *Orion*

(September–October 2006), thinks they are. He argues that the energy needs of civilization can be met without adding to global warming if we both conserve energy and deploy large numbers of wind turbines. Acceptance of wind farms, he says, could be "our generation's way of avowing our love for the next." The statistics suggest that we are well on our way.

Despite the rapid growth of wind power, the debate over its benefits is far from over. Yet, the debate centers more on industrial-scale wind farms, which deploy large numbers of large wind turbines, than on small home-scale systems. According to Jennifer Alsever, "Wind Power, the Home Edition," *Business 2.0* (January/February 2007), a dozen U.S. companies are offering wind turbines to be installed next to private homes. The Skystream costs about $10,000. Glen Salas, "A Great Wind Is Blowing," *Professional Builder* (January 2007), says that home builders can cater to "a lot of people [who] are willing to spend the extra money to generate their own electricity" and notes that "Over its life, a small residential wind turbine can offset about 1.2 tons of air pollutants and 200 tons of greenhouse gases, according to the American Wind Energy Association" (which offers fact sheets and more at www.awea.org).

A more novel proposal is to use kites that could lift wind turbines into the jet stream, where the wind is very fast; see "Getting Wind Farms off the Ground," *Economist* (June 9, 2007). Another proposal, described by Todd Woody in "Tower of Power," *Business 2.0* (August 2006), calls for a 1600-foot-tall tower that will exploit the temperature difference between the bottom and the top to set a powerful wind flowing up the tower. EnviroMission (www.enviromission.com.au/EVM/content/home.html), the Australian company behind the idea, says the tower will yield 50 MW of electricity, and larger versions are in the planning stages.

Like standard wind farms, both kites and towers will surely provoke protests. It is worth bearing in mind that no matter how humanity generates energy in the future, there will be environmental impacts. The only way to avoid that is to stop using energy, which is not a practical proposal. On the other hand, reducing energy use through conservation and efficiency is practical, and should surely be an essential part of the solution. One of the primary exponents of efficiency is Amory B. Lovins, whose "More Profit with Less Carbon," *Scientific American* (September 2005), urges that "Using energy more efficiently offers an economic bonanza—not because of the benefits of stopping global warming but because saving fossil fuel is a lot cheaper than buying it." Lovins is profiled by Elizabeth Kolbert in "Mr. Green," *New Yorker* (January 22, 2007).

In the YES selection, Xi Lu, Michael McElroy, and Juha Kiviluoma argue that a network of land-based 2.5 MW wind turbines operating at as little as 25 percent of rated capacity would be more than enough to meet total current and anticipated future global demand for electricity. In the contiguous United States, the potential is enough to supply more than 16 times current consumption. Offshore turbines add to the potential. In the NO selection, John Etherington argues that wind power has been vastly oversold. It cannot provide a predictable electrical supply, saves remarkably little on carbon emissions, is not cheap, and has a huge landscape footprint. We would be better off without it.

Xi Lu, Michael B. McElroy, and Juha Kiviluoma

Global Potential for Wind-Generated Electricity

Wind power accounted for 42% of all new electrical capacity added to the United States electrical system in 2008 although wind continues to account for a relatively small fraction of the total electricity-generating capacity [25.4 gigawatts (GW) of a total of 1,075 GW]. The Global Wind Energy Council projected the possibility of a 17-fold increase in wind-powered generation of electricity globally by 2030. Short et al., using the National Renewable Energy Laboratory's WinDs model, concluded that wind could account for as much as 25% of U.S. electricity by 2050 (corresponding to an installed wind capacity of ≈300 GW).

Archer and Jacobson estimated that 20% of the global total wind power potential could account for as much as 123 petawatt-hours (PWh) of electricity annually [corresponding to annually averaged power production of 14 terawatts (TW)] equal to 7 times the total current global consumption of electricity (comparable to present global use of energy in all forms). Their study was based on an analysis of data for the year 2000 from 7,753 surface meteorological stations complemented by data from 446 stations for which vertical soundings were available. They restricted their attention to power that could be generated by using a network of 1.5-megawatt (MW) turbines tapping wind resources from regions with annually averaged wind speeds in excess of 6.9 m/s (wind class 3 or better) at an elevation of 80 m. The meteorological stations used in their analysis were heavily concentrated in the United States, Europe, and Southeastern Asia. Results inferred for other regions of the world are subject as a consequence to considerable uncertainty.

The present study is based on a simulation of global wind fields from version 5 of the Goddard Earth Observing System Data Assimilation System (GEOS-5 DAS). Winds included in this compilation were obtained by retrospective analysis of global meteorological data using a state-of-the-art weather/climate model incorporating inputs from a wide variety of observational sources, including not only surface and sounding measurements as used by Archer and Jacobson but also results from a diverse suite of measurements and observations from a combination of aircraft, balloons, ships, buoys, dropsondes and satellites, in short the gamut of observational data used to provide the world with the best possible meteorological forecasts enhanced by application of these data in a retrospective analysis. The GEOS-5 wind field is currently available for the period 2004 to the present (March 20, 2009) with plans to extend the analysis 30 years back in time. The GEOS-5 assimilation was adopted in the present analysis to take advantage of the relatively high spatial resolution available with this product as compared with the lower spatial resolutions available with alternative products such as ERA-40, NECP II, and JRA-25. It is used here in a detailed study of the potential for globally distributed wind-generated electricity in 2006.

. . . The land-based turbines envisaged here are assumed to have a rated capacity of 2.5 MW with somewhat larger turbines, 3.6 MW, deployed offshore, reflecting the greater cost of construction and the economic incentive to deploy larger turbines to capture the higher wind speeds available in these regions. In siting turbines over land, we specifically excluded densely populated regions and areas occupied by forests and environments distinguished by permanent snow and ice cover (notably Greenland and Antarctica). Turbines located offshore were restricted to water depths <200 m and to distances within 92.6 km (50 nautical miles) of shore.

These constraints are then discussed, and results from the global analysis are presented followed by a more detailed discussion of results for the United States. . . .

Geographic Constraints

. . . Wind speeds are generally lower over forested areas, reflecting additional surface roughness. Consequently, turbines would have to be raised to a higher level in these environments to provide an acceptable economic return. Although it might be reasonable for some regions and some forest types, we elected for these reasons to exclude forested areas in the present analysis.

The exclusion of water-covered areas is more problematic. Wind speeds are generally higher over water as compared with land. However, it is more expensive to site turbines in aquatic as compared with terrestrial environments. Public pressures in opposition to the former are also generally more intense, at least in the U.S.

Topographic relief data for both land and ocean areas were derived from the Global Digital Elevation Model (GTOPO30) of the Earth Resources Observation and Science Data Center of the U.S. Geological Survey. The spatial resolution of this data source for offshore environments (bottom topography) is $\approx$1 km $\times$ 1 km. A number of factors conspire to limit the development of offshore wind farms. Aesthetic considerations, for example, have limited development of wind resources in the near-shore environment in the U.S. although objections to near-shore development in Europe appear to have been less influential. There is a need to also accommodate requirements for shipping, fishing, and wildlife reserves and to minimize potential interference with radio and radar installations. Accounting for these limitations, Musial and Butterfield and Musial, in a study of offshore wind power potential for the contiguous U.S., chose to exclude development of wind farms within 5 nautical miles (nm) (9.3 km) of shore and restrict development to 33% of the available area between 5 and 20 nm (9.3–37 km) offshore, expanding the potential area available to 67% between 20 and 50 nm (37–92.6 km).

For purposes of this study, following Dvorak et al., we consider 3 possible regimes for offshore development of wind power defined by water depths of 0–20, 20–50, and 50–200 m. Somewhat arbitrarily, we limit potential deployment of wind farms to distances within 50 nm (92.6 km) of the nearest shoreline, assuming that 100% of the area occupied by these waters is available for development.

Wind Power Potential Worldwide

Approximately 1% of the total solar energy absorbed by the Earth is converted to kinetic energy in the atmosphere, dissipated ultimately by friction at the Earth's surface. If we assume that this energy is dissipated uniformly over the entire surface area of the Earth (it is not), this would imply an average power source for the land area of the Earth of $\approx 3.4 \times 10^{14}$ W equivalent to an annual supply of energy equal to 10,200 quad [10,800 exajoules (EJ)], $\approx$22 times total current global annual consumption of commercial energy. Doing the same calculation for the lower 48 states of the U.S. would indicate a potential power source of 1.76×10^{13} W corresponding to an annual yield of 527 quad (555 EJ), some 5.3 times greater than the total current annual consumption of commercial energy in all forms in the U.S. Wind energy is not, however, uniformly distributed over the Earth and regional patterns of dissipation depend not only on the wind source available in the free troposphere but also on the frictional properties of the underlying surface.

We focus here on the potential energy that could be intercepted and converted to electricity by a globally distributed array of wind turbines, the distribution and properties of which were described above. Accounting for land areas we judge to be inappropriate for their placement (forested and urban regions and areas covered either by water or by permanent ice), the potential power source is estimated at 2,350 quad (2,470 EJ). . . . We restricted attention in this analysis to turbines that could function with capacity factors at or >20%.

. . . Table 1 presents a summary of results for the 10 countries identified as the largest national emitters of CO_2. The data included here refer to national reporting of CO_2 emissions and electricity consumption for these countries in 2005. An updated version of the table would indicate that China is now the world's largest emitter of CO_2, having surpassed the U.S. in the early months of 2006. Wind power potential for the world as a whole and the contiguous U.S. is summarized in Table 2.

The results in Table 1 indicate that large-scale development of wind power in China could allow for close to an 18-fold increase in electricity supply relative to consumption reported for 2005. The bulk of this wind power, 89%, could be derived from onshore installations. The potential for wind power in the U.S. is even greater, 23 times larger than current electricity consumption, the bulk of which, 84%, could be supplied onshore. Results for the contiguous U.S. will be discussed in more detail in the next section. If the top 10 CO_2 emitting countries were ordered in terms of wind power potential, Russia would rank number 1, followed by Canada with the U.S. in the third position. There

Table 1

Annual Wind Energy Potential, CO_2 Emissions, and Current Electricity Consumption for the Top 10 CO_2-Emitting Countries

Country	CO_2 Emission, Million Tonnes	Electricity Consumption, TWh	Potential Wind Energy, TWh	
			Onshore	Offshore
U.S.	5,956.98	3,815.9	74,000	14,000
China	5,607.09	2,398.5	39,000	4,600
Russia	1,696.00	779.6	120,000	23,000
Japan	1,230.36	974.1	570	2,700
India	1,165.72	488.8	2,900	1,100
Germany	844.17	545.7	3,200	940
Canada	631.26	540.5	78,000	21,000
U.K.	577.17	348.6	4,400	6,200
S. Korea	499.63	352.2	130	990
Italy	466.64	307.5	250	160

CO_2 emission and electricity consumption are for 2005; data are from the Energy Information Administration (http://tonto.eia.doe.gov/country/index.cfm).

Table 2

Annual Wind Energy Potential Onshore and Offshore for the World and the Contiguous U.S.

Areas	Worldwide, PWh		Contiguous U.S., PWh	
	No CF Limitation	20% CF Limitation	No CF Limitation	20% CF Limitation
Onshore	1,100	690	84	62
Offshore 0–20 m	47	42	1.9	1.2
Offshore 20–50 m	46	40	2.6	2.1
Offshore 50–200 m	87	75	2.4	2.2
Total	1,300	840	91	68

Analysis assumes loss of 20% and 10% of potential power for onshore and offshore, respectively, caused by inter-turbine interference. Analysis assumes offshore siting distance within 50 nm (92.6 km) of the nearest shoreline.

is an important difference to be emphasized, however, between wind power potential in the abstract and the fraction of the resource that is likely to be developed when subjected to realistic economic constraints. Much of the potential for wind power in Russia and Canada is located at large distances from population centers. Given the inevitably greater expense of establishing wind farms in remote locations and potential public opposition to such initiatives, it would appear unlikely that these resources will be developed in the near term. Despite these limitations, it is clear that wind power could make a significant contribution to the demand for electricity for the majority of the countries listed in Table 1, in particular for the 4 largest CO_2 emitters, China, the U.S., Russia, and Japan. It should be noted, however, the resource for Japan is largely confined to the offshore area, 82% of the national total. To fully exploit these global resources will require inevitably significant investment in transmission systems capable of delivering this power to regions of high load demand. . . .

Wind Power Potential for the United States

An estimate of the electricity that could be generated for the contiguous U.S. on a monthly basis (subject to the siting and capacity limitations noted above) . . . were computed by using wind data for 2006. Not surprisingly, the wind power potential for both environments is greatest in winter, peaking in January, lowest in summer, with a minimum in August. Onshore potential for January . . . exceeds that for August by a factor of 2.5: the corresponding ratio computed for offshore locations is slightly larger, 2.9.

. . . Demand for electricity exhibits a bimodal variation over the course of a year with peaks in summer and winter, minima in spring and fall. Demand is greatest in summer during the air-conditioning season. Summer demand exceeds the minimum in spring/fall demand typically by between 25% and 35% on a U.S. national basis depending on whether summers are unusually warm or relatively mild. The correlation between the monthly averages of wind power production and electricity consumption is negative. Very large wind power penetration can produce excess electricity during large parts of the year. This situation could allow options for the conversion of electricity to other energy forms. Plug-in hybrid electric vehicles, for example, could take advantage of short-term excesses in electricity system, while energy-rich chemical species such as H_2 could provide a means for longer-term storage.

Potential wind-generated electricity available from onshore facilities on an annually averaged state-by-state basis . . . in the central plains region extending northward from Texas to the Dakotas, westward to Montana and Wyoming, and eastward to Minnesota and Iowa . . . is significantly greater than current local demand. Important exploitation of this resource will require, however, significant extension of the existing power transmission grid. Expansion and upgrading of the grid will be required in any event to meet anticipated future growth in electricity demand. It will be important in planning for this expansion to recognize from the outset the need to accommodate contributions of power from regions rich in potential renewable resources, not only wind but also solar. The additional costs need not, however, be prohibitive. The Electric Reliability Council of Texas, the operator responsible for the bulk of electricity transmission in Texas, estimates the extra cost for transmission of up to 4.6 GW of wind-generated electricity at $\approx$\$180 per kW, $\approx$10% of the capital cost for installation of the wind power-generating equipment.

An important issue relating to the integration of electricity derived from wind into a grid incorporating contributions from a variety of sources relates to the challenge of matching supply with load demand, incorporating a contribution to supply that is intrinsically variable both in time and space and subject to prediction errors. This challenge can be mitigated to some extent if the variations of wind sources contributing to an integrated transmission grid from different regions are largely uncorrected. An anomalously high contribution from one region can be compensated in this case by an anomalously low contribution from another.

. . . [Our] analysis suggests that wind power could make a relatively reliable contribution to anticipated base load demand in winter. It may be more difficult to incorporate wind power resources into projections of base load demand for other seasons, particularly for summer.

Concluding Remarks

The GEOS-5 winds used here were obtained through assimilation of meteorological data from a variety of sources, in combination with results from an atmospheric general circulation model. Transport in the boundary layer was treated by using 2 different formalisms, one applied under conditions when the boundary layer was stable, the other under conditions when the boundary layer was either unstable or capped by clouds. The variation of wind speed with altitude was calculated in the present study by using a cubic spline fit to the 3 lowest layers (central heights of 71, 201, and 332 m) of the GEOS-5 output to estimate wind speeds at the rotor heights of the turbines considered here (100 m). Wind speeds so calculated were used in deriving all of the results presented above.

The rotors of the turbines modeled in this study are of sufficient size that as the blades rotate they traverse significant portions of the 2 lowest layers of the GEOS-5-simulated atmosphere. Use of wind speed for a single level (100 m) must be consequently subject to some uncertainty. To assess this uncertainty we explored results derived with an alternate approach. The power intercepted by the blades of the rotors passing through the separate layers was calculated initially on the basis of the reported average wind speeds for the involved layers. Adopting a typical value of ≈135 m for the height of the boundary between the first 2 layers, given a rotor diameter of 100 m as appropriate for the assumed onshore turbines, it follows that 99% of the area swept out by the rotors would intercept air from the first layer, with only 1% encountered in the second layer. The power intercepted by the rotors may be calculated in this case by averaging appropriately the power intercepted in the 2 layers. Implementing this approach yielded results that differed typically slightly lower, by $<15\%$ for the onshore results presented above, by $<7\%$ for the offshore results.

The GEOS-5 data had a spatial resolution of ≈66.7 km × 50.0 km. It is clear that wind speeds can vary significantly over distances much smaller than the resolution of the present model in response to changes in topography and land cover (affected in both cases by variations in surface roughness). In general, we expect the electricity yield computed with a low-resolution model to underestimate rather than overestimate what would be calculated by using a higher-resolution model. The GEOS-5 data are expected to provide a useful representation of winds on a synoptic scale as required for example to describe the transport between adjacent grid elements. They would not be expected to account for subgrid scale variations in wind speeds even though the latter might be expected, at least under some circumstances, to make a significant contribution to the potentially available wind power. To test this hypothesis we explored the implications of a high-resolution wind atlas available for an altitude of 100 m for Minnesota. Wind speeds indicated by the high-resolution database are higher than the wind speeds indicated by GEOS-5, supporting our hypothesis. The close association of wind speed with surface land classification implied by the high-resolution Minnesota wind atlas suggests that land classification data could provide a useful basis for at least a preliminary downscaling of the relatively coarse spatial resolution of the potential wind resources in the present study.

We elected in this study to exclude forested, urban, permanently ice covered, and inland water regions. Given the relatively coarse spatial resolution of the GEOS-5 database, it is possible that this approach may have failed to identify localized environments where wind resources may be unusually favorable and where investments in wind power could provide an acceptable economic return. To explore this possibility, we developed a global land-based map of the efficiencies with which turbines with rotors centered at 100 m might be capable of converting wind energy to electricity. We included all land areas with the exception of regions identified as permanently ice-covered (notably Greenland and Antarctica). . . . Regions with particularly favorable capacity factors, even though forested, urban, or occupied by extensive bodies of inland waters, might be considered as potential additional targets for development.

It is apparent, for example, that the low-resolution GEOS-5 record underestimates the wind resource available in Spain and Portugal (a consequence most likely of the complex terrains present in these regions). Sweden is another example where wind resources indicated with an available high-resolution wind atlas are significantly higher than those implied by GEOS-5. The discrepancy in this case may be attributed to the extensive forest cover of the region and the a priori decision to neglect such regions in the present global study. Assessment of the potential of mountainous or hilly regions is also problematic. On average, wind speeds in these regions may be relatively low. Particularly favorable conditions may exist, however, on mountain ridges or in passes through mountainous regions. The Appalachian mountain range in the U.S. offers a case in point. In general the low-resolution results tend to slightly overestimate wind resources in regions of flat terrain, while underestimating the potential for regions defined by more complex topography.

The analysis in this article suggests that a network of land-based 2.5-MW turbines operating at as little as 20% of rated capacity, confined to nonforested, ice-free regions would be more than sufficient to account for total current and anticipated future global demand for electricity. The potential for the contiguous U.S. could amount to >16 times current consumption. Important additional sources of electricity could be obtained by deploying wind farms in near-shore shallow water environments.

An extensive deployment of wind farms may be considered as introducing an additional source of atmospheric friction. For example, if the entire current demand for electricity in the U.S. were to be supplied by wind, the sink for kinetic energy associated with the related turbines would amount to ≈6% of the sink caused by surface friction over the entire contiguous U.S. land area, 11% for the region identified as most favorable for wind farm development. . . . The potential impact of major wind electricity development on the circulation of the atmosphere has been investigated in a number of recent studies. Those studies suggest that high levels of wind development as contemplated here could result in significant changes in atmospheric circulation even in regions remote from locations where the turbines are deployed. They indicate that global dissipation of kinetic energy is regulated largely by physical processes controlling the source rather than the sink. An increase in friction caused by the presence of the turbines is likely to be compensated by a decrease in frictional dissipation elsewhere. Global average surface temperatures are not expected to change significantly although temperatures at higher latitudes may be expected to decrease to a modest extent because of a reduction in the efficiency of meridional heat transport (offsetting the additional warming anticipated for this environment caused by the build-up of greenhouse gases). In ramping up exploitation of wind resources in the future it will be important to consider the changes in wind resources that might result from the deployment of a large number of turbines, in addition to changes that might arise as a result of human-induced climate change, to more reliably predict the economic return expected from a specific deployment of turbines.

John Etherington

The Wind Farm Scam: An Ecologist's Evaluation

Introduction

"He who refuses to do arithmetic is doomed to talk nonsense."

(John McCarthy, computer pioneer, Stanford University)

Electricity generated by modern industrial wind turbines has many failings, all of which can be traced back to the physical laws which govern air movement and limit the energy that can be extracted from the wind. The huge size of the machines which makes them so inappropriate in the countryside is also a consequence of those same physical laws.

Some of the biggest wind turbines on land are now almost 600 feet tall—about one and a half times the height of Salisbury Cathedral spire, the tallest in Britain, which Constable painted so many times. It is their huge size and constant movement which raises the ire of landscape protection groups and it is their visual impact on the countryside which influences the average person, if only because they see a potential effect on the value of their homes.

One of the Frequently Asked Questions on the British Wind Energy Association (BWEA)'s website is

> Why don't they make turbines that look like old fashioned windmills?

Well, that would need wind turbines to be a quarter or less of the present height and somehow avoid the mass-produced "plastic and steel" look, not to mention dispensing with the "aircraft propeller"! That the question has been asked at all is acknowledgement of a real visual problem, despite a page heading elsewhere on the BWEA website:

> Wind Farms and Tourism. A valuable addition to the landscape.

The answer is that if the wind is to give significant amounts of power there is no option to these huge machines, because so little energy can be carried by moving air which is very light. The available energy from wind or water power is a product of the mass of air or water passing through the rotor per unit time. Doubling the circular area of a rotor doubles the amount of air or water which can pass through it. Air is about one 800th the density of water which is why a tiny old fashioned waterwheel mill could grind as much grain as a 100 foot traditional windmill. For the same power output, a windmill's sails need to encompass almost a thousand times the circular area of the equivalent waterwheel—a similar relationship exists between a modern hydroelectric turbine rotor and that of a wind turbine. This is because the energy harvested is, at a given air or water flow-rate, proportional to the mass of air passing through the rotor. It is an inevitable outcome of physical laws and powerful wind turbines can never grow smaller.

Really gigantic wind turbines like the 600 foot behemoths can generate 6 megawatts (MW) of power but only when it is very windy—a Beaufort scale "Near Gale" and about the wind speed at which our TV weather ladies start warning us about gusts and driving conditions. Six megawatts by the way is about half of one per cent of the power output of a large "real" power station—a pretty poor return for effort! If this optimum wind speed is halved to a comfortable breeze, the output will fall to much less than half of the maximum 6.0 MW—less than 1.0 MW. This is the consequence of another unchangeable physical law. The energy carried by a moving fluid such as water or air is proportional to the flow-rate cubed—that is halving the flow rate gives one eighth the available energy (½ × ½ × ½). . . .

Weather forecasting as we all know to our cost is a thankless task and yet electricity supply systems have to be predictable so that generation can be kept instantaneously in balance with consumption (if not there is a loud "bang" and the lights go out). We have all heard that "tomorrow's winds will be near gale force" only to wake to a gentle breeze. Pity the owner of the much subsidised wind turbine whose income has swooped with the falling wind! Indeed because the owner cannot predict how much electricity he can generate tomorrow, he would not be able to sell at all in a competitive free market. It is only the compulsion to purchase renewable electricity imposed by governments on the distribution networks, and the consumer-sourced subsidies, which allow the wind power industry to exist at all.

The problem of unpredictable wind and unpredictable wind electricity supply arises from the chaotically turbulent nature of the atmosphere and the longer-term swirls of global weather systems. In many places, particularly the windward shores of large oceans, it is possible for weather fronts and large shifts of wind velocity to pass through the few hundred miles of ocean fringe in less than half a day and at other times anticyclones of colossal size may become stationary for days on end bringing cloudless, wind free hot weather in summer and similar but freezing cold conditions in winter. The North Atlantic coast of Europe is just such a place and it is here that the great wind farms of Denmark, Germany and the UK are situated.

Over the past three decades, wind power seems to have appeared from nowhere, to dominate the world of power generation. This is hardly because it provides a lot of electricity and yet it has become a source of vigorous support by extremists, public dissent and political coercion. Entering "wind power" in a Google search gives about 20 million hits (as I write)! And yet worldwide, wind provides just one or two per cent of power. Try a search for "electricity generation" and there are 1.5 million hits—a "spin disproportion" I think.

In the UK during 2007 wind power gave just 1.3% of our electricity compared with a total of 91% from fossil fuels and nuclear. Of the other renewable sources of power, hydroelectricity gave 1.2% and various forms of biomass combustion about 2.3%. Wind power has however grown very rapidly—more than sixfold in the UK, from less than 0.2% of supply in the late 1990s. With the exceptions of a few countries which have a huge hydroelectric resource or with unusually high reliance on nuclear power, these percentages and growth rates are more or less mirrored in many other western developed countries.

Why has there been such a growth in renewables and why should wind power have so outstripped other sources? The average person in the street could probably answer part of this question if they have absorbed the mantra which is repeated *ad nauseam* in the media which claim that renewable electricity:

1. can be generated without burning fossil fuel which is rapidly running out.
2. emits little or no carbon dioxide (CO_2) and so ameliorates man-made global warming.
3. will allow us to dispense with nuclear power which is supposedly dangerous.
4. will give us very cheap electricity because nature provides the power with no fuel-cost.
5. increases security of supply by avoiding reliance on imported fuel.

These reasons for exploiting renewables are in some cases untrue or are serious exaggerations of the truth. They camouflage the fact that the rush to renewables is more a matter of quick profit than about saving the world from self-inflicted disaster. The public at large is beginning to suspect this, but there is huge political inertia based on the fear that individuals or parties may be branded as "climate change deniers" with all the horrific undertones that phrase carries. Indeed extremists in the media have already proposed "Nuremberg type climate trials" for the deniers! There is little reason for politicians to fear such accusations for, as we shall see, even if every claim concerning global warming were correct there is not a hope that wind power could alter CO_2 emissions sufficiently to measurably alter the progress of warming. The huge growth of wind power compared with other renewables has happened because it is a relatively mature technology compared with more recent developments such as solar photovoltaic or wave-power, and also because several renewables are already near their limits of exploitation. For example the most productive hydroelectric sites have already been utilised, waste biomass is limited in availability and the land for dedicated biomass cultivation is needed for agriculture, thus its exploitation is forcing-up food prices or destroying biodiversity by taking virgin land. Biomass combustion does of course liberate CO_2 in similar quantities to fossil fuel and is thus only "carbon neutral" in the context of photosynthetic recycling. By comparison, wind and water power effectively produce no CO_2.

Nuclear power also produces next to no CO_2 but until very recently "green" activity has precluded even discussion of it as a greenhouse-free power source even though it can provide continuous huge output of base-load supply which wind cannot, and also with new PWR reactors will be able to load-follow which, again, wind cannot do.

The intention of this book is to highlight and justify reasons for rejecting wind power as a large scale source of electricity and of reducing carbon emission. In using the word "scam" in the title I shall draw down the anger of many environmentalists and green campaigners. However, a "scam" is often defined as "an attempt to swindle people which involves gaining their confidence." Exactly! The repeated reference to "tackling climate change" as a motivation for wind power, and the constant citation of the 1997 Kyoto treaty "targets" as justification is just such an attempt to gain public confidence. The spectre of climate change is being used as a scaretactic to get people to buy wind power. This is the old quack-doctor trick—"Scare 'em to death and they'll buy anything." It will certainly be seen by history as a swindle supported by untruths and half-truths. I hope without too much exaggeration to "tell it as it is."

One of the first attempts to bring really large scale wind power to the UK was made in the late 1990s by the US Enron Corporation which collapsed in financial scandal in 2001. By that time the Kyoto Treaty was already coming into effect. A senior Enron employee gave a damning pre-Kyoto assessment in an internal memo, in which he wrote:

> If implemented, this agreement will do more to promote Enron's business than will almost any other regulatory initiative outside of restructuring of the energy and natural gas industries in Europe and the United States.

The truth of this statement survived longer than the company which made it!

It does not escape the more cynically inclined "doubters" of wind power that some of the loudest voices in the development of the industry come from the same large companies which are responsible for excessive release of CO_2, the supposed reason for needing renewables in the first place, among them RWE-npower and E.ON UK. Indeed, the Prince erf Wales's UK Corporate Leaders Group on Climate Change, which has said "deep and rapid" cuts were needed in greenhouse gas emissions, includes the executive officers of Tesco, BAA, Shell, and energy group E.ON—a fact which has prompted Greenpeace to describe the comment as "hypocrisy of a previously unknown magnitude". Now, with the world in the midst of the worst economic slump for decades, some of the world's leading investment banks have come together to discuss "cashing-in" on the carbon trading business.

Pre-Industrial Wind Power

Man first noticed and then harnessed the mechanical forces of nature as wind and water currents. A floating log in a flooded stream could not convey a clearer message, "This is how to move me". Carry a large leafy branch in a strong breeze and it practically shouts at you "I want to go this way—with the wind".

Paddled or poled boats appeared as very early human technology and by the British Bronze Age, sophisticated sewn plank boats existed, such as the seagoing Dover boat dating from 1500 BC, but long before this the Egyptian archaeological record preserved evidence of square-sailed boats as old as 3000 BC, and China may have used sail some centuries before this.

Such is human ingenuity that it did not take long to use the propulsive force of a sail to provide a stationary power source, driving a mill, a water-lift or perhaps a hammer. We have no clue how this happened but the running lines on a sail boat, coiled on a cylindrical belaying pin could have revealed the means of generating a turning force and at an informed guess, led to the first wooden pulleys. From these it is but a small leap of imagination to the sail-driven wind-shaft, but a leap which nevertheless took a millennium or more as the first vertical shaft windmills date from around the seventh century AD in Persia but the much commoner design today, horizontal shaft machines—the familiar windmill—were not developed until the tenth century.

Alongside the development of wind power, water also played its part and, where it was available, was preferred to wind. The power available from a moving liquid or gas is a direct function of its density, so, being 800 times denser than air, a relatively small waterwheel will do the same work as an enormous windmill of almost 1,000 times the circular area, as we saw above. By the seventeenth century much of the world's work was done by wind and water. The windmill on land had reached a high degree of sophistication and at sea, sailing vessels were a complex triumph of man's control of the elements. Some thousands of watermills are listed in the Domesday Book, but the windmill did not appear in Britain until the twelfth or thirteenth century.

By the time of Cobbett's *Rural Rides* in the early 1800s there were some 10,000 British windmills, but by the end of World War II almost all had been destroyed, dismantled or just abandoned. I remember one or two working mills in the 1940s and '50s but the stronger memory is of the ruined stumps of derelict windmills, two within a couple of miles of my childhood home in Kent.

The loss of the mills happened almost as if a contagious disease had swept the country clear of its former inhabitants, but it was not contagion, simply a logical succession to the convenience and cheapness of the invading steam engine. The engines won so quickly because wind does not blow all the time, and when it does blow it may not be hard enough, or it may be so hard that sailing vessels are wrecked or their rigging carried away while onshore, the sails of mills are destroyed or fire started by friction in "runaway" accidents. This unpredictable variability of wind and water, which killed the traditional mills, is a key fact. . . .

The modern wind power industry repeatedly cites the aesthetic appeal of the traditional mills as justification for the new generation of airfoil wind turbines—for example the turbine which has inexplicably been allowed above the iconic Glyndebourne Opera House in West Sussex will stand on an former site of a windmill, and much was made of this fact prior to the planning application. This totally misses the point that the comparison is of magnificent hand-crafted structures with mass-produced plastic, steel and concrete machines, not to mention that the average windmill was quarter the height of a large wind turbine and equivalent in landscape scale to churches, large barns and the trees of a semi-wooded landscape, a fact celebrated by several generations of landscape artists. . . .

Epilogue

Wind power . . . fails in a whole list of criteria.

It cannot provide a predictable electricity supply at the click of a switch. Consequently it would not be marketable without an obligation to purchase—a bitter pill which is sweetened by an enormous subsidy.

It was initially sold to the public as preventing emission of CO_2. In fact it saves a remarkably small amount, which has been grossly exaggerated, as the industry has been humiliatingly forced to admit. The need for backup has been ignored in the carbon-accounting by all but a few multinational energy companies with experience of very large-scale wind power operation.

As a supplier of electricity and a mitigation of carbon emission it is remarkably expensive, in particular because each megawatt-worth of installed capacity and transmission line produces or carries not much more than a quarter of that output because of the wind-limited load factor. In due course, dedicated backup will be required but will be unable to cover its own costs as by definition it must be standing-by and not generating electricity for much of the time.

It has a huge landscape footprint for a tiny but costly contribution to electricity supply—often 300 to 400 km^2 of visual impact for no more than a few megawatts of wobbly power.

The RO subsidy in the UK has however made wind power an attractive investment for multinationals with the financial backing to carry the initial capital expenditure and indeed, in all countries the unprecedented subsidy has driven a culture of bullying, misrepresentation and occasionally law-breaking. For example whilst writing these final words, the following appeared:

> Italy police arrest 8 in Mafia wind farms plot. Operation "Aeolus," named after the ancient Greek god of winds, netted eight suspects, arrested in the Trapani area of western Sicily [and on the Italian mainland]. Police in Trapani said the local Mafia bribed city officials in nearby Mazara del Vallo so the town would invest in wind farms to produce energy. (Associated Press, 2009 [Google hosted])

Truth and even legality have come to grief in this industry. It is very apparent that with unlimited sums of money for a nearly invisible product, what else shall we grow but a mafia-culture?

This damning compendium of failure suggests that we would be better off without wind power at all—that is certainly my view, but it has been part of the successful strategy of the wind power industry to represent any opposition as being both negative and misled. Thus I feel no more compelling need to suggest an "alternative" to wind than I would offer in exchange for the fairies at the bottom of my garden, pedalling their infinite power generator! . . .

EXPLORING THE ISSUE

Is Wind Enough?

Critical Thinking and Reflection

1. Do we need to develop all available sites for wind power to meet society's need for electricity?
2. What are the essential features of an "ideal" wind farm?
3. Should wind farms be built on land or at sea?
4. If we would be better off not using wind power, what can we use instead?
5. John Etherington says one of the problems with wind power is that electricity supply systems must be predictable, and wind is not. What lifestyle changes would be necessary to live with an unpredictable electricity supply?
6. What might the drawbacks be of wind turbines held aloft by giant kites?

Is There Common Ground?

One of the objections to both wind and solar power is that they are not available all the time. Night and clouds interfere with solar power, and wind power is available only when the wind is blowing. But even though the sun is not shining or the wind is not blowing *here,* they surely are somewhere else. If we have enough solar and wind power installations, spread widely enough, all we have to do is connect them up. This is one of the ideas behind calls for a "green grid" to replace the existing and admittedly antiquated electrical grid and deliver electricity from far-flung, scattered locations to wherever it is needed. See Mike Martin, "The Great Green Grid," *E Magazine* (July/August 2010).

1. What is meant by a "green grid"?
2. How can a "green grid" help to prevent electrical blackouts?
3. Would a "green grid" be essential to making the enormous number of wind turbines environed by Lu et al. workable?
4. Would a "green grid" answer any of Etherington's objections to wind power?

ISSUE 17

Are Biofuels a Reasonable Substitute for Fossil Fuels?

YES: Keith Kline, Virginia H. Dale, Russell Lee, and Paul Leiby, from "In Defense of Biofuels, Done Right," *Issues in Science and Technology* (Spring 2009)

NO: Donald Mitchell, from "A Note on Rising Food Prices," The World Bank Development Prospects Group (July 2008)

Learning Outcomes

After studying this issue, students will be able to:

- Explain why there is so much interest in using biofuels to replace fossil fuels for powering vehicles.
- Describe how biofuels production affects food supply and prices.
- Describe how biofuels production affects the environment.
- Explain why some kinds of biofuels are more desirable than others.

ISSUE SUMMARY

YES: Keith Kline, et al. argue that the impact of biofuels production on food prices is much less than alarmists claim. If biofuels development focused on converting biowastes and fast-growing trees and grasses into fuels, the overall impact would be even better with a host of benefits.

NO: Donald Mitchell argues that although many factors contributed to the increase in internationally traded food prices from January 2002 to June 2008, the most important single factor—accounting for as much as 70 percent of the rise in food prices—was the large increase in biofuels production from grains and oilseeds in the United States and European Union.

The threat of global warming has spurred a great deal of interest in finding new sources of energy that do not add to the amount of carbon dioxide in the air. Among other things, this has meant a search for alternatives to fossil fuels, which modern civilization uses to generate electricity, heat homes, and power transportation. Finding alternatives for electricity generation (which relies much more on coal than on oil or natural gas) or home heating (which relies more on oil and natural gas) is easier than finding alternatives for transportation (which relies on oil, refined into gasoline and diesel oil). In addition, the transportation infrastructure, consisting of refineries, pipelines, tank trucks, gas stations, and an immense number of cars and trucks that will be on the road for many years, is well designed for handling liquid fuels. It is not surprising that industry and government would like to find nonfossil liquid fuels for cars and trucks (as well as ships and airplanes).

There are many suitably flammable liquids. Among them are the so-called biofuels or renewable fuels, plant oils, and alcohols that can be distilled from plant sugars. According to Daniel M. Kammen, "The Rise of Renewable Energy," *Scientific American* (September 2006), the chief biofuel in the United States so far is ethanol, distilled from corn and blended with gasoline. Production is subsidized with $2 billion of federal funds, and "when all the inputs and outputs were correctly factored in, we found that ethanol" contains about 25 percent more energy (to be used when it is burned as fuel) than was used to produce it. At least one study says the "net energy" is actually less than the energy used to produce ethanol from corn; see Dan Charles, "Corn-Based Ethanol Flunks Key Test," *Science* (May 1, 2009). If other sources, such as cellulose-rich switchgrass or cornstalks, can be used, the "net energy" is much better. However, generating ethanol requires first converting cellulose to fermentable sugars, which is so far an expensive process (although many people are working on making the process cheaper; see Jennifer Chu, "Reinventing Cellulosic Ethanol Production," *Technology Review* (online) (June 10, 2009, www.technologyreview.com/energy/22774/?nlid=2091), George W. Huber and Bruce E. Dale, "Grassoline at the Pump," *Scientific American* (July 2009), and Rachel Ehrenberg, "The Biofuel Future," *Science News* (August 1, 2009). A significant additional concern is the amount of land needed for growing crops to be turned into biofuels; in a world where hunger is widespread, this means land is taken out of food production. If additional land is cleared to grow biofuel crops, this must mean loss of forests and wildlife habitat, increased erosion, and other environmental problems.

Under the Energy Policy Act of 2005, the U.S. Environmental Protection Agency (EPA) requires that gasoline sold in the United States contains a minimum volume of renewable fuel. Under the Renewable Fuel Program (also known as the Renewable Fuel Standard Program, or RFS Program), this volume will increase over the years, reaching 36 billion gallons by 2022. According to the EPA (www.epa.gov/otaq/fuels/renewablefuels/index.htm), "the RFS program was developed in collaboration with refiners, renewable fuel producers, and many other stakeholders." However, some think it is premature and even dangerous to put so much emphasis on biofuels. Robbin S. Johnson and C. Ford Runge, "Ethanol: Train Wreck Ahead," *Issues in Science and Technology* (Fall 2007), argue that the U.S. government's bias in favor of corn-based ethanol rigs the market against more efficient alternatives. It also leads to rising food prices, which particularly affects the world's poor. Pat Thomas, introducing

The Ecologist's special report on biofuels in the March 2007 issue (www.theecologist.org/archive_detail.asp?content_id=838), wrote that "the science is far from complete, the energy savings far from convincing and, although many see biofuels as a way to avoid the kind of resource wars currently raging in the Middle East and elsewhere, going down that road may in the end provoke a wider series of resource wars—this time over food, water and habitable land." William Tucker, "Food Riots Made in the USA," *The Weekly Standard* (April 28, 2008), blames the food riots seen in many countries in 2008 on price increases due in large part to the shift in agricultural production from food grains to biofuels crops. See also Robert Bryce, "A Promise Not Kept: The Case against Ethanol," *The Conference Board Review* (May/June 2008).

In March 2007, President Bush visited Brazil, which meets much of its need for vehicle fuel with ethanol from sugarcane, and agreed to work with Brazil in developing and promoting biofuels. According to the U.S. State Department, the agreement "reassures small countries in Central America and the Caribbean that they can reduce their dependence on foreign oil." In both Europe and the United States, governments are rushing to encourage the production and use of biofuels. L. Pelmans et al., "European Biofuels Strategy," *International Journal of Environmental Studies* (June 2007), attempts to classify nations according to their strategies so that "the formulation of a strategy to support the advancement of biofuels and alternative motor fuels in general should become more manageable." Corporations and investors see huge potential for profit, and many environmentalists see benefits for the environment.

However, there *are* problems with biofuels, and those problems are getting a great deal of attention. Robin Maynard, "Against the Grain," *The Ecologist* (March 2007), stresses that when food and fuel compete for farmland, food prices will rise, perhaps drastically. The poor will suffer, as will rainforests. Renton Righelato, "Forests or Fuel," *The Ecologist* (March 2007), reminds us that when forests are cleared, they no longer serve as "carbon sinks"; deforestation thus adds to the global warming problem, and it may take a century for the benefit of biofuels to show itself. David Pimentel, "Biofuel Food Disasters and Cellulosic Ethanol Problems," *Bulletin of Science, Technology & Society* (June 2009), says that because using 20 percent of the U.S. corn crop displaces a mere 1 percent of oil consumption, corn ethanol is a disaster, while using crop wastes and other biological materials poses its own problems. Heather Augustyn, "A Burning Issue," *World Watch* (July/August 2007), describes the impact of forest fires to clear land for oil palm plantations in Indonesia. Palm oil holds great promise as a biofuel, but the plantations displace natural ecosystems and destroy habitat for numerous species, as well as for indigenous peoples. Man Kee Lam et al., "Malaysian Palm Oil: Surviving the Food versus Fuel Dispute for a Sustainable Future," *Renewable & Sustainable Energy Reviews* (August 2009), are more optimistic, saying that with care, palm oil may be able to satisfy needs for both edible oils and fuel oils. Robert F. Service, "Another Biofuels Drawback: The Demand for Irrigation," *Science* (October 23, 2009), notes that biofuel crops can compete for irrigation water, leading to both water supply and water quality problems. Gernot Stoeglehner and Michael Narodoslawsky, "How Sustainable Are Biofuels? Answers and Further Questions Arising from an Ecological Footprint Perspective," *Bioresource Technology* (August 2009), note that the sustainability of biofuels production depends very much on regional context.

Laura Venderkam, "Biofuels or Bio-Fools?" *American: A Magazine of Ideas* (May/June 2007), describes the huge amounts of money being invested in companies planning to bring biofuels to market. Research is on going, including efforts to use genetic engineering to produce enzymes that can cheaply and efficiently break cellulose into its component sugars (see Matthew L. Wald, "Is Ethanol for the Long Haul?" *Scientific American* (January 2007) and Michael E. Himmel et al., "Biomass Recalcitrance: Engineering Plants and Enzymes for Biofuels Production," *Science* (February 9, 2007)), make bacteria or yeast that can turn a greater proportion of sugar into alcohol (see Francois Torney et al., "Genetic Engineering Approaches to Improve Bioethanol Production from Maize," *Current Opinion in Biotechnology* (June 2007)), and even make bacteria that can convert sugar or cellulose into hydrocarbons that can easily be turned into gasoline or diesel fuel (see Neil Savage, "Building Better Biofuels," *Technology Review* (July/August 2007)). If these efforts succeed, the price of biofuels may drop drastically, leading investors to abandon the field. Such a price drop would, of course, benefit the consumer and lead to wider use of biofuels.

D. A. Walker, "Biofuels—For Better or Worse?" *Annals of Applied Biology* (May 2010), views much biofuels advocacy as based on misinformation. Peter Rosset, "Agrofuels, Food Sovereignty, and the Contemporary Food Crisis," *Bulletin of Science, Technology & Society* (June 2009), says that biofuels are not a prime cause of the 2008 food crisis but they are "clearly contraindicated." On the other hand, Jose C. Escobar et al., "Biofuels: Environment, Technology and Food Security," *Renewable & Sustainable Energy Reviews* (August 2009), considers that increased reliance on biofuels is inevitable due to "the imminent decline of the world's oil production, its high market prices and environmental impacts."

Suzanne Hunt, "Biofuels, Neither Saviour Nor Scam: The Case for a Selective Strategy," *World Policy Journal* (Spring 2008), argues that all the concerns are real, but the larger problem is that "our current agricultural, energy, and transport systems are failing." John Ohlrogge et al., "Driving on Biomass," *Science* (May 22, 2009), point to a potential fix for the failure of the transportation system when they show that converting biomass to biofuels is much less efficient than burning it to produce electricity and using the electricity to power electric cars, which are in fact projected to "gain substantial market share in the coming years."

In the YES selection, Keith Kline, et al. argue that the impact of biofuels production on food prices is much less than alarmists claim, and the overall impact would come with a host of benefits. In the NO selection, Donald Mitchell argues that although many factors contributed to the increase in internationally traded food prices from January 2002 to June 2008, the most important single factor—accounting for as much as 70 percent of the rise in food prices—was the large increase in biofuels production from grains and oilseeds in the United States and European Union. Without these increases, global wheat and maize stocks would not have declined appreciably and price increases due to other factors would have been moderate.

YES Keith Kline, Virginia H. Dale, Russell Lee, and Paul Leiby

In Defense of Biofuels, Done Right

Biofuels have been getting bad press, not always for good reasons. Certainly important concerns have been raised, but preliminary studies have been misinterpreted as a definitive condemnation of biofuels. One recent magazine article, for example, illustrated what it called "Ethanol USA" with a photo of a car wreck in a corn field. In particular, many criticisms converge around grain-based biofuel, traditional farming practices, and claims of a causal link between U.S. land use and land-use changes elsewhere, including tropical deforestation.

Focusing only on such issues, however, distracts attention from a promising opportunity to invest in domestic energy production using biowastes, fast-growing trees, and grasses. When biofuel crops are grown in appropriate places and under sustainable conditions, they offer a host of benefits: reduced fossil fuel use; diversified fuel supplies; increased employment; decreased greenhouse gas emissions; enhanced habitat for wildlife; improved soil and water quality; and more stable global land use, thereby reducing pressure to clear new land.

Not only have many criticisms of biofuels been alarmist, many have been simply inaccurate. In 2007 and early 2008, for example, a bumper crop of media articles blamed sharply higher food prices worldwide on the production of biofuels, particularly ethanol from corn, in the United States. Subsequent studies, however, have shown that the increases in food prices were primarily due to many other interacting factors: increased demand in emerging economies, soaring energy prices, drought in food-exporting countries, cut-offs in grain exports by major suppliers, market-distorting subsidies, a tumbling U.S. dollar, and speculation in commodities markets.

Although ethanol production indeed contributes to higher corn prices, it is not a major factor in world food costs. The U.S. Department of Agriculture (USDA) calculated that biofuel production contributed only 5% of the 45% increase in global food costs that occurred between April 2007 and April 2008. A Texas A&M University study concluded that energy prices were the primary cause of food price increases, noting that between January 2006 and January 2008, the prices of fuel and fertilizer, both major inputs to agricultural production, increased by 37% and 45%, respectively. And the International Monetary Fund has documented that since their peak in July 2008, oil prices declined by 69% as of December 2008, and global food prices declined by

33% during the same period, while U.S. corn production has remained at 12 billion bushels a month, one-third of which is still used for ethanol production.

In another line of critique, some argue that the potential benefits of biofuel might be offset by indirect effects. But large uncertainties and postulations underlie the debate about the indirect land-use effects of biofuels on tropical deforestation, the critical implication being that use of U.S. farmland for energy crops necessarily causes new land-clearing elsewhere. Concerns are particularly strong about the loss of tropical forests and natural grasslands. The basic argument is that biofuel production in the United States sets in motion a necessary scenario of deforestation.

According to this argument, if U.S. farm production is used for fuel instead of food, food prices rise and farmers in developing countries respond by growing more food. This response requires clearing new land and burning native vegetation and, hence, releasing carbon. This "induced deforestation" hypothesis is based on questionable data and modeling assumptions about available land and yields, rather than on empirical evidence. The argument assumes that the supply of previously cleared land is inelastic (that is, agricultural land for expansion is unavailable without new deforestation). It also assumes that agricultural commodity prices are a major driving force behind deforestation and that yields decline with expansion. The calculations for carbon emissions assume that land in a stable, natural state is suddenly converted to agriculture as a result of biofuels. Finally, the assertions assume that it is possible to measure with some precision the areas that will be cleared in response to these price signals.

A review of the issues reveals, however, that these assumptions about the availability of land, the role of biofuels in causing deforestation, and the ability to relate crop prices to areas of land clearance are unsound. Among our findings:

First, sufficient suitably productive land is available for multiple uses, including the production of biofuels. Assertions that U.S. biofuel production will cause large indirect land-use changes rely on limited data sets and unverified assumptions about global land cover and land use. Calculations of land-use change begin by assuming that global land falls into discrete classes suitable for agriculture—cropland, pastures and grasslands, and forests—and results depend on estimates of the extent, use, and productivity of these lands, as well as presumed future interactions among land-use classes. But several major organizations, including the Food and Agriculture Organization (FAO), a primary data clearinghouse, have documented significant inconsistencies surrounding global land-cover estimates. For example, the three most recent FAO Forest Resource Assessments, for periods ending in 1990, 2000, and 2005, provide estimates of the world's total forest cover in 1990 that vary by as much as 470 million acres, or 21% of the original estimate.

Cropland data face similar discrepancies, and even more challenging issues arise when pasture areas are considered. Estimates for land used for crop production range from 3.8 billion acres (calculated by the FAO) to 9 billion acres (calculated by the Millennium Ecosystem Assessment, an international effort spearheaded by the United Nations). In a recent study attempting to reconcile cropland use circa 2000, scientists at the University of Wisconsin-Madison and McGill University estimated that there were 3.7 billion acres of cropland, of which 3.2 billion were actively cropped or harvested. Land-use studies consistently acknowledge serious data limitations and uncertainties, noting that a majority of global crop lands are constantly shifting the location of cultivation, leaving at any time large areas fallow or idle that may not be captured in statistics. Estimates of idle croplands, prone to confusion with pasture and grassland, range from 520 million acres to 4.9 billion acres globally. The differences illustrate one of many uncertainties that hamper global land-use change calculations. To put these numbers in perspective, USDA has estimated that in 2007, about 21 million acres were used worldwide to produce biofuel feedstocks, an area that would occupy somewhere between 0.4% and 4% of the world's estimated idle cropland.

Diverse studies of global land cover and potential productivity suggest that anywhere from 600 million to more than 7 billion additional acres of underutilized rural lands are available for expanding rain-fed crop production around the world, after excluding the 4 billion acres of cropland currently in use, as well as the world's supply of closed forests, nature reserves, and urban lands. Hence, on a global scale, land per se is not an immediate limitation for agriculture and biofuels.

In the United States, the federal government, through the multiagency Biomass Research and Development Initiative (BRDI), has examined the land and market implications of reaching the nation's biofuel target, which calls for producing 36 billion gallons by 2022. BRDI estimated that a slight net reduction in total U.S. active cropland area would result by 2022 in most scenarios, when compared with a scenario developed from USDA's so-called "baseline" projections. BRDI also found that growing biofuel crops efficiently in the United States would require shifts in the intensity of use of about 5% of pasture lands to more intensive hay, forage, and bioenergy crops (25 million out of 456 million acres) in order to accommodate dedicated energy crops, along with using a combination of wastes, forest residues, and crop residues. BRDI's estimate assumes that the total area allocated to USDA's Conservation Reserve Program (CRP) remains constant at about 33 million acres but allows about 3 million acres of the CRP land on high-quality soils in the Midwest to be offset by new CRP additions in other regions. In practice, additional areas of former cropland that are now in the CRP could be managed for biofuel feedstock production in a way that maintains positive impacts on wildlife, water, and land conservation goals, but this option was not included among the scenarios considered.

Yields are important. They vary widely from place to place within the United States and around the world. USDA projects that corn yields will rise by 20 bushels per acre by 2017; this represents an increase in corn output equivalent to adding 12.5 million acres as compared with 2006, and over triple that area as compared with average yields in many less-developed nations. And there is the possibility that yields will increase more quickly than projected in the USDA baseline, as seed companies aim to exceed 200 bushels per acre by 2020. The potential to increase yields in developing countries offers tremendous opportunities to improve welfare and expand production while reducing or maintaining the area harvested. These improvements are consistent with U.S. trends during the past half century showing agricultural output growth averaging 2% per year while cropland use fell by an average of 0.7% per year. Even without large yield increases, cropland requirements to meet biofuel production targets may not be nearly as great as assumed.

Concerns over induced deforestation are based on a theory of land displacement that is not supported by data. U.S. ethanol production shot up by more than 3 billion gallons (150%) between 2001 and 2006, and corn production increased 11%, while total U.S. harvested cropland fell by about 2% in the same period. Indeed, the harvested area for "coarse grains" fell by 4% as corn, with an average yield of 150 bushels per acre, replaced other feed grains such as sorghum (averaging 60 bushels per acre). Such statistics defy modeling projections by demonstrating an ability to supply feedstock to a burgeoning ethanol industry while simultaneously maintaining exports and using substantially less land. So although models may assume that increased use of U.S. land for biofuels will lead to more land being cleared for agriculture in other parts of the world, evidence is lacking to support those claims.

Second, there is little evidence that biofuels cause deforestation, and much evidence for alternative causes. Recent scientific papers that blame biofuels for deforestation are based on models that presume that new land conversion can be simulated as a predominantly market-driven choice. The models assume that land is a privately owned asset managed in response to global price signals within a stable rule-based economy—perhaps a reasonable assumption for developed nations.

However, this scenario is far from the reality in the smoke-filled frontier zones of deforestation in less-developed countries, where the models assume biofuel-induced land conversion takes place. The regions of the world that are experiencing first-time land conversion are characterized by market isolation, lawlessness, insecurity, instability, and lack of land tenure. And nearly all of the forests are publicly owned. Indeed, land-clearing is a key step in a long process of trying to stake a claim for eventual tenure. A cycle involving incremental degradation, repeated and extensive fires, and shifting small plots for subsistence tends to occur long before any consideration of crop choices influenced by global market prices.

The causes of deforestation have been extensively studied, and it is clear from the empirical evidence that forces other than biofuel use are responsible for the trends of increasing forest loss in the tropics. Numerous case studies document that the factors driving deforestation are a complex expression of cultural, technological, biophysical, political, economic, and demographic interactions. Solutions and measures to slow deforestation have also been analyzed and tested, and the results show that it is critical to improve governance, land tenure, incomes, and security to slow the pace of new land conversion in these frontier regions.

Selected studies based on interpretations of satellite imagery have been used to support the claims that U.S. biofuels induce deforestation in the

Amazon, but satellite images cannot be used to determine causes of land-use change. In practice, deforestation is a site-specific process. How it is perceived will vary greatly by site and also by the temporal and spatial lens through which it is observed. Cause-and-effect relationships are complex, and the many small changes that enable larger future conversion cannot be captured by satellite imagery. Although it is possible to classify an image to show that forest in one period changed to cropland in another, cataloguing changes in discrete classes over time does not explain why these changes occur. Most studies asserting that the production and use of biofuels cause tropical deforestation point to land cover at some point after large-scale forest degradation and clearing have taken place. But the key events leading to the primary conversion of forests often proceed for decades before they can be detected by satellite imagery. The imagery does not show how the forest was used to sustain livelihoods before conversion, nor the degrees of continual degradation that occurred over time before the classification changed. When remote sensing is supported by a ground-truth process, it typically attempts to narrow the uncertainties of land-cover classifications rather than research the history of occupation, prior and current use, and the forces behind the land-use decisions that led to the current land cover.

First-time conversion is enabled by political, as well as physical, access. Southeast Asia provides one example where forest conversion has been facilitated by political access, which can include such diverse things as government-sponsored development and colonization programs in previously undisturbed areas and the distribution of large timber and mineral concessions and land allotments to friends, families, and sponsors of people in power. Critics have raised valid concerns about high rates of deforestation in the region, and they often point an accusing finger at palm oil and biofuels.

Palm oil has been produced in the region since 1911, and plantation expansion boomed in the 1970s with growth rates of more than 20% per year. Biodiesel represents a tiny fraction of palm oil consumption. In 2008, less than 2% of crude palm oil output was processed for biofuel in Indonesia and Malaysia, the world's largest producers and exporters. Based on land-cover statistics alone, it is impossible to determine the degree of attribution that oil palm may share with other causes of forest conversion in Southeast Asia. What is clear is that oil palm is not the only factor and that palm plantations are established after a process of degradation and deforestation has transpired. Deforestation data may offer a tool for estimating the ceiling for attribution, however. In Indonesia, for example, 28.1 million hectares were deforested between 1990 and 2005, and oil palm expansion in those areas was estimated to be between 1.7 million and 3 million hectares, or between 6% and 10% of the forest loss, during the same period.

Initial clearing in the tropics is often driven more by waves of illegitimate land speculation than agricultural production. In many Latin American frontier zones, if there is native forest on the land, it is up for grabs, as there is no legal tenure of the land. The majority of land-clearing in the Amazon has been blamed on livestock because, in part, there is no alternative for classifying the recent clearings and, in part, because land holders must keep it "in production" to maintain claims and avoid invasions. The result has been the frequent burning and the creation of extensive cattle ranches. For centuries, disenfranchised groups have been pushed into the forests and marginal lands where they do what they can to survive. This settlement process often includes serving as low-cost labor to clear land for the next wave of better-connected colonists. Unless significant structural changes occur to remove or modify enabling factors, the forest-clearing that was occurring before this decade is expected to continue along predictable paths.

Testing the hypothesis that U.S. biofuel policy causes deforestation elsewhere depends on models that can incorporate the processes underlying initial land-use change. Current models attempt to predict future land-use change based on changes in commodity prices. As conceived thus far, the computational general equilibrium models designed for economic trade do not adequately incorporate the processes of land-use change. Although crop prices may influence short-term land-use decisions, they are not a dominant factor in global patterns of first-time conversion, the land-clearing of chief concern in relating biofuels to deforestation. The highest deforestation rates observed and estimated globally occurred in the 1990s. During that period, there was a surplus of commodities on world markets and consistently depressed prices.

Third, many studies omit the larger problem of widespread global mismanagement of land. The recent arguments focusing on the possible deforestation attributable to biofuels use idealized representations of crop and land markets, omitting what may be larger issues of concern. Clearly, the causes of global deforestation are complex and are not driven merely by a single crop market. Additionally, land mismanagement, involving both initial clearing and maintaining previously cleared land, is widespread and leads to a process of soil degradation and environmental damage that is especially prevalent in the frontier zones. Reports by the FAO and the Millennium Ecosystem Assessment describe the environmental consequences of repeated fires in these areas. Estimates of global burning vary annually, ranging from 490 million to 980 million acres per year between 2000 and 2004. The vast majority of fires in the tropics occur in Africa and the Amazon in what were previously cleared, nonforest lands. In a detailed study, the Amazon Institute of Environmental Research and Woods Hole Research Center found that 73% of burned area in the Amazon was on previously cleared land, and that was during the 1990s, when overall deforestation rates were high.

Fire is the cheapest and easiest tool supporting shifting subsistence cultivation. Repeated and extensive burning is a manifestation of the lack of tenure, lack of access to markets, and severe poverty in these areas. When people or communities have few or no assets to protect from fire and no incentive to invest in more sustainable production, they also have no reason to limit the extent of burning. The repeated fires modify ecosystem structure, penetrate ever deeper into forest margins, affect large areas of understory vegetation (which is not detected by remote sensing), and take an ever greater cumulative toil on soil quality and its ability to sequester carbon. Profitable biofuel markets, by contributing to improved incentives to grow cash crops, could reduce the use of fire and the pressures on the agricultural frontier. Biofuels done right, with attention to best practices for sustained production, can make significant contributions to social and economic development as well as environmental protection.

Furthermore, current literature calculates the impacts from an assumed agricultural expansion by attributing the carbon emissions from clearing intact ecosystems to biofuels. If emission analyses consider empirical data reflecting the progressive degradation that occurs (often over decades) before and independently of agriculture market signals for land use, as well as changes in the frequency and extent of fire in areas that biofuels help bring into more stable market economies, then the resulting carbon emission estimates would be worlds apart.

Brazil provides a good case in point, because it holds the globe's largest remaining area of tropical forests, is the world's second-largest producer of biofuel (after the United States), and is the world's leading supplier of biofuel for global trade. Brazil also has relatively low production costs and a growing focus on environmental stewardship. As a matter of policy, the Brazilian government has supported the development of biofuels since launching a National Ethanol Program called Proálcool in 1975. Brazil's ethanol industry began its current phase of growth after Proálcool was phased out in 1999 and the government's role shifted from subsidies and regulations toward increased collaboration with the private sector in R&D. The government helps stabilize markets by supporting variable rates of blending ethanol with gasoline and planning for industry expansion, pipelines, ports, and logistics. The government also facilitates access to global markets; develops improved varieties of sugarcane, harvest equipment, and conversion; and supports improvements in environmental performance.

New sugarcane fields in Brazil nearly always replace pasture land or less valuable crops and are concentrated around production facilities in the developed southeastern region, far from the Amazon. Nearly all production is rainfed and relies on low input rates of fertilizers and agrochemicals, as compared with other major crops. New projects are reviewed under the Brazilian legal framework of Environmental Impact Assessment and Environmental Licensing. Together, these policies have contributed to the restoration or protection of reserves and riparian areas and increased forest cover, in tandem with an expansion of sugarcane production in the most important producing state, Sao Paulo.

Yet natural forest in Brazil is being lost, with nearly 37 million acres lost between May 2000 and August 2006, and a total of 150 million acres lost since 1970. Some observers have suggested that the increase in U.S. corn production for biofuel led to reduced soybean output and higher soybean prices, and that these changes led, in turn, to new deforestation in Brazil. However, total deforestation rates in Brazil appear to fall in tandem with rising soybean prices. This co-occurrence illustrates a lack of connection between commodity prices and initial land clearing. This phenomenon has been observed around the globe and suggests an alternate hypothesis: Higher global commodity prices focus production and investment where it can be used most efficiently, in the plentiful previously cleared and underutilized lands around the world. In times of falling prices and incomes, people return to forest frontiers, with all of their characteristic tribulations, for lack of better options.

Biofuels Done Right

With the right policy framework, cellulosic biofuel crops could offer an alternative that diversifies and boosts rural incomes based on perennials. Such a scenario would create incentives to reduce intentional burning that currently affects millions of acres worldwide each year. Perennial biofuel crops can help stabilize land cover, enhance soil carbon sequestration, provide habitat to support biodiversity, and improve soil and water quality. Furthermore, they can reduce pressure to clear new land via improved incomes and yields. Developing countries have huge opportunities to increase crop yield and thereby grow more food on less land, given that cereal yields in less developed nations are 30% of those in North America. Hence, policies supporting biofuel production may actually help stop the extensive slash-and-burn agricultural cycle that contributes to greenhouse gas emissions, deforestation, land degradation, and a lifestyle that fails to support farmers and their families.

Biofuels alone are not the solution, however. Governments in the United States and elsewhere will have to develop and support a number of programs designed to support sustainable development. The operation and rules of such programs must be transparent, so that everyone can understand them and see that fair play is ensured. Among other attributes, the programs must offer economic incentives for sustainable production, and they must provide for secure land tenure and participatory land-use planning. In this regard, pilot biofuel projects in Africa and Brazil are showing promise in addressing the vexing and difficult challenges of sustainable land use and development. Biofuels also are uniting diverse stakeholders in a global movement to develop sustainability metrics and certification methods applicable to the broader agricultural sector.

Given a priority to protect biodiversity and ecosystem services, it is important to further explore the drivers for the conversion of land at the frontier and to consider the effects, positive and negative, that U.S. biofuel policies could have in these areas. This means it is critical to distinguish between valid concerns calling for caution and alarmist criticisms that attribute complex problems solely to biofuels.

Still, based on the analyses that we and others have done, we believe that biofuels, developed in an economically and environmentally sensible way, can contribute significantly to the nation's—indeed, the world's—energy security while providing a host of benefits for many people in many regions.

Donald Mitchell

A Note on Rising Food Prices

I. Introduction

Internationally traded food commodities prices have increased sharply since 2002 and especially since late-2006, and prices of major staples, such as grains and oilseeds, have doubled in just the past two years. Rising prices have caused food riots in several countries and led to policy actions such as the banning of grain and other food exports by a number of countries and tariff reductions on imported foods in others. The policy actions reflect the concern of governments about the impact of food price increases on the poor in developing countries who, on average, spend half of their household incomes on food. This paper examines how internationally traded food commodities prices (maize, wheat, rice, soybeans, etc.) have changed, and analyzes the factors contributing to these increases. In particular, it looks at the contribution of biofuels production to food price increases. In this paper biofuels refer to ethanol and biodiesel.

II. The Rise in Global Food Prices

The IMF's index of internationally traded food commodities prices increased 130 percent from January 2002 to June 2008 and 56 percent from January 2007 to June 2008. Prior to that, food commodities prices had been relatively stable after reaching lows in 2000 and 2001 following the Asia financial crisis. The low levels of global grain stocks had been identified as a cause for concern in a number of fora and the risk of higher food prices was highlighted in a recent World Bank publication and online.

The increase in food commodities prices was led by grains which began sustained price increases in 2005 despite a record global crop in the 2004/05 crop year that was 10.2 percent larger than the average of the three previous years and a near record crop in 2005/06 that was still 8.9 percent larger. Global stocks of grain increased in 2004/05 but declined in 2005/06 as demand increased more than production. From January 2005 until June 2008, maize prices almost tripled, wheat prices increased 127 percent and rice prices increased 170 percent. The increase in grain prices was followed by increases in fats & oils prices in mid-2006, and that also followed a record 2004/05 global oilseed crop that was 13 percent larger than in the previous year and an even larger crop in 2005/06. Fats & oils prices have shown similar increases to grains, with palm oil prices up 200 percent from January 2005 until June 2008, soybean oil prices up 192 percent, and other vegetable oils prices increasing by similar amounts. Other foods prices (sugar, citrus, bananas, shrimp and meats) increased 48 percent from January 2005 to June 2008.

III. Recent Estimates of the Contribution of Biofuels Production to Food Prices

Estimates of the contribution of biofuels production to food price increases are difficult, if not impossible to compare. Estimates can differ widely due to different time periods considered, different prices (export, import, wholesale, retail) considered, and different coverage of food products. . . .

[But] [d]espite all the differences in approach, many studies recognize biofuels production as a major driver of food prices. The USDA's chief economist in testimony before the Joint Economic Committee of Congress on May 1, attributed much of the increase in farm prices of maize and soybeans to biofuels production. The IMF estimated that the increased demand for biofuels accounted for 70 percent of the increase in maize prices and 40 percent of the increase in soybean. Collins used a mathematical simulation to estimate that about 60 percent of the increase in maize prices from 2006 to 2008 may have been due to the increase in maize used in ethanol. Rosegrant, et al., using a general equilibrium model, calculated the long-term impact on weighted cereal prices of the acceleration in biofuel production from 2000 to 2007 to be 30 percent in real terms. Maize prices were estimated to have increased 39 percent in real terms, wheat prices increased 22 percent and rice prices increased 21 percent. During this period, the U.S. CPI increased by 20.4 percent, which would imply nominal prices increases of 47, 26, and 25, respectively, for maize, wheat and rice prices. This is the same order of magnitude as was calculated with the World Bank's linkages model. Differences in the estimates of the impact of biofuels on the price index of all food depend largely on how broadly the food basket is defined and what is assumed about the interaction between prices of maize and vegetable oils (directly influenced by demand for biofuels) to prices of other crops such as rice through substitution on the supply or demand side. For example, the Council of Economic Advisors estimated that retail food prices increased only about 3 percent over the past 12 months due to ethanol production, in part because they only considered the impact of maize prices, directly and indirectly, on retail prices.

Many other potential drivers of the escalating food prices are mentioned in discussions, but there are few quantitative estimates of their impacts. For example, a recent USDA report attributed the increase in world market prices for major food commodities such as grains and vegetable oils to many factors including biofuels as well as other factors including the declining dollar, rising energy prices, increasing agricultural costs of production, growing foreign exchange holdings by major food-importing countries, and recent policies by some exporting countries to mitigate their own food-price inflation.

IV. Estimates of Factors Contributing to the Rise in Food Commodities Prices

There are a number of factors that have contributed to the rise in food prices. Among these are the increase in energy prices and the related increases in prices of fertilizer and chemicals, which are either produced from energy or are heavy users of energy in their production process. This has increased the cost of production, which ultimately gets reflected in higher food prices. Higher energy prices have also increased the cost of transportation, and increased the incentive to produce biofuels and encouraged policy support for biofuels production. The increase in biofuels production has not only increased demand for food commodities, but also led to large land use changes which reduced supplies of wheat and crops that compete with food commodities used for biofuels. Drought in Australia in 2006 and 2007 and poor crops in Europe in 2007 added to the grain and oilseed price increases, and rapid import demand increases for oilseeds by China to feed its growing livestock and poultry industry contributed to oilseed price increases. Other factors, including the decline of the dollar, and the increased investment in commodities by institutional investors to hedge against inflation and diversify portfolios may have also contributed to the price increases. The remainder of this section will examine these factors.

High energy prices have contributed about 15–20 percent to higher U.S. food commodities production and transport costs. Production costs per acre for U.S. corn, soybeans and wheat increased 32.3, 25.6 and 31.4 percent, respectively, from 2002 to 2007, according to the USDA's cost-of-production surveys and forecasts. However, yield increases during this period reduced the per bushel cost increases to 17.0, 24.1 and 6.7 percent, respectively. The contribution of the energy-intensive components of production costs—fertilizer, chemicals, fuel, lubricants and electricity—were 13.4 percent for corn, 6.7 percent for soybeans and 9.4 percent for wheat per bushel. The production-weighted average increase in the cost of production due to these energy-intensive inputs for these crops was 11.5 percent between 2002 and 2007. In addition to the increase in production costs, transport costs also increased due to higher fuel costs and the margin between domestic and export prices reflect this cost. However, these margins also include handling and other charges, such as insurance, which increase with crop prices. The margin for corn between central Illinois cash and the Gulf Ports barge increased from $0.36 to $0.72 per bushel for an increase of 15.5 percent, while the margin between Kansas City and the Gulf Ports wheat increased only $1 per metric ton. An export weighted average of these prices suggests that transport costs could have added as much as 10.2 percent to the export prices of corn and wheat. Comparable data was not available for soybeans. Thus, the combined increase in production costs and transport costs for the major U.S. food commodities—corn, soybeans and wheat—was at most 21.7 percent, and this amount likely overstates the increase, because transport costs are not estimated separately. It therefore seems reasonable to conclude that higher energy and related costs increased export prices of major U.S. food commodities by about 15–20 percent between 2002 and 2007.

Increased biofuel production has increased the demand for food commodities. The use of maize for ethanol grew especially rapidly from 2004 to 2007 and used 70 percent of the increase in global maize production. In contrast, feed use of maize, which accounts for 65 percent of global maize use, grew by only 1.5 percent per year from 2004 to 2007 while ethanol use grew by 36 percent per year. The share of global feed use of total use declined in response to maize price rises from 69 to 64 percent from 2004 to 2007, and from 70 to 67 percent when the feed by-products from biofuel production are included in feed use.

The United States is the largest producer of ethanol from maize and is expected to use about 81 million tons for ethanol in the 2007/08 crop year. Canada, China, and the European Union used roughly an additional 5 million tons of maize for ethanol in 2007, bringing the total use of maize for ethanol to 86 million tons, which was about 11 percent of global maize production. The large use of maize for ethanol in the U.S. has important global implications, because the U.S. accounts for about one-third of global maize production and two-thirds of global exports and used 25 percent of its production for ethanol in 2007/08.

About 7 percent of global vegetable oil supplies were used for biodiesel production in 2007 and about one-third of the increase in consumption from 2004 to 2007 was due to biodiesel. The largest biodiesel producers were the European Union, the United States, Argentina, Australia, and Brazil, with a combined use of vegetable oils for biodiesel of about 8.6 million tons in 2007 compared with global vegetable oils production of 132 million tons according to the USDA (2008f). From 2004 to 2007, global consumption of vegetable oils for all uses increased by 20.8 million tons, with food use accounting for 80 percent of total use and 60 percent of the increase. Industrial uses of vegetable oils (which include biodiesel) grew by 15 percent per annum from 2004 to 2007, compared with 4.2 percent per annum for food use. The share of industrial use of total use rose from 14.4 percent in 2004 to 18.7 percent in 2007.

Imports of vegetable oils by the EU and U.S. have increased substantially, with the EU-27 increasing imports from 4.4 to 6.9 million tons from 2000 to 2007 and the U.S. increasing imports from 1.7 to 2.9 million tons. The large imports coincided with the increase in biodiesel production in the EU-27 from .45 billion gallons in 2004 to 1.9 billion gallons in 2007 and from .03 billion gallons in the U.S. in 2004 to an estimated .44 billion gallons in 2007.

Brazilian ethanol production from sugar cane has not contributed appreciably to the recent increase in food commodities prices, because Brazilian sugar cane production has increased rapidly and sugar exports have nearly tripled since 2000. Brazil uses approximately half of its sugar cane to produce ethanol for domestic consumption and exports and the other half to produce sugar. The increase in cane production has been large enough to allow sugar production to increase from 17.1 million tons in 2000 to 32.1 million tons in 2007 and exports to increase from 7.7 million tons to 20.6 million tons. Brazil's share of global sugar exports increased from 20 percent in 2000 to 40 percent in 2007, and that was sufficient to keep sugar price increases small except for 2005 and early 2006 when Brazil and Thailand had poor crops due to drought.

The increases in biofuels production in the EU, U.S., and most other biofuel-producing countries have been driven by subsidies and mandates. The U.S. has a tax credit available to blenders of ethanol of $0.51 per gallon and an import tariff of $0.54 per gallon, as well as a biodiesel blenders tax credit $1.00 per gallon. The U.S. mandated 7.5 billion gallons of renewable fuels by 2012 in its 2005 legislation and raised the mandate to 15 billion gallons of ethanol from conventional sources (maize) by 2022 and 1.0 billion gallons of biodiesel by 2012 in energy legislation passed in late-2007. The new U.S. mandates will require ethanol production to more than double and biodiesel production to triple if they are met from domestic production. The EU has a specific tariff of €0.192/liter of ethanol (€0.727 or about $1.10 per gallon) and an ad valorem duty of 6.5 percent on biodiesel. EU member states are permitted to exempt or reduce excise taxes on biofuels, and several EU member states have introduced mandatory blending requirements. Individual member states have also provided generous excise tax concessions without limit, and Germany for example, has provided tax exemptions of €0.4704/ ($0.64) per liter of biodiesel and €0.6545 ($0.88) per liter of ethanol prior to new legislation in 2006. These strong incentives and mandates encouraged the rapid expansion of biofuels in both the EU and U.S.

The EU began to rapidly expand biodiesel production after the EU directive on biofuels entered into effect in October 2001 stipulating that national measures must be taken by EU countries aimed at replacing 5.75 percent of all transport fossil fuels with biofuels by 2010. This led to an increase in biodiesel production from 0.28 billion gallons in 2001 to 1.78 billion gallons in 2007 (FAPRI 2008). Rapeseed was the primary feedstock, followed by soybean oil and sunflower oil. The combined use of vegetable oils for biodiesel was 6.1 million tons in 2007 compared with about 1.0 million tons in 2001.

The U.S. expanded its biodiesel production following legislation passed in 2004 which took effect in January 2005, providing an excise tax credit of US$1.00 per gallon of biodiesel made from agricultural products. This contributed to an increase in biodiesel production in the U.S. from 0.03 billion gallons in 2005 to .44 billion gallons in 2007 and used 3.0 million tons of soybean oil and 0.3 million tons of other fats and oils. These two policies encouraged the rapid expansion of oilseeds production for biodiesel and contributed to the surge in vegetable oils prices, with annual average soybean oil prices rising from $354/ton in 2001 to $881 per ton in 2007. Monthly soybean oil prices rose to $1,522/ton in June 2008. Since oilseeds are close substitutes and prices highly correlated, this led to similar increases in other oilseeds prices.

Land use changes due to expanded biofuel's feedstock production have been large and have led to reduced production of other crops. The U.S. expanded maize area 23 percent in 2007 in response to high maize prices and rapid demand growth for maize for ethanol production. This expansion resulted in a 16 percent decline in soybean area, which reduced soybean production and contributed to a 75 percent rise in soybean prices between April 2007 and April 2008.

While maize displaced soybeans in the U.S., other oilseeds displaced wheat in the EU and other wheat exporting countries. The expansion of biodiesel production in the EU diverted land from wheat and slowed the increase in wheat production, which would have otherwise kept wheat stocks higher. In response to the increased demand and rising prices for oilseeds, land planted to oilseeds increased, especially rapeseed and to a lesser extent sunflower. The increase was primarily in the countries that are also major wheat exporters such as Argentina, Canada, the EU, Russia, and Ukraine. Oilseeds and wheat are grown under similar climatic conditions and in similar areas and most of the expansion of rapeseed and sunflower displaced wheat or was on land that could have grown wheat. The 8 largest wheat exporting countries expanded area in rapeseed and sunflower by 36 percent (8.4 million hectares) between 2001 and 2007 while wheat area fell by 1.0 percent. The wheat production potential of this land was 26 million tons in 2007 based on average wheat yields in each country, and the cumulative wheat production potential of that land totaled 92 million tons from 2002 to 2007. To illustrate the impact of this land shift on wheat stocks, . . . [simulations show that] if the land planted to rapeseed and sunflower had been planted to wheat and if wheat stocks had increased by the same amounts . . . wheat stocks would have been almost as large in 2007 as in 2001 rather than lower by almost half.

Export bans and restrictions fueled the price increases by restricting access to supplies. A number of countries have imposed export restrictions or bans on grain exports to contain domestic price increases. These include Argentina, India, Kazakhstan, Pakistan, Ukraine, Russia, and Vietnam. . . . According to the USDA and the International Grains Council, there were no other important market developments at that time that could account for the subsequent rice price increases. The USDA had projected India to export 4.1 million tons in the month prior to the ban and that was revised to 3.4 million tons in the month following the ban. The ban on exports led to a steady increase in prices over the following weeks. While it is probably not correct to say that all of the price increases were due to the ban, it likely focused attention on the market fundamentals and the rise in wheat prices and caused market participants to reconsider their imports and exports.

Rice is not used for biofuels, but the increase in prices of other commodities contributed to the rapid rise in rice prices. Rice prices almost tripled from January to April 2008 despite little change in production or stocks. This increase was mostly in response to the surge in wheat prices in 2007 (up 88 percent from January to December) which raised concerns about the adequacy of global grain supplies and encouraged several countries to ban rice exports to protect consumers from international price increases, and caused others to increase imports.

Weather-related production shortfalls have been identified as a major factor underpinning world cereals prices, especially in Australia, U.S., EU, Canada, Russia, and Ukraine. The back-to-back droughts in Australia in 2006 and 2007 reduced grain exports by an average of 9.2 million tons per year compared with 2005, and poor crops in the EU and Ukraine reduced their exports by an additional 10 million tons in 2007. However, these declines were more than offset by large crops in Argentina, Kazakhstan, Russia, and the U.S. Total grain exports from these countries in 2007 increased by about 22 million tons compared with 2006. Global grain production did decline by 1.3 percent in 2006, but it then increased 4.7 percent in 2007. Thus, the production shortfall in grains would not, by itself, have been a major contributor to the increase in grain prices. But when combined with large increases in biofuels production, land use changes, and stock declines it undoubtedly contributed to higher prices. The production shortfall was most significant in wheat, where global production declined 4.5 percent in 2006 and then increased only 2 percent in 2007. Global oilseed production rose 5.4 percent in 2006/07 and declined 3.4 percent in 2007/08.

Rapid income growth in developing countries has not led to large increases in global grain consumption and was not a major factor responsible for the large grain price increases. However, it has contributed to increased oilseed demand and higher oilseed prices as China increased soybean imports for its livestock and poultry industry. Both China and India have been net grain exporters since 2000, although exports have declined as consumption has increased. Global consumption of wheat and rice grew by only 0.8 and 1.0 percent per annum, respectively, from 2000 to 2007 while maize consumption grew by 2.1 percent (excluding the demand for biofuels in the U.S.). This was slower than demand growth during 1995–2000 when wheat, rice, and maize consumption increased by 1.4, 1.4, and 2.6 percent per annum, respectively.

Other factors, such as the decline of the dollar, contributed to food commodity price increases. . . .

Speculative and investor activity has also increased and could have contributed to food price increases. A reflection of this increased activity was the quadrupling of the number of wheat futures contacts traded on the Chicago Board of Trade from 2002 to 2006. However, the increase in futures contracts does not coincide closely with the increase in wheat prices, which raises doubts about the impact on prices. The impact on prices is hard to quantify and most studies do not find that such activity changes prices from the levels which would have prevailed without such activity, however, they may change the rate of adjustment to a new equilibrium when fundamental factors change.

Summary and Conclusions

The increase in internationally traded food prices from January 2002 to June 2008 was caused by a confluence of factors, but the most important was the large increase in biofuels production from grains and oilseeds in the U.S. and EU. Without these increases, global wheat and maize stocks would not have declined appreciably, and price increases due to other factors would have been moderate. Land use changes in wheat exporting countries in response to increased plantings of oilseeds for biodiesel production limited expansion of wheat production that could have otherwise prevented the large declines in global wheat stocks and the resulting rise in wheat prices. The rapid rise in oilseed prices was caused mostly by demand for biodiesel production in response to incentives provided by policy changes in the EU beginning in 2001 and in the U.S. beginning in 2004. The large increase in rice prices was largely a response to the increase in wheat prices rather than to changes in rice production or stocks, and was thus indirectly related to the increase in biofuels. Recent export bans on grains and speculative activity would probably not have occurred without the large price increases due to biofuels production because they were largely responses to rising prices. Higher energy and fertilizer prices would have still increased crop production costs by about 15–20 percentage points in the U.S. and lesser amounts in countries with less intensive production practices. The back-to-back droughts in Australia would not have had a large impact because they only reduced global grain exports by about 4 percent and other exporters would normally have been able to offset this loss. The decline of the dollar has contributed about 20 percentage points to the rise in dollar food prices.

Thus, the combination of higher energy prices and related increases in fertilizer prices and transport costs, and dollar weakness caused food prices to rise by about 35–40 percentage points from January 2002 until June 2008. These factors explain 25–30 percent of the total price increase, and most of the remaining 70–75 percent increase in food commodities prices was due to biofuels and the related consequences of low grain stocks, large land use shifts, speculative activity and export bans. It is difficult, if not impossible, to compare these estimates with estimates from other studies because of different methodologies, widely different time periods considered, different prices compared, and different food products examined, however, most other studies have also recognized biofuels production as a major factor driving food prices. The increase in grain consumption in developing countries has been moderate and did not lead to large price increases. Growth in global grain consumption (excluding biofuels) was only 1.7 percent per annum from 2000 to 2007, while yields grew by 1.3 percent and area grew by 0.4 percent, which would have kept global demand and supply roughly in balance. This was slower than growth during 1995–2000 when wheat, rice and maize consumption increased by 1.4, 1.4, and 2.6 percent per annum, respectively.

The large increases in biofuels production in the U.S. and EU were supported by subsidies, mandates, and tariffs on imports. Without these policies, biofuels production would have been lower, and food commodity price increases would have been smaller. Biofuels production from sugar cane in Brazil is lower-cost than biofuels production in the U.S. or EU and has not raised sugar prices significantly because sugar cane production has grown fast enough to meet both the demand for sugar and ethanol. Removing tariffs on ethanol imports in the U.S. and EU would allow more efficient producers such as Brazil and other developing countries, including many African countries, to produce ethanol profitably for export to meet the mandates in the U.S. and EU. Biofuels policies which subsidize production need to be reconsidered in light of their impact on food prices.

EXPLORING THE ISSUE

Are Biofuels a Reasonable Substitute for Fossil Fuels?

Critical Thinking and Reflection

1. Why is there so much interest in using ethanol as a portion of automobile fuel?
2. Why does the production of biofuels affect the food supply and food prices?
3. Is using biofuels to replace fossil fuels compatible with feeding the world's population?
4. What alternatives exist to using liquid fuels to power automobiles?
5. Which biofuels are likely to have the least undesirable impact on the environment and food supply?

Is There Common Ground?

There is agreement that using prime agricultural land and expensive fertilizers and processing to grow biofuel crops necessarily reduces resources needed to grow food and unsurprisingly has effects on food supply and prices. However, there are biofuel crops that are less dependent on agricultural resources. Investigate what is happening with "cellulosic" biofuels, algae-based biofuels, and bacteria-based biofuels and answer the following questions.

1. Can these alternatives be implemented in the near future?
2. Will their prices be competitive?
3. Can they meet demand?
4. Will their production compete with food production?

ISSUE 18

Can Hydropower Play a Role in Preventing Climate Change?

YES: Alain Tremblay, Louis Varfalvy, Charlotte Roehm, and Michelle Garneau, from "The Issue of Greenhouse Gases from Hydroelectric Reservoirs: From Boreal to Tropical Regions," United Nations Symposium on Hydropower and Sustainable Development, Beijing, China (October 27–29, 2004)

NO: American Rivers, from "Hydropower: Not the Answer to Preventing Climate Change" (www.americanrivers.org) (2007)

Learning Outcomes

After studying this issue, students will be able to:

- Explain why greenhouse gas emissions may increase after the land is flooded by a reservoir.
- Describe the environmental benefits of hydroelectric power.
- Describe the environmental drawbacks of hydroelectric power.
- Explain how energy can be obtained from water without building dams.

ISSUE SUMMARY

YES: Alain Tremblay, Louis Varfalvy, Charlotte Roehm, and Michelle Garneau, researchers with Hydro-Quebec and the University of Quebec in Montreal, argue that hydropower is a very efficient way to produce electricity, with emissions of greenhouse gases between a tenth and a hundredth of the emissions associated with using fossil fuels.

NO: American Rivers, a nonprofit organization dedicated to the protection and restoration of North America's rivers, argues that suggesting that hydropower is the answer to global warming hurts opportunities for alternative renewable energy technologies such as solar and wind and distracts from the most promising solution, energy efficiency.

Dams have long been an icon of civilization. Building dams of all sizes, from those that hold back village mill ponds to the giant Hoover Dam, was a crucial step in the settling and development of America. They supplied mills with the mechanical power generated when flowing water spun waterwheels. Combined with locks and canals, dams improved the navigability of waterways in the days before railroads. They provided water for irrigation, reduced flooding, and generated electricity. In other parts of the world, building dams for these benefits has been an important step in moving from "undeveloped" to "developing" to "developed" status. See Ibrahim Yuksel, "Hydropower in Turkey for a Clean and Sustainable Energy Future," *Renewable & Sustainable Energy Reviews* (August 2008).

In the United States, the Army Corps of Engineers (ACE) built almost all hydroelectric power systems until the late 1970s. Today, according to the ACE pamphlet, "Hydropower: Value to the Nation" (www.vtn.iwr.usace.army.mil/pdfs/Hydropower.pdf), the ACE is the single largest owner and operator of hydroelectric power plants in the country and one of the largest in the world. Its 75 power plants produce nearly 100 billion kilowatt-hours of electricity per year. Hydroelectric power is renewable, efficient, and clean. It does not generate air or water pollution, and it emits no greenhouse gases to contribute to global warming.

Does this mean that building more hydroelectric power plants could help solve the global warming crisis? In the United States, most good sites for large hydropower plants have already been developed. In the rest of the world, the best sites, such as China's Three Gorges project, are rapidly being developed. Unfortunately, many people see problems in these projects. Mara Hvistendahl's "China's Three Gorges Dam: An Environmental Catastrophe?" *Scientific American* (online) (March 25, 2008) (www.sciam.com/article.cfm?id=chinas-three-gorges-dam-disaster), is indicative. In Chile, the need for electricity versus the risks to the environment is still being debated; see Gaia Vice, "Dams for Patagonia," *Science* (July 23, 2010). Because dams flood the land behind them, they destroy forests, farmland, and villages and displace thousands—even millions—of people. Species decline and vanish. Sediment trapped behind the dam no longer reaches the sea to nourish fisheries. Diseases change their patterns. When a dam breaks, the resulting sudden, immense flood can do colossal damage downstream.

Fortunately, it is possible to get energy from water without building lakes. Tidal power requires dams across estuaries; at high tide, water flows upstream through turbines; at low tide, the turbines are reversed to extract energy as the water flows back to the sea. Turbines can also be placed on the sea floor to take advantage of tidal (and other) currents. See Jonathon Porritt, "Catch the Tide," *Green Futures* (January 2008), and David Kerr, "Marine Energy," *Philosophical Transactions: Mathematical, Physical & Engineering Sciences* (April 2007). In North America, a number of tidal power projects are being started in the Bay of Fundy; see Colin Woodard, "On US Border, a Surge in Tidal-Power Projects," *Christian Science Monitor* (August 15, 2007).

There is also wave power, which uses the motion of waves to work special arrangements of pistons, levers, and air chambers to extract energy. See Ewen Callaway, "To Catch a Wave," *Nature* (November 8, 2007), David C. Holzman, "Blue Power: Turning Tides into Electricity," *Environmental Health Perspectives* (December 2007), and Elisabeth Jeffries, "Ocean Motion Power," *World Watch* (July/August 2008). Yet, even this can draw attention from environmentalists, who fear that the equipment may interfere with marine animals and that associated electromagnetic fields may harm sensitive species such as sharks and salmon; see Stiv J. Wilson, "Wave Power," *E Magazine* (May/June 2008). However, Urban Henfridsson et al., "Wave Energy Potential in the Baltic Sea and the Danish Part of the North Sea, with Reflections on the Skagerrak," *Renewable Energy: An International Journal* (October 2007), note that the potential contribution of wave energy to civilization's needs is large. Taking advantage of that potential requires "Sound engineering, in combination with producer, consumer and broad societal perspective."

One subject of debate has been whether dams really help with reducing the addition of greenhouse gases to the atmosphere and the consequent global warming. According to the International Hydropower Association (http://hydropower.org), the lakes behind dams do emit greenhouse gases, and the land covered by the lake no longer grows trees that absorb carbon dioxide from the atmosphere. But before the lake existed, the land emitted greenhouse gases, although not all such lands grew trees, and lake sediments could store large amounts of organic matter. The real issue is the *net* effect, which is ignored by those who wish to discredit hydroelectric power, who also focus on the worst cases, which tend to be in tropical environments.

The net emissions of greenhouse gases from hydropower impoundments are an area of ongoing research. In Canada, one such research program focuses on the large Eastmain-1 impoundment in northwestern Quebec, near James Bay. Commissioned in April 2007, it can generate 480 megawatts of electricity. Annual output is expected to average 2700 gigawatt-hours (2.7 billion kilowatt-hours). Alain Tremblay is the coordinator of the Reservoirs' Net Greenhouse Gas Emissions project at the site. In the YES selection, he and coauthors Louis Varfalvy, Charlotte Roehm, and Michelle Garneau argue that hydropower is a very efficient way to produce electricity, with emissions of greenhouse gases between a tenth and a hundredth of the emissions associated with using fossil fuels. In the NO selection, American Rivers, a nonprofit organization dedicated to the protection and restoration of North America's rivers, argues that suggesting that hydropower is the answer to global warming hurts opportunities for alternative renewable energy technologies such as solar and wind and distracts from the most promising solution, energy efficiency.

A year after Tremblay et al. gave their talk at the United Nations Symposium on Hydropower and Sustainable Development in Beijing, China, they published *Greenhouse Gas Emissions—Fluxes and Processes: Hydroelectric Reservoirs and Natural Environments* (Springer, 2005). The research continues today (see Suzanne Pritchard, "Emitting the Truth," *International Water Power and Dam Construction* (January 16, 2008)), but even though hydropower seems a relatively benign energy source, the objections also continue. Sara Phelan,

"Bubbling Waters," *Earth Island Journal* (Autumn 2007), notes reports that suggest that any gains in regard to carbon emissions are offset by methane emissions. Hydropower also has genuine undesirable environmental effects. There is thus a tension between real benefits and real drawbacks, as discussed by R. Sternberg in "Hydropower: Dimensions of Social and Environmental Coexistence," *Renewable & Sustainable Energy Reviews* (August 2008). Aviva Imhof and Guy R. Lanza, "Greenwashing Hydropower," *World Watch* (January/February 2010), stress the drawbacks and argue that "the industry's attempt to repackage hydropower as a green, renewable technology is both misleading and unsupported by the facts, and alternatives are often preferable. In general, the cheapest, cleanest, and fastest solution is to invest in energy efficiency."

Alain Tremblay et al.

The Issue of Greenhouse Gases from Hydroelectric Reservoirs: From Boreal to Tropical Regions

Abstract

The role of greenhouse gas emissions (GHG) from freshwater reservoirs and their contribution in increasing atmospheric GHG concentrations is actually well discussed worldwide. The amount of GHGs emitted at the air-water interface of reservoirs varies over time. The maximum is attained within 3 to 5 years after impoundment. In reservoirs older >10 years in boreal and semi-arid regions, GHG emissions are similar than those of natural lakes. In tropical regions, the time to return to natural values may be longer depending on the water quality conditions. Hydropower is a very efficient way to produce electricity, showing emission factors between one and two orders of magnitude lower than the thermal alternatives.

1.0 Introduction

The major greenhouse gases are carbon dioxide (CO_2), methane (CH_4) and nitrous oxide (N_2O). These gases are emitted from both natural aquatic (lakes, rivers, estuaries, wetlands) and terrestrial ecosystems (forest, soils) as well as from anthropogenic sources. According to both the European Environment Agency and the United States Environmental Protection Agency, CO_2 emissions account for the largest share of GHGs equivalent of ±80–85% of the emissions. Fossil fuel combustion for transportation and electricity generation are the main source of CO_2 contributing to more than 50% of the emissions. Thermal power plants represents 66% of the world's electric generation capacity. Hydropower represents about 20% of the world's electricity generation capacity and emits 35 to 70 times less GHGs per TWh than thermal power plants. Nevertheless, for the last few years GHG emissions from freshwater reservoirs and their contribution to the increase of GHGs in the atmosphere are actually at the heart of a worldwide debate concerning the electricity generating sector. However, to our knowledge, there are few emission measurements available from these environments although they are at the heart of the debate concerning methods of energy production.

From *United Nations Symposium on Hydropower and Sustainable Development,* by Alain Tremblay, Louis Varfalvy, Charlotte Roehm, and Michelle Garneau (UN Economic & Social Affairs, 27–29 October, 2004). Reprinted by permission of the author. www.un.org/esa/sustdev/sdissues/energy/op/hydro_tremblaypaper.pdf.

In this context, Hydro-Québec and its partners have adapted a technique to measure gross GHG emissions at the air-water interface that allows for a high rate of sampling in a short period of time, while increasing the accuracy of the results and decreasing the confidence intervals of the average flux measured. These measurements were done in order to compare the results with those obtained by other methods, and to better assess the gross emissions of GHG from aquatic ecosystems in boreal (northern climate), semi-arid and tropical regions. This was done also to adequately estimate the contribution of reservoirs compared to natural water bodies in these regions, as well as to compare properly various options of electricity generation.

2.0 Adapting a Simple CO2 and CH4 Emission Measuring Technique

Measuring equipment consists of a one piece Rubbermaid box (polyethylene container) with a surface area of 0.2 m^2 and a height of 15 cm, of which 2 cm is placed below the water surface. The air is sampled through an opening at the top of the floating chamber and it is returned at the opposite end of the chamber. The inlet and the outlet at opposite ends of the chamber allows the trapped air to be continuously mixed. This brings for a more representative measurement of gas concentrations. Air is analysed with a NDIR (Non-Dispersive Infrared) instrument (PP-System model Ciras-SC) or a FTIR (Fourier Transform Infrared) instrument. The accuracy of the instrument in the 350–500 ppm range is of 0.2 to 0.5 ppm for the NDIR and 3 to 5 ppm for the FTIR. The instrument takes continuous reading and the data logger stores a value every 20 seconds over a period of 5 to 10 minutes. All samples are plotted on a graph to obtain a slope and calculate the flux of CO_2 or CH_4 per m^2.

3. Results and Discussion

The results come from a sampling program that was conducted since 2001 in many boreal Canadian provinces, in the semi-arid western region of the United States of America and in tropical region of Panama and French Guiana. The results and conclusions also benefit from the Synthesis "Greenhouse Gas Emissions: Fluxes and Processes, Hydroelectric Reservoirs and Natural Environments." One must keep in mind, that most of the data on GHG from hydroelectric reservoirs come from research and measurements in boreal regions and, to a lesser extent, from a follow-up environmental program of Petit Saut reservoir in tropical French Guiana and a few Brazilian reservoirs.

The chemical, morphological and biological processes determining the fate of carbon in reservoirs are similar to those occurring in natural aquatic ecosystems. However, some of these processes might be temporally modified in reservoirs due to the flooding of terrestrial ecosystems which results from the creation of reservoirs. In boreal reservoirs, environmental follow-up programs have clearly shown that these changes generally last less than 10 years.

However in tropical reservoirs, these changes can extend over a longer period of time according to the conditions of impoundment.

In the case of reservoirs, it is known that the amount of GHGs emitted at the air-water interface varies over time. In fact, there is an initial peak which occurs immediately after impoundment.

Fluxes of GHG in boreal reservoirs are usually 3 to 6 times higher than those from natural lakes when they reach their maximum at 3 to 5 years after impoundment. In boreal reservoirs older than 10 years (10 years for CO_2 and 4 years for CH_4), CO_2 fluxes ranged between –1800 to 11200 mg $CO_2 \cdot m^{-2} \cdot d^{-1}$ and are similar to those of natural systems with flux ranged from –460 to 10800 mg $CO_2 \cdot m^{-2} \cdot d^{-1}$. Generally, degassing and ebullition emissions are not reported for boreal regions because diffusive emissions are considered the major pathway. Methane emissions are very low in these ecosystems; however, they can be substantial in some tropical areas where the ebullition pathway is important.

Despite fewer data available, similar patterns are observed in most of the studied reservoirs in boreal (Finland, British-Columbia, Manitoba, New Foundland–Labrador), semi-arid (Arizona, New Mexico, Utah) and tropical regions (Panama, Brazil, French Guiana). In tropical regions, the time to return to natural values is sometimes longer depending on the water quality conditions. For example, when anoxic conditions occur CH_4 production decreases slowly and might be maintained for longer periods by carbon input from the drainage basin. However, such situations are rare in most of the studied reservoirs.

The increase of GHG emissions in reservoirs shortly after flooding is related to the release of nutrients, enhanced bacterial activity and decomposition of labile carbon. Magnitude of emissions for both reservoirs and natural aquatic systems depend on physico-chemical characteristics of the water body and on the incoming carbon from the watershed.

There is a convergence in the results that clearly illustrates that, in both boreal and tropical reservoirs, the contribution of flooded soils of the reservoir carbon pool is important in the first few years following impoundment. After this period, terrestrial allochthonous (drainage basin level) input of dissolved organic matter (DOC) can exceed, by several times, the amount of particulate organic matter and DOC produced within the reservoir through soil leaching or primary production. In reservoirs, this is particularly important since the water residence time is generally shorter, from a few weeks to a few months, in comparison to several years to many decades in the case of lakes.

Advantages of Hydroelectricity

There is a convergence in the results, from both boreal and tropical reservoirs, that clearly illustrates that reservoirs do emit GHGs for a period of about 10–15 years. Therefore, according to the GHG emission factors reported for hydro reservoirs both by IAEA and by various studies performed during the last decade on a variety and a great number of reservoirs, it can be concluded that the energy produced with the force of water is very efficient, showing emission factors between one and two orders of magnitude lower than the

Table 1

Full Energy Chain Greenhouse Gas Emission Factors in g CO_2 equiv./kWh(e) h^{-1} (modified from IAEA 1996).

Energy Source	Emission Factor* g CO_2 equiv./kWh(e).h
Coal (lignite and hard coal)	940–1340
Oil	690–890
Gas (natural and LNG)	650–770
Nuclear Power	8–27
Solar (photovoltaic)	81–260
Wind Power	16–120
Hydro Power	4–18
Boreal reservoirs (La Grande Complexe)[1]	~33
Average boreal reservoirs[2]	~15
Tropical reservoirs (Petit-Saut)[3]	~455 (gross)/~327 (net)
Tropical reservoirs (Brazil)[4]	~6 to 2100 (average: ~160)

*: Rounded to the next unit or the next tenth respectively for values < or > 100 g CO_2 equiv./kWh(e) h^{-1}.
1: La Grande Complexe, Quebec, 9 reservoirs, 15 784 MW, ~ 174 km^2/TWh, (Hydro-Québec 2000).
2: According to average reservoir characteristics of 63 km^2/TWh, (Hydro-Québec 2000).
3: Petit-Saut, French Guiana, 115 MW, 0.315 MW/km^2 (Chap. 12).
4: 9 Brazilian reservoirs from 216 to 12 600 MW, total power = 23 518 MW, total surface = 7867 km^2.

thermal alternatives (Table 1). However in some cases, tropical reservoirs, such as the Petit Saut reservoir in French Guiana or some Brazilian reservoirs, GHG emissions could, during a certain time period, significantly exceed emissions from thermal alternatives. Similar values have been reported in the literature. With respect to GHG emissions from hydropower, we can provide the following general observations:

- GHG emission factors from hydroelectricity generated in boreal regions are significantly smaller than corresponding emission factors from thermal power plants alternatives (i.e. from < 2% to 8% of any kind of conventional thermal generation alternative);
- GHG emission factors from hydroelectricity generated in tropical regions cover a much wider range of values (for example, a range of more than 2 order of magnitude for the 9 Brazilian reservoirs). Based on a 100 year lifetime, these emissions factors could either reach very low or very high values, varying from less than 1% to more than 200% of the emission factors reported for thermal power plant generation;
- Net GHG emission factors for hydro power, should be at first sight 30% to 50% lower than the emission factors currently reported.

The vast majority of hydroelectric reservoirs built in boreal regions are emitting very small amounts of GHGs, and represent therefore one of the cleanest way to generate electricity. In some tropical reservoirs, anoxic conditions could lead to larger emissions. Therefore, GHG emissions from tropical reservoirs should be considered on a case by case base level.

References

Cole JJ, Caraco FC, Kling GW, Kratz TK (1994). Carbon dioxide supersaturation in the surface waters of lakes. Science 265:1568–1570.

Fearnside PM (1996). Hydroelectric dams in Brazilian Amazonia: response to Rosa, Schaeffer and dos Santos. Environ Conserv 23:105–108.

Gagnon L, Chamberland A (1993). Emissions from hydroelectric reservoirs and comparison of hydroelectric, natural gas and oil. Ambio 22:568–569.

Gagnon L, van de Vate JF (1997). Greenhouse gas emissions from hydropower, the state of research in 1996. Ener Pol 25:7–13.

Hope D, Billet MF, Cressner MS (1994). A review of the export of carbon in river water: fluxes and processes. Environ Pollut 84:301–324.

IAEA, Hydro-Québec (1996). Assessment of greenhouse gas emissions from the full energy chain for hydropower, nuclear power and other energy sources. Working material: papers presented at IAEA advisory group meeting jointly organized by Hydro-Québec and IAEA (Montréal, 12–14 March, 1996), Montréal, International Atomic Energy Agency et Hydro-Québec.

St. Louis VL, Kelly CA, Duchemin E, Rudd JWM, Rosenberg DM (2000). Reservoirs surfaces as sources of greenhouse gases to the atmosphere: a global estimate. BioScience 50:766–775.

Tremblay, A., Lambert, M and Gagnon, L., 2004a. CO_2 Fluxes from Natural Lakes and Hydroelectric Reservoirs in Canada. Environmental Management, Volume 33, Supplément 1: S509–S517.

Tremblay, A., Varfalvy, L., Roehm, C., and Garneau, M. (Eds.), 2004b. Greenhouse gas Emissions: Fluxes and Processes, Hydroelectric Reservoirs and Natural Environments. Environmental Science Series, Springer, Berlin, Heidelberg, New York, 731 pages.

Hydropower: Not the Answer to Preventing Climate Change

Scientists from around the world have sounded the alarm that climate change (a.k.a. global warming) is one of the greatest long-term threats facing the natural and human environment today.

Although it is difficult to predict precisely how and when the impacts of human caused climate change will manifest, likely consequences could include catastrophic storms, severe droughts, wide-spread epidemics, loss of crops, and massive species extinctions.

The river conservation community is equally concerned about climate change. The impacts that this phenomenon will have on the planet are not something that we take lightly and we adamantly support efforts to reduce this pressing threat. However, we need not sacrifice the health of our nation's rivers to prevent climate change.

The leading cause of climate change is the accumulation of atmospheric carbon dioxide which causes an increase in global temperatures through a "greenhouse effect."

The greatest contributor of human made carbon dioxide is the burning of fossil fuels, such as coal and oil, for the generation of electricity. Hydropower is often promoted as the answer to the world's greenhouse woes due to its "emissions free" generation.

It is true that generation of hydropower does not create significant air pollution or greenhouse gases but the climate benefits of hydropower do not come without significant environmental costs. Boosting U.S. hydropower generation—either through building new dams or relaxing environmental controls at existing dams—would not generate a significant amount of extra power, and would come at a terrible cost to rivers, fish, and wildlife and recreation.

Suggesting that hydropower is the answer to the climate change problem not only hurts opportunities for emerging renewable technologies such as solar and wind, but it also distracts us from the most promising solutions to this dilemma—energy efficiency.

Is Hydropower Renewable?

No. Moving Water May Be Renewable But Endangered River Species Are Not.

It is true that generation of hydropower does not create significant air pollution or greenhouse gases but this power generation definitely has environmental impacts. Of the 3.5 million river miles in the United States, almost 600,000 or 20 percent are impounded by dams, with thousands more miles indirectly affected downstream. Only recently have we begun to recognize the price that these dams have wrought on our environment.

According to The Nature Conservancy, fresh water aquatic species are up to 10 times more endangered than their terrestrial or avian cousins, in no small measure due to dams. In addition, valuable fisheries throughout the country have been adversely affected by dams. Dams create physical barriers, corrupt natural flows, and alter water chemistry and temperature as evidenced in the Columbia, Colorado, Mississippi, and Missouri River basins. On the Columbia River, salmon runs are only 2% of their historic levels.

Do We Need More Hydropower to Meet the Kyoto Treaty Targets? No!

According to the treaty on Climate Change signed in December 1997 in Kyoto, Japan, the United States is obligated to reduce its emissions to 7% below 1990 levels. The Energy Information Administration's 1997 preliminary estimates for carbon emissions show we are currently emitting close to 10% above 1990 levels, meaning that the U.S. must reduce emissions by 17% from today's current output.

Data from the Energy Information Administration suggests that hydropower currently makes up only 10% of the nation's annual electric generation, with approximately half of that coming from the nation's 2,000 non-federal owned hydropower dams. Existing hydropower can continue to make a contribution to the nation's energy needs as a source of electricity with very limited climate impact, we must not—and have no need to—sacrifice our free-flowing rivers, fish and wildlife in the process.

Damming more rivers to produce more hydropower, however, is not a realistic option for meeting the Kyoto targets. For the most part, the good dam sites in the U.S. are already dammed. Those river reaches that remain undammed are either prohibitively expensive, protected lands and waterways, or have a far greater value as recreational, aesthetic, or environmental resources.

Are Environmental Regulations Severely Reducing Hydropower's Generating Capacity? No.

The operation of these dams is regulated by the Federal Energy Regulatory Commission (FERC), which grants 30 to 50 year licenses to dam owners to operate and profit from the public's waterways. When a dam owner's license expires, the operator must apply for a new license from FERC. As part of this relicensing process, FERC can impose new conditions to protect fish and wildlife as well as accommodate other uses of the river such as recreation. A dam owner can also decide not to seek a new license, or FERC could deny the application for a new license and order the dam removed.

Some in the hydropower industry argue that by imposing environmental conditions, the government is cutting hydropower production and therefore leading to further emissions of greenhouse gases. According to FERC however, the changed license conditions of the 273 dam licenses which expired in 1993 resulted in an average loss of only 1% of annual generation. In fact, according to statistics kept by the Energy Information Administration, hydropower generation is at an all-time high despite environmental regulation imposed during the last several years of relicensing.

Over the next 12 years, more than 500 dams or approximately half of FERC-regulated hydropower capacity (2.5% of the annual generation of the United States), will come up for relicensing. Based on FERC's experience from their 1993 class of dams, we can expect about a 1% reduction in the annual generation of these projects and therefore a 0.025% reduction in the nation's over-all annual generation.

That is only enough energy to meet the country's needs for 2.2 hours of an entire year!!! This 0.025% reduction of annual generation is significantly less than the 5% average fluctuation of energy demand caused by factors such as weather, fuel prices, and technology. This small loss could be made up by retrofitting one out of every 30 homes in the nation with just one energy efficient compact fluorescent light bulb!

Is This Small Loss in Power Really Worth It? Yes.

The licensing requirements that lead to losses in power generation at hydropower dams put water back into stream channels, restoring valuable fisheries, improving flood control, and ensuring compliance with state water quality standards. Given the perilous state of rivers and riverine species, and the ability of rivers to heal themselves once natural flows are restored, the cost of regulation is not only a small price to pay, it is a necessary one.

Relicensing of FERC regulated hydropower dams can and has rehabilitated rivers. For example, on the Deerfield River, a settlement signed in October 1994, between the New England Power Company and state, federal and resource agencies enhanced thirty-three river miles and returned water to twelve miles of river, which had been previously de-watered by diversions. This same project was recently sold at a price well above what was expected by market analysts. Along the Au Sable, Manistee, and Muskegon Rivers, license conditions calling for restoration of a more natural flow regime have already resulted in significant increases in natural fish reproduction.

Operational changes can also guarantee equal use of the public's rivers for what are often more profitable recreational interests. In 1996, 29 million anglers spent nearly $31 billion on fishing related products and services. In 1992, the economic value from power production of hydropower

Project #3 on the Kern River in California was $17,000 per day. By contrast, the total economic benefit from commercial rafting on the Kern was in excess of $35,000 per day. Mitigation efforts at each of these dams will continue to improve fish and wildlife habitat and water quality while maintaining profitable operations for the dam owner.

But Won't These Reductions in Hydropower Generation Make a Significant Contribution to Climate Change? No.

Even assuming that this small amount of power would need to be replaced—and it is unclear that it would—it is likely that, given the current market, most new generation will consist of natural gas turbines. While these facilities do emit greenhouse gases and use non-renewable fossil fuels, they are 35% more efficient than coal and emit 40% less carbon dioxide.

Emerging technologies such as solar, wind, and cogeneration are becoming increasingly cost competitive and are likely to meet growing demand. But conservation and efficiency are perhaps the greatest untapped resource in meeting the energy needs of the future. None of these substitutes will add significantly to greenhouse gas emissions. Thus, any fear that lost generation would be made up from the dirtiest coal-fired power plant is unrealistic and unfounded.

Won't Interest in Dam Removal Lead to Dramatic Decreases in Hydropower Capacity? No.

No one is advocating the decommissioning and removal of all hydropower dams in this country. The vast majority of dam removal in this country will affect small, non-generating dams that no longer serve their original purpose. Beyond those, the Hydropower Reform Coalition supports the removal of a small number of economically unviable hydropower dams and a handful of projects that cause acute environmental harm, which cannot be mitigated.

The generating capacity of those non-federal projects that the Coalition has identified add up to less than 100 MW and accounts for less than 0.2% of total non-federal hydropower capacity. Only 0.006% of carbon emissions from natural gas or coal generation would be saved if the projects remained in production. These dam removals will result in big river benefits with virtually no negative climate impacts.

How Will We Meet the Kyoto Targets? Efficiency and Renewables.

Cutting back environmental protection at existing hydropower dams will not get us where we need to be to meet the Kyoto targets; neither will development of the few marginal sites left in this country. Natural gas provides an interim solution, but is not sustainable in the long run and can also have adverse impacts on rivers through water withdrawals and thermal pollution. During this interim period the U.S. should focus research and development on truly renewable sources of energy, such as solar, wind, and new technologies like fuel cells. There is still an enormous amount of growth potential in these technologies.

However, the best and most immediate answers lie not with new generation sources but instead with more efficient use of existing resources. Rather than continue pouring water into a leaky bucket, we need to patch the holes! According to the Rocky Mountain Institute, an energy think-tank, we can easily halve what we currently use while improving our standard of living through additional use of more efficient lighting, heating, cooling, building, and industrial processes.This is not futuristic talk of technologies on the horizon. We have the capability—all we need are the right incentives.

Buildings—Energy and carbon savings can be made at no additional cost by implementing energy efficient building practices, such as those advocated by the 1995 modern energy codes, which include selecting appropriate building materials and strategically placing windows and doors to reduce the need for heating and cooling systems and light fixtures.

Lighting—The initial costs of compact fluorescent light bulbs and Energy Star light fixtures are balanced by long-term economic and environmental benefits. Already this program is saving more than 70 billion kilowatt-hours annually (about a quarter of the conventional hydroelectric generation in the U.S. or 4% of annual coal generation).

Insulation—Ceiling, wall, and foundation insulation in existing buildings is saving the United States about 12 quads—that's 15 percent of the U.S. national energy bill—and another 2.2 quads could be saved insulating all new and existing buildings to the levels recommended by modern energy codes.

Investing in wind and photovoltaic technologies—Wind generation and photovoltaics are two efficient and consumer-friendly alternatives to hydropower generation. Since the 1980's, wind power prices have dropped to near competitive rates (44/kWh) and are in high demand in markets across the country. Thirty-seven U.S. states have a wind resource that could support development of utility-scale wind plants—13 of these exceed the wind resource available in California, which is the world's leader in wind energy development. Photovoltaics are advantageous because of their easy installation, adaptability to consumers needs, and their independence from electric transmission systems. Manufacturing photovoltaic cells could provide an excellent opportunity for high-tech companies adapting to shrinking military contracts.

Does Reducing the Threat of Global Climate Change Have to Harm Rivers? No.

The nation's solution to the climate change problem need not involve trading jobs or economic prosperity for the environment. Nor should it mean that we trade healthy rivers for clean air and reduced threat of climate change. We can meet and surpass our climate treaty obligations by 2010 at low or no cost per unit saved by building better buildings, using better lights, using insulation, improving industrial practices, constructing combined cycle gas turbines, and investing in wind and photovoltaic technologies.

When people incorporate energy-efficient appliances and heating and cooling systems into their homes and businesses they save money and the environment. The future economic and environmental benefits outweigh the initial costs of making the transition to more energy-efficient systems and living practices. In our effort to reduce the threat of climate change we must continue to protect the health of our nation's rivers.

EXPLORING THE ISSUE

Can Hydropower Play a Role in Preventing Climate Change?

Critical Thinking and Reflection

1. Why may greenhouse gas emissions increase after the land is flooded by a reservoir?
2. In some parts of the world, much of the water that rivers carry comes from melting snowpack and glaciers. In a "global warming" world, can a hydroelectric dam on such a river be said to provide "renewable" electricity?
3. Can improved energy efficiency meet the needs of developing nations?
4. The Introduction mentions ways of extracting energy from waves and ocean currents. Might any of these techniques work in rivers?

Is There Common Ground?

Many of the drawbacks of hydroelectric power have to do with constructing dams. Instream turbines, such as those proposed for generating electricity from ocean currents, can also be used in rivers (although they may pose navigational problems). Research this topic (begin at Hydro Green Energy, www.hgenergy.com/current.html) and answer the following questions:

1. Are there environmental drawbacks?
2. What can turbine operators do if a river dries up?
3. Are they affordable and practical in developing countries?

ISSUE 19

Is It Time to Put Geothermal Energy Development on the Fast Track?

YES: Susan Petty, from "Testimony on the National Geothermal Initiative Act of 2007 Before the Senate Committee on Energy and Natural Resources" (September 26, 2007)

NO: Alexander Karsner, from "Testimony on the National Geothermal Initiative Act of 2007 Before the Senate Committee on Energy and Natural Resources" (September 26, 2007)

Learning Outcomes

After studying this issue, students will be able to:

- Explain how geothermal energy is produced.
- Explain why geothermal energy is considered "renewable."
- Explain how increased governmental funding will increase development of geothermal power.
- Describe the hazards associated with geothermal energy.
- Describe the benefits of geothermal energy.

ISSUE SUMMARY

YES: Susan Petty, president of AltaRock Energy, Inc., argues that the technology already exists to greatly increase the production and use of geothermal energy. Supplying 20 percent of U.S. electricity from geothermal energy by 2030 is a very realistic goal.

NO: Alexander Karsner, Assistant Secretary for Energy Efficiency and Renewable Energy at the U.S. Department of Energy, argues that it is not feasible to supply 20 percent of U.S. electricity from geothermal energy by 2030.

In June 2007, Senator Jeff Bingaman (D-NM) introduced the National Geothermal Initiative Act of 2007. Hearings were held in September of that year. The point of the bill was to establish a national goal of achieving "20 percent of total electrical energy production in the United States from geothermal resources by not later than 2030." To accomplish that goal, the Department of Energy and the Department of the Interior would characterize the complete U.S. geothermal resource base by 2010 and, among other things, develop policies and programs to sustain a 10 percent annual growth rate in the use of geothermal energy. A similar bill in the House of Representatives—the Advanced Geothermal Energy Research and Development Act of 2007—had a hearing on May 17, 2007, before the Subcommittee on Energy and Environment of the House Committee on Science and Technology. Both bills failed to leave committee. The reason may lie in the fact that, as Mark D. Myers, director of the U.S. Geological Survey, said in his Senate testimony, although increasing the use of geothermal energy is a laudable goal, funds may not be available to achieve the goals of the proposed legislation in the specified time frame. The shortage of funds is due to competing demands, a record federal deficit, and a weakened national economy.

However, the Geothermal Technologies Program of the U.S. Department of Energy's Energy Efficiency and Renewable Energy office (www.eere.energy.gov/) announced in July 2008 a new $30.5 billion loan guarantee program for renewable energy (including geothermal energy) projects. On June 2, 2009, the Department of Energy announced that $50 million of Recovery Act funding would go to helping the deployment of geothermal heat pumps. Enhanced Geothermal Systems (EGS) would receive $80 million. How quickly and how extensively geothermal energy will be developed remains to be seen.

In February 2010, the Natural Resources Committee of the U.S. House of Representatives held a hearing on the Geothermal Energy Production Expansion Act. A Senate version of the bill was introduced in December 2010. Geothermal energy production increased 26 percent in 2009, and the U.S. Department of Energy sees a near-term potential of about 5 percent of U.S. power needs, with longer term potential more than five times as great; see Leslie Blodgett, "U.S. Geothermal Energy Growth," *Electric Light & Power* (May/June 2010). It is clear that geothermal energy is an important alternative energy source. Unlike solar and wind power, it is available 24 hours a day, 12 months a year. Geothermal heat pumps have immense potential for home and commercial heating, according to Joanna R. Turpin, "Commercial Geothermal: Bright Spot in a Gloomy Economy," *Air Conditioning, Heating & Refrigeration News* (May 4, 2009), and the federal government is using the Federal Economic Stimulus Bill to provide investment tax credits to businesses that install them.

Geothermal energy can also be used to meet demand for electricity. "Conventional" geothermal sources are those where the Earth's heat comes fairly close to the surface (as in places like Yellowstone National Park) and the rock of the heated zone has numerous cracks. Groundwater in the cracks may be heated to the point of boiling and erupt as geysers. Surface water may also be pumped into hot rock zones where it is heated. In both cases, steam can be captured and used to generate electricity.

Hot rock may also exist close to the surface but without cracks. In such cases, it may be "enhanced" by such measures as drilling and the use of explosives to create cracks in the rock. Where the hot rock is further from the surface—and there is hot rock everywhere in the world, if one goes deep enough—enhancement must involve drilling more deeply. The more enhancement that is needed, the more expensive the geothermal resource is to tap (which affects feasibility), but even conventional geothermal sources are not cheap. The hot water and steam coursing through the pipes, pumps, and turbines of geothermal plants can be highly corrosive, which makes maintenance expensive. Even so, however, geothermal energy is highly renewable (it is constantly regenerated by the slow decay of radioactive substances deep within the Earth), it does not depend on foreign sources, it uses no fuel, and it does not contribute to global warming.

The first geothermal energy plant was established in Lardarello, Italy, in 1904. In the United States, the first plant started operations in 1922, at The Geysers, California. Today, The Geysers has the largest complex of geothermal power plants in the world and produces more than 850 megawatts of electricity—enough to power a million homes.

When Olafur Ragnar Grimsson, president of Iceland, testified before the Senate Committee on Energy and Natural Resources on September 26, 2007, he said that Iceland meets more than half of its total energy needs with geothermal power and that

> For the United States of America, geothermal energy can become a major energy resource, contributing to the security of the country, limiting dependence on the import of fossil fuels, reducing the risks caused by fluctuating oil prices and providing opportunities for new infrastructures supporting the cities and regions where the resources are located. . . .
>
> [G]eothermal energy is a reliable, flexible and green energy resource which can supply significant amounts of power to households and industry. Furthermore, it uses land economically, gives social returns and it is cost-effective.

MIT's 2006 report, "The Future of Geothermal Energy: Impact of Enhanced Geothermal Systems (EGS) on the United States in the 21st Century" (www1.eere.energy.gov/geothermal/pdfs/future_geo_energy.pdf), notes that:

> Geothermal energy from EGS represents a large, indigenous resource that can provide baseload electric power and heat at a level that can have a major impact on the United States, while incurring minimal environmental impacts. With a reasonable investment in R&D, EGS could provide 100 GWe or more of cost competitive generating capacity in the next 50 years. Further, EGS provides a secure source of power for the long term that would help protect America against economic instabilities resulting from fuel price fluctuations or supply disruptions. Most of the key technical requirements to make EGS work economically over a wide area of the country are in effect, with remaining goals easily within reach . . . within a 10 to 15 year period nationwide.
>
> In spite of its enormous potential, the geothermal option for the United States has been largely ignored . . .

Despite the great potential benefits—large amounts available, continuous availability, zero use of fossil fuels, and hence no impact on global warming—

some researchers and activists do worry about unforeseen consequences. For instance, Michael N. Bates et al., "Cancer Incidence, Morbidity and Geothermal Air Pollution in Rotorua, New Zealand," *International Journal of Epidemiology* (vol. 27, no. 1, 1998), reported elevated incidences of some cancers and diseases of the nervous system and eye, consistent with exposure to hydrogen sulfide and mercury, both of which can be found in the vapors given off by geothermal systems. Water pumped from below ground may be contaminated with arsenic and other minerals, requiring that the water be reinjected into the ground or otherwise kept out of waterways; see Alper Baba and Halldor Armannsson, "Environmental Impact of the Utilization of Geothermal Areas," *Energy Sources Part B: Economics, Planning & Policy* (July–September 2006). Ernest L. Majer et al., "Induced Seismicity Associated with Enhanced Geothermal Systems," *Geothermics* (June 2007), note that concerns that the drilling and fluid injection associated with EGS may cause earthquakes "has been the cause of delays and threatened cancellation of at least two EGS projects worldwide"; see also Sonal Patel, "Assessing the Earthquake Risk of Enhanced Geothermal Systems," *Power* (December 2009). There is a need for careful site selection and attention to public acceptance.

It sounds enticing, but even though tapping geothermal energy is almost entirely an engineering problem, engineering requires time, money, and political commitment. In the YES selection, Susan Petty, president of AltaRock Energy, Inc., a company that hopes to profit from efforts to increase the use of geothermal energy, argues that the technology already exists to greatly increase the production and use of geothermal energy. Immediate federal commitment and investment will improve that technology and lower costs. Supplying 20 percent of U.S. electricity from geothermal energy by 2030 is a very realistic goal. In the NO selection, Alexander Karsner, assistant secretary for Energy Efficiency and Renewable Energy at the U.S. Department of Energy, argues that it is not feasible to supply 20 percent of U.S. electricity from geothermal energy by 2030.

YES  **Susan Petty**

Testimony on the National Geothermal Initiative Act of 2007 Before the Senate Committee on Energy and Natural Resources

One of the goals of S. 1543 is to achieve 20% of electric power generation from geothermal energy by 2030. You may be asking yourself if this a realistic goal? In the fall of 2004, I was included in a 12 member panel led by Dr. Jefferson Tester of the Massachusetts Institute of Technology that looked at the Future of Geothermal Energy. Our group consisted of members from both industry and academia. While some of us started the study convinced that it was possible to engineer or enhance geothermal systems (EGS) with today's technology, many of us, including myself, were skeptical. As we reviewed data, and listened to experts who were actively researching new methods, testing them in the field, and starting commercial enterprises to develop power projects from geothermal energy using this emerging technology, I believe all of us became convinced that a way had been found to tap into the vast geothermal resource under our feet.

Everywhere on Earth, the deeper you go, the hotter it gets. In some places, high temperatures are closer to the surface than others. We have all heard of the "Ring of Fire," characterized by volcanoes, hot springs and fumaroles around the rim of the Pacific Ocean, including the Cascades, the Aleutian Islands, Japan, the Philippines and Indonesia. We know that along the tectonic rifts such as the Mid-Atlantic Ridge including Iceland and the Azores, the East African Rift Valley, the East Pacific Rise, the Rio Grande Rift running up through New Mexico and Colorado and the Juan de Fuca Ridge the earth's heat is right at the surface. But other geologic settings allow high temperatures to occur at shallow depths, such as the faulted mountains and valleys of the Basin and Range, the deep faults in the Rocky Mountains and the Colorado Plateau. In addition, the sedimentary basins that insulate granites heated by radioactive decay along the Gulf Coast, in the Midwest, along the Chesapeake Bay and just west of the Appalachians can not only provide oil and gas, but hot water as well. [. . .]

The heat contained in this vast resource is so large that it is really difficult to contemplate. Even with very conservative calculations, the MIT study panel found that the amount of heat that could be realistically recovered in the U.S. from rocks at depths of 3 km to 10 km (about 2 miles to 6 miles) is

U.S. Senate, September 26, 2007.

almost 3,000 times the current energy consumption of the country. [. . .] Listening to the experience of those developing the Soultz project in France, the Rosemanowes project in the UK and the Cooper Basin project in Australia, the panel members began to understand that the technology to recover this heat was here today. We can drill wells into high temperature rocks at depths greater than 3 km. We can fracture large volumes of hot rock. We can target wells into these man-made fractures and intersect them. We can circulate water through these created fractures, picking up heat and produce it at the other side heated to the temperature of reservoir rocks. We can produce what we inject without having to add more water. Long term tests have been conducted at fairly modest flow rates on these created reservoirs without change in temperature over time. No power plants have yet been built, but several are in progress in Europe.

Does this mean that we can build economic geothermal power plants based on EGS technology right now? At the best sites, where high temperatures occur at shallow depths in large rock masses with similar properties, geothermal power production from EGS technology is economic today. But to bring on line the huge resource stretching across the country from coast to coast, we need to do some work.

I'd like to talk about the economics of geothermal power production so you can better understand what needs to happen to enable widespread development of power projects using EGS.

At some places in the Earth's crust, faults and fractures allow water to circulate in contact with hot rock naturally. These are hydrothermal systems where natural fractures and high permeability allow high production rates. Even low temperature systems can be economic if the flow rates produced are high enough. The capital cost for the wells and wellfield-related equipment generally is between 25%–50% of the total capital cost of the power project. The capital cost for hydrothermal projects can range from around $2,500/installed kW to over $5,000/kW, largely depending on the flow rate per well and the depth of the wells. The levelized break-even cost of energy for commercially viable hydrothermal projects currently ranges from $35/MWh to over $80/MWh. Of this, about $15–25/MWh is operating cost. The rest is the cost to amortize the power generation equipment and the wellfield.

Hydrothermal power is a good deal: Clean, small foot print, cost-effective. So why isn't more power from hydrothermal sources on line? The issue for hydrothermal power is risk. Because the risk related to finding the resource and successfully drilling and completing wells into the resource is high, development by utilities is unlikely. In order to accept this risk, independent power producers need a long-term contract at a guaranteed price and a high return on their investment. Utilities are loath to give a long-term contract because the payments to the generator will be treated as debt in determining their debt-to-equity ratio for credit and bond ratings.

Hydrothermal projects also tend to be small in size. While some of the potential future hydrothermal projects might be large, many of these are associated with scenic volcanic features protected as national parks or revered by Native Americans. A large scale project might mitigate the risk by spreading it over a much larger number of MW. In addition, there is a true economy of scale for geothermal power projects. For instance, the same number of people are needed to operate a 10 MW geothermal project as operate a 120 MW, or even a 250 MW, project.

Most of the really good (i.e. economic) hydrothermal systems are in the arid West. Not only is cooling water, which improves project economics by improving plant efficiency, an issue in this part of the country, but also the wide open spaces mean high-potential sites are often far from transmission, operators, supplies and large population centers with a high demand for power. Little potential for producing power from conventional geothermal, i.e., hydrothermal, sources exists in the Midwest, Southeast or East Coast.

Still, hydrothermal power has the potential to supply the country with more than 20,000 MW, or about 2% of our current installed capacity. However, the very high reliability of geothermal power means that this would be about 4% of our current annual generation. And this power is baseload or power that is available night and day.

Over the years, the cost of generating electricity from hydrothermal sources has dropped from around $130/MWh to less than $50/MWh. This was facilitated by incentives provided both by the market during the mid-1980s oil crisis, and by the government in the form of tax subsidies encourage the construction of over 2,000 MW of geothermal power that went on line from 1986–1995. Some of this drop in cost is due to research conducted by the U.S. Department of Energy (DOE). For instance, in 1980 the DOE completed the first demonstration binary power plant at Raft River. This plant enabled the use of fluids at temperatures much lower than had been developed in the past. Industry commercialized this technology, and now most of the new geothermal power plants being built today are binary plants. DOE research, together with industry, developed high-temperature tools that are now essential to the evaluation of geothermal wells. A combination of DOE-supported research and industry effort has improved binary power plant efficiency by almost 50% from the earliest commercial plants in the 1980s, and flash power-plant efficiency by almost 35% over the same time period. This translates directly into reduction in overall project cost and power prices because fewer wells and less equipment is needed to generate the same amount of energy.

The MIT study started with the current state of the geothermal industry. The first task we realized we needed to undertake was a realistic look at the size and potential cost of developing geothermal power across the continent. It has long been realized by scientists that a vast geothermal resource exists everywhere as long as technology allows us to drill deep enough, develop a reservoir by creating fractures or enhancing natural fractures, and connect wells to circulate fluid through that reservoir. The U.S. Geological Survey has been tasked with a detailed evaluation of the U.S. geothermal resource, but this could not be finished in time for our study. The MIT panel, therefore, undertook a preliminary assessment of the geothermal resource in the U.S.

Using data collected over the years with DOE support, maps of the temperature at depth were developed by Dr. David Blackwell's group at SMU. Temperature at the midpoint of 1 km thick slices was projected at 1 km intervals starting at a depth of 3 km and extending down to 10 km, a reasonable limit for drilling using today's technology. The heat resource contained in each cubic kilometer of rock at these temperatures at each depth was then calculated. The amount of energy stored in this volume of rock is so enormous that it is really impossible to comprehend. . . . We then looked at the studies that had estimated what fraction of this heat might be recovered, and at what efficiency this recovered heat might be turned into electric power. Studies showed that for economic systems, 40% or more of the total heat stored in the rock is recoverable. We also considered the more conservative recoverable estimates of 2% and 20%. Even at 2%, the amount of energy that could be realistically recovered, leaving economics and cost considerations aside, is more than 3,000 times the current total energy consumption of the U.S., including transportation uses.

In order to understand the technology needed to recover this energy, we turned to the published literature on the experiments done in the past at Fenton Hill, Rosemanowes, Hijiori, Ogachi and Soultz. We also brought in experts who are currently working on the Soultz project and on commercial engineered and enhanced geothermal projects in Europe and in Australia to tell us about the status of their work and their future efforts and needs. By the end of the study, we had concluded that EGS technology is technically feasible today. We can:

- Drill wells deep enough and successfully using standard geothermal and oil-and-gas drilling technology with existing infrastructure to tap the geothermal resource across the U.S., including areas in the Midwest, East and Southeast
- Consistently fracture large rock volumes of rock
- Monitor and map these created or enhanced fractures
- Drill production wells into the fractured rock
- Circulate cold water into the injection well and produce heated water from the production wells
- Operate the system without having to add significant amounts of water over time
- Operate the circulation system over extended test periods without measurable drop in temperature
- Generate power from the circulating water at Fenton Hill and Ogachi

In addition, EGS power projects are scalable. Once the first demonstration unit has been tested at a site, the potential exists to develop a really large scale project of 250 to 1000 MW. Combined with the fact that good EGS sites where large bodies of hot rock with fairly uniform properties can be found across the U.S., that the sites are so many that they can be selected to avoid places with no transmission capacity or those located near areas of scenic beauty or environmental sensitivity, generating power from EGS technology looks like a winning proposition.

The real question then becomes, not is it realistic to anticipate generating 20% of our nation's electric power from geothermal energy, but can we make it cost effective?

The MIT panel included members from industry and research who are experts in the economics of power generation. The panel developed a list of key technologies that could help reduce the cost of generating power from EGS. They considered the changes in the cost of power generation from hydrothermal systems over the last 20 years, and the current state of EGS technology. They also considered research currently underway, not only that sponsored by DOE through universities and the national laboratories, but that being done by industry. Using models developed by both DOE and MIT, the cost of power and the impact on that cost of these possible technology improvements was examined. In addition, the panel looked at the impact of "learning by doing" on the cost of power.

We concluded that at the best sites, those with very high temperatures at depths of around 3–4 km in areas with low permeability natural fractures, EGS is economic today. . . . With current technology power from [a 300°C site at a depth of 3 km] this site could be generated for a levelized cost of power of about $74/MWh. This isn't the price that power could be sold for, since it doesn't include profit. It does, however, include financing charges at higher than utility rates, operating costs and the cost of amortizing the capital investment in the welfield and power plant. At deeper depths and lower temperatures, the cost of generating power using EGS technology is much higher, about $192/MWh. . . .

With incremental technology improvement, the cost of power could be cut in half or more, particularly for the deeper high temperature systems. These incremental technology improvements include things like improving conversion cycle efficiency, being able to isolate the part of the wellbore that has been treated so that untreated parts can be fractured, redesigning wells to reduce the number of casing strings and improved understanding of rock/fluid interaction to prevent or repair short circuiting through the reservoir. None of these technology improvements require game changing strategies, just the kind of advancement that comes from persisting in extending our knowledge to the next level. Looking at the high temperature example . . ., the levelized cost of power could be cut to $54/MWh or about 27% with these technology improvements implemented. The moderate temperature site could see a much larger reduction of over 60% to $74/MWh. . . .

We concluded that at the best sites, those with very high temperatures at depths of around 3–4 km in areas with low-permeability natural fractures, EGS is economic today. With incremental technology improvement, the cost of power could be cut in half or more, particularly for the deeper high temperature systems. These incremental technology improvements include things such as improving conversion cycle efficiency, being able to isolate the part of the wellbore that has been treated so that untreated parts can be fractured, redesigning wells to reduce the number of casing strings and improved understanding of rock/fluid interaction to prevent or repair short circuiting through the reservoir. None of these technology improvements require gamechanging or revolutionary strategies, just the kind of advancement that comes from persisting in extending our knowledge to the next level.

The cost of this type of technology improvement is not high. The panel felt that an investment of ~$368,000,000 over a period of about 8–10 years combined with industry involvement could result in 100,000 MW on line by 2030. This would be 10% of the current installed capacity and over 20% of the current electric generation of the country. Combined with the hydrothermal resource, it is a very realistic goal to have geothermal energy provide 20% of the nation's electricity by 2030. However, the effort would require federal support, university, laboratory and industry research, and development and a real commitment to renewable energy use.

Currently more than eight companies are developing EGS power projects in Europe and more than 20 companies are working to get power on line using this technology in Australia. AltaRock Energy Inc. is the only company focused on commercializing power generation from EGS technology in the U.S. In Europe, price subsidies and European Union-sponsored research are helping to start more than 50 EGS projects. In Australia, government grants, help with transmission access, research, and legislation requiring generation from renewable energy sources are driving EGS technology to commercialization. Other countries with fewer economic geothermal resources are planning to include geothermal energy in their generation portfolio. The U.S. needs to commit to this clean, baseload, renewable power source for our own energy future. . . .

Alexander Karsner

Testimony on the National Geothermal Initiative Act of 2007 Before the Senate Committee on Energy and Natural Resources

S. 1543 establishes a national goal of achieving "20 percent of total electrical energy production in the United States from geothermal resources by not later than 2030." To accomplish that goal, the legislation requires the Department of Energy and the Department of the Interior to characterize the complete U.S. geothermal resource base by 2010; develop policies and programs to sustain an annual growth rate in geothermal power, heat, and heat pump applications of at least 10 percent, and to achieve new power or commercial heat production from geothermal resources in at least 25 states; demonstrate state-of-the-art geothermal energy production; and develop tools and techniques to construct an engineered geothermal system power plant. Additionally, the legislation directs the Secretary to establish a geothermal research, development, demonstration, commercialization, outreach and education program in support of the 20 percent national goal.

The Department has significant concerns with the feasibility of the national goal established in this legislation. Generating 20 percent of our nation's electricity from geothermal resources would require more than 165,000 megawatts of geothermal power plant capacity by 2030, in Energy Information Administration's (EIA) reference case electricity demand forecast.[1] The 1978 USGS National Geothermal Resource Assessment estimated 23,000 megawatts of identified conventional geothermal resources, also called hydrothermal technology, that can be developed for electricity. The difference, more than 142,000 megawatts, would have to come from new discoveries, conventional resources that were not viable at the time of the 1978 assessment, and unconventional means such as Enhanced Geothermal Systems (EGS), coproduced fluid from oil and gas wells, and geopressured-geothermal resources, as well as and avoided electricity use from heat, and heat pump applications. With the exception of one small co-production generator, none of these unconventional resources are being used currently to generate commercial power. A recent report by the Massachusetts Institute of Technology (MIT), The Future of Geothermal Energy, estimates that 100,000 megawatts of electricity could be installed by 2050 using EGS technology. The MIT projection

U.S. Senate, September 26, 2007.

assumes a 15-year technology development program is conducted by the public and private sector prior to wide-scale installations.

While the Department shares the Committee's interest in rapidly accelerating market penetration of all renewable energy technologies, including geothermal, this particular goal may be technically unattainable within the timeframe specified. The Department looks forward to working with the Committee to resolve these and other technical concerns with S. 1543.

Since the founding of the Department of Energy, the agency has supported geothermal research and development. Over that period, a number of key accomplishments have contributed to increased commercial development of hydrothermal resources—to a point where it has reached market maturity. The Department's investment contributed to the identification of those resources, accurate characterization and modeling of hydrothermal reservoirs, improved drilling techniques, and advanced means of converting the energy for productive uses. The Federal government has realized many successes in hydrothermal technology development, as evidenced by winning eight R&D 100 Awards in the past ten years. I would like to share with the Committee the Department's current assessment of the geothermal industry, and discuss briefly the future potential for geothermal development as a part of a diversified, domestic clean energy portfolio.

Geothermal Industry

Geothermal energy is the heat from deep inside the earth, coming in large part from the decay of radioactive elements. Geothermal heat is considered a base load renewable energy source, and can be used for electricity generation and direct use (space heating, district heating, snow melting, aquaculture, etc.). While geothermal energy is available at some depth everywhere, in the U.S., it is most accessible in western states such as California, Nevada, Utah, and Hawaii, where it is found at shallow depths as hydrothermal resources. This is where the bulk of conventional, commercial geothermal development is taking place, but a number of other states, notably Idaho, Oregon, Arizona and New Mexico, could see new power projects coming online in the very near future.

Geothermal resources can be subdivided into four categories: 1. hydrothermal; 2. deep geothermal (Enhanced Geothermal Systems or EGS); 3. geopressured; and 4. fluid co-produced with oil and gas. Of these, hydrothermal resources, which are characterized by ample heat, fluid, and permeability, have been developed commercially around the world. The other resource categories have not reached commercial maturity and are less accessible through conventional geothermal processes. The United States has been and continues to be the world leader in online capacity of hydrothermal resources for electric power generation.

Currently, the U.S. has approximately 2850 MWe of installed capacity and about 2,900 MWe of new geothermal power plants under development in 74 projects in the Western U.S., according to industry estimates. In 2006, EIA estimates that geothermal energy generated approximately 14,842 gigawatt-hours (GWh) of electricity. The geothermal industry presently accounts for approximately 5% of renewable energy-based electricity consumption in the U.S. Most of the balance is split between hydropower and biomass, with wind and solar contributing a small portion.

In general, conventional hydrothermal technology is sufficiently mature, based on the following:

- The Western Governors Association geothermal task force recently identified over 140 sites with an estimated 13,000 MWe of power with near-term development potential.
- Hydrothermal reservoirs discovered at shallow depths using existing drilling technology, based upon similar available oil and gas practices used in the industry, are cost-effective.
- Power plant technology is based on standard cycles and can be bought off-the-shelf. Major development of binary-cycle power plant technology has enabled the development of increasingly lower temperature hydrothermal resources.
- Hydrothermal-generated electricity is cost competitive in certain regions of the country, where the resource can be maximized.

Favorable provisions of the Energy Policy Act of 2005 (EPACT 2005) and other federal and local incentives encourage industry to develop hydrothermal resources. EPACT 2005 contains significant provisions to promote the installation of geothermal power plants and geothermal heat pumps. These include:

- Resource Assessment—USGS has been directed to update its 1978 assessment of geothermal resources (Circular 790). EPACT 2005 mandates that USGS complete the Resource Assessment report by September 2008. To date, the Department of Energy has contributed over $1 million in financial support as well as technical support through its national laboratories and the Department's Geothermal Resources Exploration and Definitions activity.
- Programmatic Environmental Impact Statement (PEIS)—A PEIS is being developed for the major geothermal areas in the Western U.S. by the Bureau of Land Management (BLM), in partnership with the U.S. Forest Service. DOE is a cooperating agency for the PEIS and the Department anticipates that completion of the PEIS will encourage geothermal production.
- Streamlined Permitting and Royalty Structure—EPACT changed the royalty structure for leasing on Federal land from a 50/50 State/Federal split to a 50/25/25 split for State/Federal/local, providing an incentive for local governments to attract geothermal resource developers. EPAct also streamlined leasing requirements, which lowers costs for potential developers.
- Federal Purchases of Renewable Energy—EPAct 2005 requires that the Secretary of Energy seek to ensure that federal consumption of electric energy during any fiscal year should include the following amounts of renewable energy; 1) not less than 3 percent in fiscal years 2007 through 2009, 2) not less than 5% in fiscal years 2010 through 2012 and 3) not less than 7.5% in fiscal year 2013 and each fiscal year thereafter.

- Loan Guarantees—EPACT 2005 authorizes the Department to issue loan guarantees to eligible projects that "avoid, reduce, or sequester air pollutants or anthropogenic emissions of greenhouse gases" and "employ new or significantly improved technologies as compared to technologies in service in the United States at the time the guarantee is issued." On May 16, 2007, the Department issued a Notice of Proposed Rulemaking to establish the loan guarantee program. The comment period for that rulemaking has closed, and the Department anticipates finalizing the rule shortly. In addition, on August 3, 2007, the Department named David G. Frantz as the Director of the Loan Guarantee Office, reporting directly to the Department's Chief Financial Officer. By providing the full faith and credit of the United States government, loan guarantees will enable the Department to share some of the financial risks of projects that employ new or significantly improved technologies. DOE is currently authorized to provide $4 billion in loan guarantees, and the 2008 President's Budget requested $9 billion in loan volume limitation.

In addition, the Tax Relief and Health Care Act of 2006 extended the production tax credit for geothermal and other renewables that are put into service through December 31, 2008. This provision has had a significant impact on encouraging new installations of conventional geothermal power facilities; as I mentioned previously, over 2,900 MWe are now under development in the U.S. An investment tax credit of 10 percent is also available to the industry, but cannot be combined with the production tax credit. Because conventional geothermal is a mature technology and favorable policy changes have clearly resulted in the growth of the industry, the FY 2008 Budget Request terminates the current Geothermal Technology program.

Enhanced Geothermal Systems (EGS)

Enhanced Geothermal Systems (EGS) involves technology that enables geothermal resources that lack sufficient water or permeability (compared to conventional hydrothermal resources) to be developed. The ultimate intent is to tap energy from hot impermeable rocks that are at a depth of between 3 and 10 kilometers in the earth's crust. Such rock formations require engineered enhancements to enable productive reservoirs.

DOE funded MIT to conduct a study of EGS potential in the U.S. MIT made the following key findings:

- EGS has the potential to produce up to approximately 100,000 MW of new electric power by 2050 based in part on an abundance of available geothermal resources.
- Elements of the technology to capture EGS are in place.
- Multiple reservoir experiments are required.
- Successful R&D could provide performance verification at a commercial scale within a 15-year period nationwide.

The Department is currently considering the findings of the MIT study. DOE is holding discussions with industry and academic experts, further defining technical barriers and gaps, and determining the technical and commercial actions that can help industry overcome the barriers and to bridge the gaps. Input has come from oil and gas companies, service companies, academia, the geothermal industry, international experts, government agencies, and the national laboratories. We expect to release this evaluation by the end of 2007.

Conclusion

In conclusion, Mr. Chairman, the Department anticipates that geothermal resources will continue to play an important and potentially growing role in our nation's energy portfolio, as we look to rapidly expand the availability of clean, secure, reliable domestic energy. The industry currently benefits from tax incentives and regulatory streamlining in EPACT 2005, and future industry investments in enhanced geothermal have the potential to significantly expand domestic geothermal energy production. The Department looks forward to working with this Committee to resolve concerns related to S. 1543, and to continue our national commitment to clean, renewable energy production. . . .

Note

1. The Energy Information Administration projects Total Electric Power Sector Capacity in 2030 to be 1159 GW. This projection is based on an assumption that geothermal power plant has a capacity factor of 80–85 percent.

EXPLORING THE ISSUE

Is It Time to Put Geothermal Energy Development on the Fast Track?

Critical Thinking and Reflection

1. In 2009, government funding for geothermal research and development increased greatly. Do some research and describe the difference this has made.
2. Explain the difference between a hydrothermal geothermal resource and an Enhanced Geothermal Systems (EGS) resource.
3. Explain how geothermal energy may cause air and water pollution.
4. How might the development of geothermal power cause earthquakes?
5. What are the advantages of geothermal energy when compared to fossil fuels? When compared to wind and solar power?

Is There Common Ground?

In this issue's essays, the greatest disagreement centers on the time needed to develop geothermal power to the point where it can supply 20 percent of U.S. electricity.

1. What methods are available to hasten the development of geothermal energy?
2. What are the major reasons for supporting the expansion of geothermal power? Look back at earlier issues in this book for material to support your answer.
3. How does the "footprint" of geothermal power compare to those of wind, solar, and hydroelectric power?

Contributors to This Volume

EDITOR

THOMAS A. EASTON is a professor of science at Thomas College in Waterville, Maine, where he has been teaching environmental science; science, technology, and society; emerging technologies; and computer science since 1983. He received a B.A. in biology from Colby College in 1966 and a Ph.D. in theoretical biology from the University of Chicago in 1971. He writes and speaks frequently on scientific and futuristic issues. His books include *Focus on Human Biology,* 2nd ed., coauthored with Carl E. Rischer (HarperCollins, 1995), *Careers in Science,* 4th ed. (VGM Career Horizons, 2004), *Taking Sides: Clashing Views in Science, Technology and Society* (McGraw-Hill, 10th ed., 2011), *Taking Sides: Clashing Views on Environmental Issues* (McGraw-Hill, 15th ed., 2012), and *Classic Editions Sources: Environmental Studies* (McGraw-Hill, 4th ed., 2011). Dr. Easton is also a well-known writer and critic of science fiction.

AUTHORS

AMERICAN RIVERS (www.americanrivers.org) is a national organization that protects and promotes rivers as vital to health, safety, and quality of life.

ROBERT F. ANDERSON is Ewing-Lamont Research Professor, Lamont-Doherty Earth Observatory, Columbia University.

JESSE H. AUSUBEL is a senior research associate at Rockefeller University in New York City, where he directs the Program for the Human Environment. He is also a Program Director for the Alfred P. Sloan Foundation.

STEPHEN L. BAIRD is a technology education teacher for the Virginia Beach City Public School system and an adjunct assistant professor at Old Dominion University.

LAKE H. BARRETT is a former official of the Nuclear Regulatory Commission and the Department of Energy. He once directed the DOE program for disposal of spent fuel and high-level waste.

KATE BRASH is assistant director of the Columbia Climate Center, Earth Institute, Columbia University.

ROBERT BRYCE is the managing editor of *Energy Tribune* and has been writing about energy for almost 20 years.

KEVIN BULLIS is the Energy Editor of *Technology Review.*

MARY-ELENA CARR is associate director of the Columbia Climate Center, Earth Institute, Columbia University.

LUTHER J. CARTER is the author of *Nuclear Imperatives and Public Trust: Dealing with Radioactive Waste* (Resources for the Future, 1987).

PAUL CICIO is president of the Industrial Energy Consumers of America (IECA).

VIRGINIA H. DALE is a scientist in the Center for BioEnergy Sustainability at the Oak Ridge National Laboratory, Oak Ridge, Tennessee.

THE DEPARTMENT OF ENERGY is the United States government's agency charged with advancing the national, economic, and energy security of the United States; promoting scientific and technological innovation in support of that mission; and ensuring the environmental cleanup of the national nuclear weapons complex.

JOHN ETHERINGTON is a Thomas Huxley Medallist at the Royal College of Science and a former coeditor of the *Journal of Ecology.*

AARON EZROJ is a Law Fellow at Adams Broadwell Joseph & Cardozzo. He studied market-based environmental mechanisms in Europe as a Fulbright Scholar.

CHARLES D. FERGUSON was the Philip D. Reed senior fellow for science and technology at the Council on Foreign Relations until January 1, 2010, when he took office as President of the Federation of American Scientists.

PHILLIP J. FINCK is deputy associate laboratory director, Applied Science and Technology and National Security, Argonne National Laboratory.

JAMES R. FLEMING is a historian of science and technology and professor of Science, Technology and Society at Colby College, Maine. His books include *Fixing the Sky: The Checkered History of Weather and Climate Control* (Columbia University Press, 2010), *The Callendar Effect* (American Meteorological Society, 2007), *Historical Perspectives on Climate Change* (Oxford University Press, 1998), and *Meteorology in America, 1800–1870* (Johns Hopkins University Press, 1990).

MICHELLE GARNEAU is a professor in the Department of Geography at the University of Quebec in Montreal. She studies the impact of climate change.

ERIC GHOLZ is assistant professor of public affairs at the University of Texas, Austin.

THE GLOBAL HUMANITARIAN FORUM was established to bring together a global community and create the conditions for dialogue and rapid action. Its mission was to inspire others and to connect them so that they can take action together. It closed in 2010 for lack of funds.

RICHARD N. HAASS is the president of the Council on Foreign Relations. From 2001 to 2003, he was the Director of Policy Planning for the United States Department of State.

DAVID G. HAWKINS is the director of the Climate Center of the Natural Resources Defense Council.

MICHAEL HORN is a retired aerospace scientist.

ALEXANDER KARSNER is the assistant secretary for Energy Efficiency and Renewable Energy at the U.S. Department of Energy.

DIANE KATZ is director of Risk, Environment and Energy Policy for the Fraser Institute, an independent nonpartisan research and educational organization based in Canada.

JUHA KIVILUOMA is a research scientist at the VTT Technical Research Centre of Finland.

KEITH KLINE is a scientist in the Center for BioEnergy Sustainability at the Oak Ridge National Laboratory, Oak Ridge, Tennessee.

ANDREA LARSON is a professor in the Darden School of Business, University of Virginia. Her research has focused on innovation and entrepreneurship, strategy, and sustainability.

RUSSELL LEE is a scientist in the Center for BioEnergy Sustainability at the Oak Ridge National Laboratory, Oak Ridge, Tennessee.

PAUL LEIBY is a scientist in the Center for BioEnergy Sustainability at the Oak Ridge National Laboratory, Oak Ridge, Tennessee.

STEVEN F. LEER is the chairman and chief executive officer of Arch Coal, Inc.

BJORN LOMBORG is an adjunct professor in the Copenhagen Business School and organizer of the Copenhagen Consensus, a conference of top economists who work to prioritize the best solutions for the world's greatest challenges. His latest book is *Cool It! The Skeptical Environmentalist's Guide to Global Warming* (Knopf, 2007).

XI LU is a Ph.D. candidate at the Harvard School of Engineering and Applied Sciences.

ALLISON MACFARLANE is an associate professor of Environmental Science and Policy at George Mason University and a member of the U.S. Energy Department's Blue Ribbon Commission on America's Nuclear Future. She is the coauthor, with Rod Ewing, of *Uncertainty Underground: Yucca Mountain and High-Level Nuclear Waste Disposal* (MIT Press, 2006).

MICHAEL B. McELROY is Gilbert Butler Professor of Environmental Studies at the Harvard School of Engineering and Applied Sciences.

DONALD MITCHELL is lead economist in the World Bank's Development Prospects Group.

SUSAN MORAN teaches magazine journalism at the University of Colorado at Boulder's School of Journalism and Mass Communication.

RICK NEWMAN is chief business correspondent for *U.S. News and World Report* and the author of *Firefight: Inside the Battle to Save the Pentagon on 9/11* (Presidio, 2008), and *Bury Us Upside Down: The Misty Pilots and the Secret Battle for the Ho Chi Minh Trail* (Presidio, 2006).

SUSAN PETTY is the president of AltaRock Energy, Inc., a developer of "geothermal technology to produce clean, renewable power."

DARYL G. PRESS is associate professor of government at Dartmouth College in Hanover, New Hampshire.

CHARLOTTE ROEHM is currently a postdoctoral researcher in the Department of Ecology and Environmental Science at Umea University in northern Sweden.

KENNETH C. ROGERS is a former president of Stevens Institute of Technology. He was a member of the NRC from 1987 to 1997.

MARY ANNETTE ROSE is an assistant professor in the Department of Technology at Ball State University in Muncie, Indiana.

CHARLES W. SCHMIDT is a freelance science writer specializing in the environment, genomics, and information technology, among other topics. In 2002, he won the National Association of Science Writers' Science-in-Society Journalism Award for his reporting on hazardous electronic waste exports to developing countries.

KRISTIN SHRADER-FRECHETTE is the O'Neill Family Professor, Department of Biological Sciences and Department of Philosophy, at the University of Notre Dame. She is the author of *Taking Action, Saving Lives: Our Duties to Protect Environmental and Public Health* (Oxford University Press, 2007).

ALAIN TREMBLAY is the coordinator of the Reservoirs' Net Greenhouse Gas Emissions project at Hydro-Quebec's Eastmain-1 site.

LOUIS VARFALVY is a researcher with Hydro-Quebec.

DEBORAH WEISBERG is an award-winning journalist whose work appears in the *Pittsburgh Post-Gazette, New York Times*, and other publications.

TOM WHIPPLE is the editor of *Peak Oil Review*, a publication of the Association for the Study of Peak Oil & Gas.